CATALYSIS OF ORGANIC REACTIONS

CHEMICAL INDUSTRIES

A Series of Reference Books and Textbooks

Consulting Editor

HEINZ HEINEMANN
Heinz Heinemann, Inc.
Berkeley, California

Additional Volumes in Preparation

CATALYSIS OF ORGANIC REACTIONS

edited by
William E. Pascoe
Eastman Kodak Company
Rochester, New York

Marcel Dekker, Inc.   New York • Basel • Hong Kong

Library of Congress Cataloging-in-Publication Data

Catalysis of organic reactions / edited by William E. Pascoe.
    p.   cm. -- (Chemical industries; v. 47)
    "Collection of papers and posters presented at the Thirteenth
  Conference on Catalysis of Organic Reactions in Boca Raton, Florida
  in May, 1990." -- Pref.
    Includes bibliographical references and index.
    ISBN 0-8247-8573-8
    1. Organic compounds--Synthesis--Congresses. 2. Catalysis-
  -Congresses. I. Pascoe, William. II. Conference on Catalysis of
  Organic Reaction (13th : 1990 : Boca Raton, Fla.) III. Series.
  QD262.C35   1992
  547.2--dc20
                                                          91-32982
                                                              CIP

This book is printed on acid-free paper.

Marcel Dekker, Inc.
270 Madison Avenue, New York, New York 10016

Current printing (last digit):
10 9 8 7 6 5 4 3 2 1

PRINTED IN THE UNITED STATES OF AMERICA

Preface

This volume of *Catalysis of Organic Reactions* is a collection of
the papers and posters presented at the Thirteenth Conference
on Catalysis of Organic Reactions in Boca Raton, Florida, in May
1990. The conference was sponsored by the Organic Reactions
Catalysis Society (ORCS), an affiliate of the Catalysis Society of
North America.

ORCS has been providing a forum for chemists from chemical
and pharmaceutical industries and from academia to present and
discuss their work on the use of catalysis in organic syntheses
since 1966. At past conferences the presentations had been limited
to papers, but at the Thirteenth Conference a poster session was
also included.

Part I includes papers and posters covering a variety of topics
involved in catalytic hydrogenations. Part II covers other uses
of catalysis in organic syntheses. Included are topics on acid catalysis, dehydrogenation, oxidation, and other selected special topics.

ORCS acknowledges the financial support of our major industrial
sponsors, Calsicat, Degussa Corporation, Engelhard Corporation,
Johnson Matthey, Parr Instrument Company, and W. R. Grace & Company. Their support, not only of this conference but over the
years, has greatly contributed to the success of the society. ORCS
would also like to acknowledge the following industries for their
financial support of this conference: Air Products and Chemicals,
Inc., Autoclave Engineers, Inc., Ethyl Corporation, E. I. du Pont
de Nemours & Co., G. D. Searle & Co., Merck Sharp & Dohme
Research Lab., Monsanto Co., The NutraSweet Co., Parke-Davis/
Warner-Lambert Co., Uniroyal Chemical Co., and United Catalyst,
Inc.

I wish to express my appreciation to all the presenters and session chairmen for their participation in the program. I also wish to thank the members of the Executive Committee, Russell Malz, Jr., Anthony King, Michael Scaros, Louis Seif, John Kosak, Robert Augustine, Paul Rylander, John Cihonski, and Dale Blackburn for their help in organizing this conference.

William E. Pascoe

Contents

Oxidation

Special Topics in Catalysis

Contributors

Robert L. Augustine   *Department of Chemistry, Seton Hall University, South Orange, New Jersey*

M. Bartók   *Department of Organic Chemistry, József Attila University, Szeged, Hungary*

James R. Behling   *Department of Chemical Development, Searle, Skokie, Illinois*

Cv. Bezouhanova   *Department of Chemistry, University of Sofia, Sofia, Bulgaria*

Thomas J. Blacklock   *Merck Sharp & Dohme Research Laboratories, Rahway, New Jersey*

Anthony P. Bonds   *First Chemical Corporation, Pascagoula, Mississippi*

Donna M. Brestensky   *Department of Chemistry, Indiana University, Bloomington, Indiana*

John W. Butcher   *Merck Sharp & Dohme Research Laboratories, Rahway, New Jersey*

Pochen Chu   *Mobil Research and Development Corporation, Paulsboro, New Jersey*

Mark D. Conner   *Air Products and Chemicals, Inc., Allentown, Pennsylvania*

John F. Daeuble   *Department of Chemistry, Indiana University, Bloomington, Indiana*

J. L. Dallons   *UCB, Drogenbos, Belgium*

K. Deller   *Degussa AG, Hanau, Germany*

B. Despeyroux   *Degussa AG, Hanau, Germany*

Joseph K. Doles   *Chemicals Development Division, Eastman Kodak Company, Rochester, New York*

Ming-Jaw Don   *Department of Chemistry, Center for Organometallic Research and Education, University of North Texas, Denton, Texas*

Jonathan S. Dordick   *Department of Chemical and Biochemical Engineering, University of Iowa, Iowa City, Iowa*

Michael P. Doyle   *Department of Chemistry, Trinity University, San Antonio, Texas*

Payman Farid   *Department of Chemical Development, Searle, Skokie, Illinois*

Michael E. Ford   *Air Products and Chemicals, Inc., Allentown, Pennsylvania*

P. Gallezot   *Institut de Recherches sur la Catalyse, C.N.R.S., Villeurbanne, France*

William E. Garwood   *Mobil Research and Development Corporation, Paulsboro, New Jersey*

A. Giroir-Fendler   *Institut de Recherches sur la Catalyse, C.N.R.S., Villeurbanne, France*

Owen J. Goodmonson*   *Department of Chemical Development, Searle, Skokie, Illinois*

Edward J. J. Grabowski   *Merck Sharp & Dohme Research Laboratories, Rahway, New Jersey*

Harold Greenfield   *First Chemical Corporation, Pascagoula, Mississippi*

**Current affiliation*: C.A.P.D., Abbott Laboratories, North Chicago, Illinois

Xiangyao Guo *Department of Chemical Engineering, University of Waterloo, Waterloo, Ontario, Canada*

Kenneth G. High *Department of Chemistry, Trinity University, San Antonio, Texas*

Hikmat Hilal *Department of Chemistry, Mississippi State University, Mississippi State, Mississippi*

James J. Huson *Eastman Kodak Company, Rochester, New York*

Brian R. James *Department of Chemistry, University of British Columbia, Vancouver, British Columbia, Canada*

G. Jannes *CERIA-ISI, Brussels, Belgium*

Makarand G. Joshi *Manufacturing Research and Engineering, Eastman Kodak Company, Rochester, New York*

J. A. Kaduk *Amoco Research Center, Amoco Chemical Company, Naperville, Illinois*

Yu. Kalvachev *Department of Chemistry, University of Sofia, Sofia, Bulgaria*

Ish Khanna *Department of Molecular and Cell Biology, Searle, Skokie, Illinois*

K. Laali *Department of Chemistry, Kent State University, Kent, Ohio*

Theresa R. Lamanec *Merck Sharp & Dohme Research Laboratories, Rahway, New Jersey*

Kevin R. Lassila *Air Products and Chemicals, Inc., Allentown, Pennsylvania*

H. Lechert *University of Hamburg, Institute of Physical Chemistry, Hamburg, Germany*

C.-Y. Lin *Uniroyal Chemical Company, Inc., Naugatuck, Connecticut*

You-Jyh Lin *Washington Research Center, W. R. Grace & Company, Columbia, Maryland*

Wayne S. Mahoney *Department of Chemistry, Indiana University, Bloomington, Indiana*

R. J. Malone   *Herzog-Hart Corporation, Boston, Massachusetts*

R. E. Malz, Jr.   *Uniroyal Chemical Company, Inc., Naugatuck, Connecticut*

John R. Medich   *Department of Chemical Development, Searle, Skokie, Illinois*

Patrick E. McMahon   *Amoco Research Center, Amoco Chemical Company, Naperville, Illinois*

H. L. Merten   *Herzog-Hart Corporation, Boston, Massachusetts*

M. Musoiu   *Molecular Science Program, Southern Illinois University at Carbondale, Carbondale, Illinois*

Victor L. Mylroie   *Chemicals Development Division, Eastman Kodak Company, Rochester, New York*

D. S. Nagvekar   *Department of Chemistry, Kent State University, Kent, Ohio*

Carey L. Nesloney   *Department of Chemistry, Trinity University, San Antonio, Texas*

F. Notheisz   *Department of Organic Chemistry, József Attila University, Szeged, Hungary*

Shaun T. O'Leary   *Department of Chemistry, Seton Hall University, South Orange, New Jersey*

D. Ostgard   *Department of Chemistry and Biochemistry, Southern Illinois University at Carbondale, Carbondale, Illinois*

Sanghamitra Parida   *Department of Chemical and Biochemical Engineering, University of Iowa, Iowa City, Iowa*

William E. Pascoe   *Manufacturing Research and Engineering, Eastman Kodak Company, Rochester, New York*

Damodar R. Patil   *Department of Chemical and Biochemical Engineering, University of Iowa, Iowa City, Iowa*

Charles U. Pittman, Jr.   *Department of Chemistry, Mississippi State University, Mississippi State, Mississippi*

Michael L. Prunier   *Department of Chemical Development, Searle, Skokie, Illinois*

P. Kanta Rao   *Catalysis Section, Indian Institute of Chemical Technology, Hyderabad, India*

V. Nageshwar Rao   *Catalysis Section, Indian Institute of Chemical Technology, Hyderabad, India*

G. L. Rempel   *Department of Chemical Engineering, University of Waterloo, Waterloo, Ontario, Canada*

David G. Rethwisch   *Department of Chemical and Biochemical Engineering, University of Iowa, Iowa City, Iowa*

Michael Reynolds   *Uniroyal Chemical Company, Naugatuck, Connecticut*

D. Richard   *Institut de Recherches sur la Catalyse, C.N.R.S., Villeurbanne, France*

Michael G. Richmond   *Department of Chemistry, Center for Organometallic Research and Education, University of North Texas, Denton, Texas*

Keungarp Ryu   *Department of Chemical and Biochemical Engineering, University of Iowa, Iowa City, Iowa*

Larry C. Satek   *Amoco Research Center, Amoco Chemical Company, Naperville, Illinois*

Mike G. Scaros   *Department of Chemical Development, Searle, Skokie, Illinois*

Willard E. Shearin   *Merck Sharp & Dohme Research Laboratories, Rahway, New Jersey*

Richard F. Shuman   *Merck Sharp & Dohme Research Laboratories, Rahway, New Jersey*

C. Sivaraj   *Catalysis Section, Indian Institute of Chemical Technology, Hyderabad, India*

G. V. Smith   *Molecular Science Program and Department of Chemistry and Biochemistry, Southern Illinois University at Carbondale, Carbondale, Illinois*

Paul Sohar Merck Sharp & Dohme Research Laboratories, Rahway, New Jersey

Ruozhi Song Department of Chemistry and Biochemistry, Southern Illinois University at Carbondale, Carbondale, Illinois

S. T. Srinivas Catalysis Section, Indian Institute of Chemical Technology, Hyderabad, India

Joseph Stieber Uniroyal Chemical Company, Naugatuck, Connecticut

J. Stoch Institute of Catalysis and Surface Chemistry, Polish Academy of Sciences, Kraków, Poland

Jeffrey M. Stryker Department of Chemistry, Indiana University, Bloomington, Indiana

S. Tjandra Department of Chemistry and Biochemistry, Southern Illinois University at Carbondale, Carbondale, Illinois

Louis F. Valente Eastman Kodak Company, Rochester, New York

A. Van Gysel UCB, Drogenbos, Belgium

Richard M. Weier Department of Molecular and Cell Biology, Searle, Skokie, Illinois

John P. Westrich[*] Department of Chemical Development, Searle, Skokie, Illinois

Geoffrey T. White Engelhard Corporation, Beachwood, Ohio

T. Wiltowski Molecular Science Program, Southern Illinois University at Carbondale, Carbondale, Illinois

---

*Current affiliation: S. C. Johnson and Son, Inc., Skokie, Illinois

1

Selectivity Control in Cinnamaldehyde Hydrogenation by Metal Catalysts of Precise Structure and Morphology

P. Gallezot, A. Giroir-Fendler, and D. Richard

Institut de Recherches sur la Catalyse, C.N.R.S., Villeurbanne, France

I. INTRODUCTION

Preparation of unsaturated alcohols by hydrogenation of α, β-unsaturated aldehydes on metal catalysts is a subject of continuous interest from both industrial and academic standpoints. The selectivity to unsaturated alcohol is governed by the nature of the metal and by the presence of additives [1]. Thus, palladium is quite unselective, osmium and iridium are intrinsically selective, and the selectivity of platinum can be tuned by promoters. Addition of metallic salts to the reaction medium has been known for a long time [2] to increase the selectivity of platinum catalysts. However, the mode of action of these additives remains obscure; thus it is still not established whether they are fixed on the base metal, and there is doubt concerning their oxidation state. The selectivity of platinum metals can also be improved by the addition of Brønsted or Lewis bases [3-5], but the reasons for these promoting effects are not yet known. Other factors that might improve selectivity, such as the nature and texture of the support or the morphology of the metal particles, have been hardly studied. To throw some light on these different points, a few years ago we started a research program to find out how much and why selectivity to unsaturated alcohols is affected by the composition, structure, and texture of metal catalysts. Only one substrate, cinnamaldehyde, was hydrogenated, under standard reaction conditions on tailor-made catalysts obtained by innovative preparation techniques and thoroughly characterized by high resolution, conventional and analytical electron microscopy. Catalysts of homogeneous composition and morphology were selected to establish correlations between structure and selectivity.

II. EXPERIMENTAL

The size and distribution of metal particles on supports were determined by high resolution transmission electron microscopy (TEM)
with a JEOL 100 CX microscope (resolution on lattice, 0.14 nm).
To obtain the distribution of metal particles in microporous supports
(charcoal and zeolite), the catalysts were embedded in an epoxy-
resin and cut into thin sections with an ultramicrotome equipped
with a diamond knife. In the case of bimetallic catalysts, the composition of the particles was determined by energy-dispersive X-ray
emission analysis (EDX) in a scanning transmission electron microscope (STEM) VGHB 501 equipped with a field emission electron
gun. The spatial resolution of analysis is 1.5 nm, which permits
the composition of individual particles to be measured.

   The hydrogenation of cinnamaldehyde was conducted in a well-
stirred autoclave. Typically, 400 mg of catalyst was suspended
in a liquid phase containing 37.5 cm^3 of isopropanol, 10 cm^3 of
water, and 2.5 cm^3 of sodium acetate (0.1 mol 1^{-1}). The catalyst
was activated by stirring the mixture for 2 hours at 333 K under
4 MPa H$_2$ pressure; then 0.1 mol of cinnamaldehyde was introduced
under H$_2$ pressure and the reaction was started by stirring at
1500 rpm under 4 MPa H$_2$ pressure at 333 K. The reaction rate
and product distribution were monitored by repetitive sampling
and gas-chromatographic analysis. Under our reaction conditions,
cinnamaldehyde (CAL) hydrogenation gives only cinnamyl alcohol
(COL), hydrocinnamaldehyde (HCAL), and hydrocinnamyl alcohol
(HCOL) as shown in Figure 1.1.

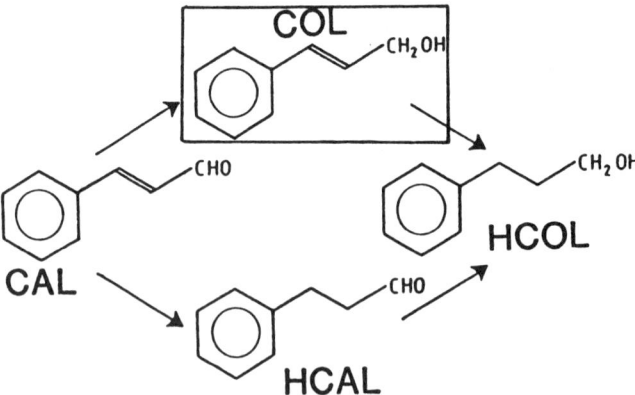

Figure 1.1 Cinnamaldehyde hydrogenation.

III. RESULTS AND DISCUSSION

A. Selectivity Control by Support-Induced Electronic Effects

Two catalysts, Pt/C (3.8 wt % Pt) and Ru/C (4.6 wt % Ru), were prepared by ion exchange with $Pt(NH_3)_4^{2+}$ and $Ru(NH_3)_6^{3+}$ cations of a charcoal support (CECA 50S, 1400 m^2g^{-1}), previously function-alized by NaClO oxidation as described elsewhere [7]. The catalysts were then reduced by flowing hydrogen at 573 K. Figure 1.2a, a TEM view through an ultramicrotome section of a charcoal grain of Pt/C, shows that the particles are in the size range 1–1.5 nm (average size, 1.3 nm) and that they are homogeneously distributed throughout the grain. It was also checked by TEM that the ruthenium in the Ru/C is in the form of homogeneously distributed particles smaller than 1 nm.

Two catalysts, Pt/G_{ex} (3.6 wt % Pt) and Ru/G_{ex} (3.6 wt % Ru) were prepared by ion exchange of a high surface area graphite (LONZA HSAG 12,300 m^2/g) oxidized by NaClO. The catalysts were then reduced at 573 K. The morphology of Pt/G_{ex} has been studied in detail [6,8]. Briefly, it can be said that the NaClO treat-ment increases the number of steps on the graphite surface and that these steps are functionalized; that is, carboxylic groups are created at the extremities of the graphite basal planes. These groups act as ion-exchange sites and, after reduction, as anchoring sites for the metal particles. This preparation results in a very homoge-neous distribution of 1–1.5 nm particles all located along the graphite steps as shown on the TEM view of Pt/G_{ex} (Figure 1.2b) and in Figure 1.3.

The catalyst Ru/G_{ex} has a similar morphology but with smaller metal particles. The catalyst Pt/G_C was obtained by decomposition of the zero-valent platinum complex $Pt(dpo)_2$ (dpo = 1,5-diphenyl-1,4-pentadiene-3-one) on graphite suspended in a CH_2Cl_2 solution of the complex, as described elsewhere [9]. The metal particles are deposited selectively on the basal plane of graphite. The dis-tributions of products as a function of the CAL conversion are given in Figure 1.4 for Pt/C, Pt/G_{ex}, Ru/C, and Ru/G_{ex}. Table 1.1 gives the selectivities to COL at 0, 25, and 50% conver-sion.

It is noteworthy that the selectivities to COL are much higher on graphite-supported than on charcoal-supported catalysts. Since the particle sizes are the same on the two supports, the higher selectivities of Pt/G_{ex} and Ru/G_{ex} are attributed to an effect of the support. Indeed it has been shown [7,8] that graphite acts as an electron-donating macroligand, increasing the electron density on platinum particles anchored on the graphite steps. This was

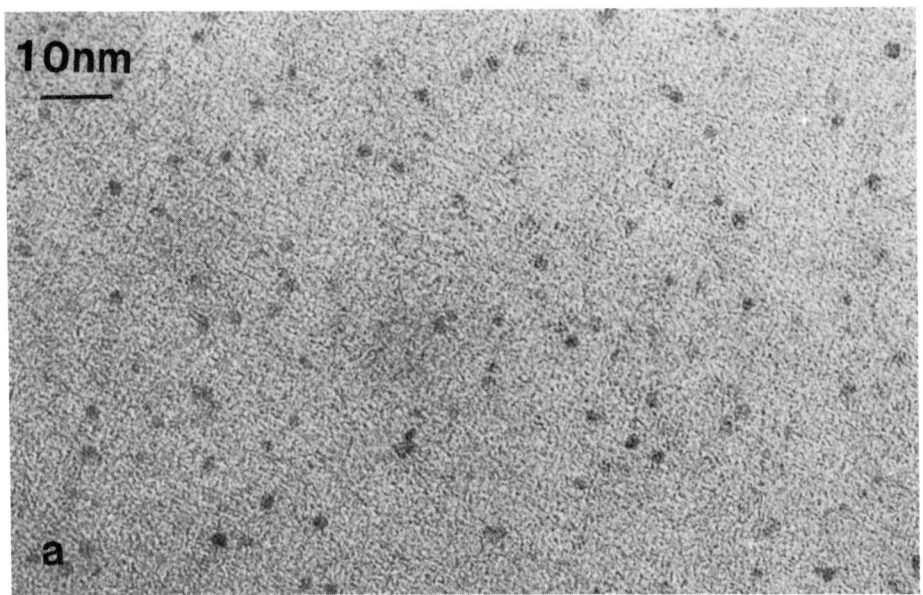

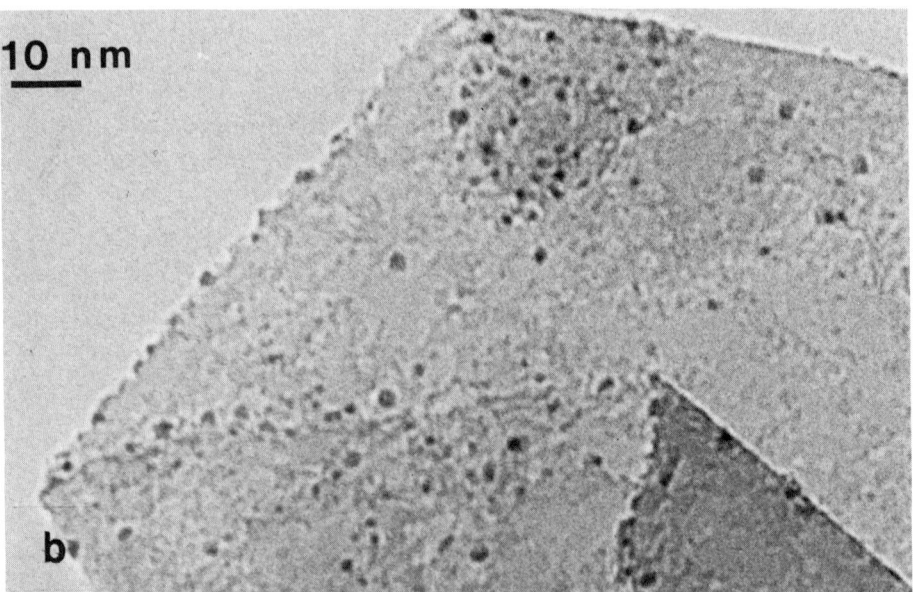

Figure 1.2   TEM view (a) through a section of a grain of Pt/C
and (b) through Pt/G_{ex}.

Figure 1.3 Particle location on graphite steps.

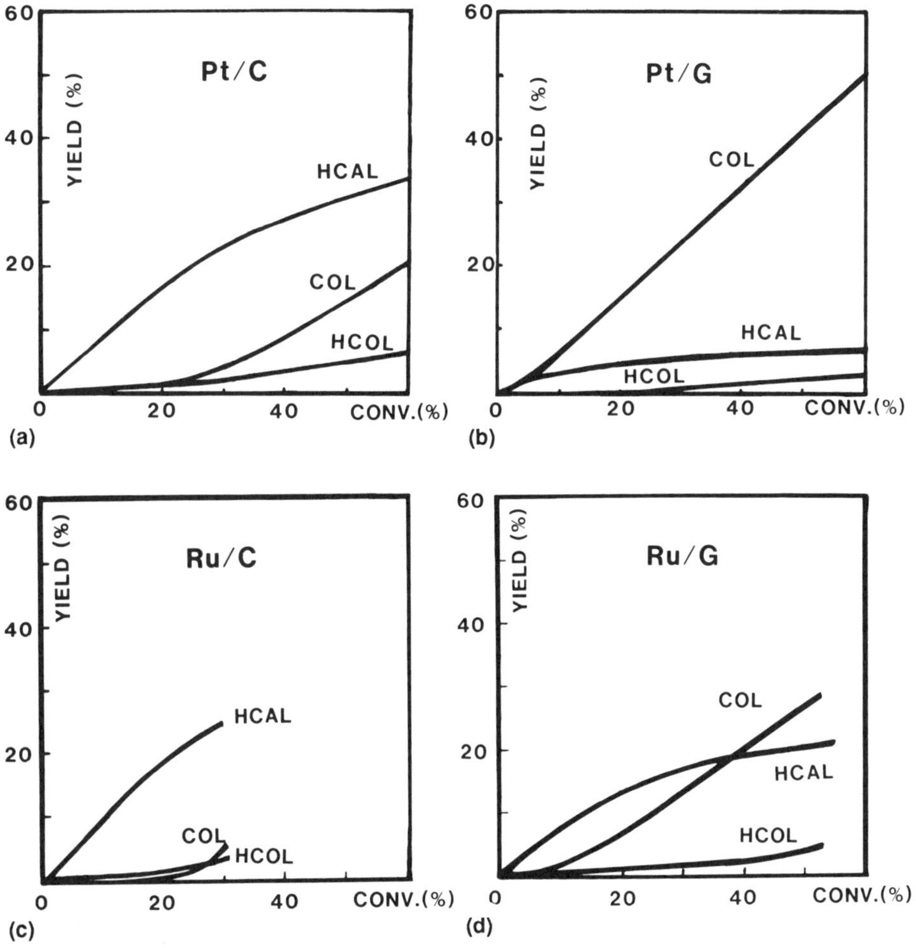

Figure 1.4 Distribution of products versus CAL conversion: (a) Pt/C, (b) Pt/G_{ex}, (c) Ru/C, and (d) Ru/G_{ex}.

Table 1.1   Characteristics and Selectivities ($S^{\% conv}$) to Cinnamyl
Alcohol of Catalysts at Different Conversions

Catalysts	x (wt %)[a]	D (nm)[b]	S^0 (%)	S^{25} (%)	S^{50} (%)
Pt/C	3.8	1-1.5 (1.3)	0	33	55
Ru/C	4.6	<1	0	5	
Rh/C	2.7	2-3 (2.5)	0	0	5
Pt/G$_{ex}$	3.6	1-1.5 (1.3)	72	78	83
Rh/G$_{ex}$	3.4	2-3 (2.5)	12	14	18
Ru/G$_{ex}$	3.6	<1	12	39	53
Pt/G$_C$	1.1	1-5	54	73	
Pt/G$_{HT}$	3.6	3-6 (5)	91	96	98
Rh/G$_{HT}$	3.4	3-9 (7)	32	36	42
PtAd$_1$	100	2-6	73	74	81
PtAd$_2$	100	20-200	90	90	93
Pt/Y	11.0	(1.3)	0	74	82
Pt/Y*	14.0	(5)	0	97	97
Rh/Y	7.0	(1)	0	16	30

[a]Wt % metal in the catalyst.
[b]Particle size, average values in parentheses.

evidenced by an expansion of the platinum lattice and by a decrease
of the ratio $K_{T/B}$ of the adsorption coefficients of toluene and benzene (because toluene is a better electron donor than benzene,
this ratio is a good probe for the electron structure of metal [10]).
The good selectivity of Pt/G$_{ex}$ can be attributed to the reduced
ability of the platinum particles to hydrogenate the C=C bond of
the CAL molecule. Indeed, because of the higher electron density
of the metal, there is a lower probability for the activation of the
C=C bond, which involves as a first step an electron transfer of
π-electrons to the metal d-band. This interpretation can also account
for the higher selectivity to COL of Ru/G$_{ex}$ with respect to Ru/C.

It is now interesting to compare the selectivities of Pt/G$_{ex}$ and
Pt/G$_C$ catalysts. Their supports are the same, but the particles are
located selectively at the extremities of basal planes in the former
and on top of basal planes in the latter. Table 1.1 shows that the
selectivities are smaller on Pt/G$_C$ than on Pt/G$_{ex}$. It has been
established [8,9] that the electron transfer from graphite to platinum
is smaller on Pt/G$_C$ than on Pt/G$_{ex}$ because the metal particles are
in different locations. Therefore the different selectivities could
again be interpreted in terms of the higher electron density on
Pt/G$_{ex}$, which decreases the probability for the C=C bond activation.

Note that whatever the catalysts (examined in this section or in forthcoming ones), the selectivity to COL increases with CAL conversion. It has been shown that HCAL added to the reaction medium acts as an inhibitor for CAL conversion. This indicates that part of the HCAL molecules produced by the hydrogenation of CAL remain adsorbed on the metal surface and modify the selectivity. The surface ligand could act by a steric effect (e.g., by forcing the oncoming CAL molecules to adsorb vertically via the C=O group) or by an electronic effect (e.g., by transferring electrons to the metal, which would decrease the probability for the C=C bond activation as discussed above).

We conclude this section by observing that a parallel can be drawn between the effect of graphite acting as an electron-donating, polydentate ligand and other ligands acting as Lewis bases. Thus phosphines added to the reaction medium improve the selectivity of platinum [4] because these electron-donating ligands could increase the electron density on the metal. Finally, it is interesting to notice that the activities of graphite-supported catalysts are little modified [7,9], unlike those of the bimetallic catalysts discussed below (Section D), for which both activity and selectivity are enhanced. This corroborates our interpretation according to which the better selectivity of metals interacting with electron-donating ligands is due more to an inhibition of the hydrogenation of the C=C bond than to an activation of the C=O bond.

B. Selectivity Control by Shape Selective Effects in Metal Zeolite Catalysts

Catalysts Pt/Y (11 wt % Pt) and Rh/Y (7 wt % Rh) were prepared by ion exchange of a NaY zeolite (linde LZY 52) with $Pt(NH_3)_4^{2+}$ and $Rh(NH_3)_5Cl^{2+}$ cations, respectively. A high dispersion of the metal in zeolite micropores is obtained by following well-documented treatment conditions [11]. Figure 1.5a is a TEM view through a ultramicrotome section of a grain of Pt/Y zeolite. The metal is uniformly distributed throughout the grain in the form of metal particles small enough (1 ± 0.3 nm) to fit into the zeolite supercages (1.3 nm).

The selectivities at different conversions are compared to those of other catalysts in Table 1.1. The initial selectivities to COL are close to zero because it takes some time before COL can be detected in the reaction medium. After this induction period, which is probably due to mass transfer in the micropores, the selectivities are much higher than on Pt/C and Rh/C catalysts taken as reference samples (Table 1.1). Thus the selectivities at 50% conversion on Rh/C and Pt/C are 5 and 55%, respectively, whereas they are as high as 30 and 82% on RhY and Pt/Y.

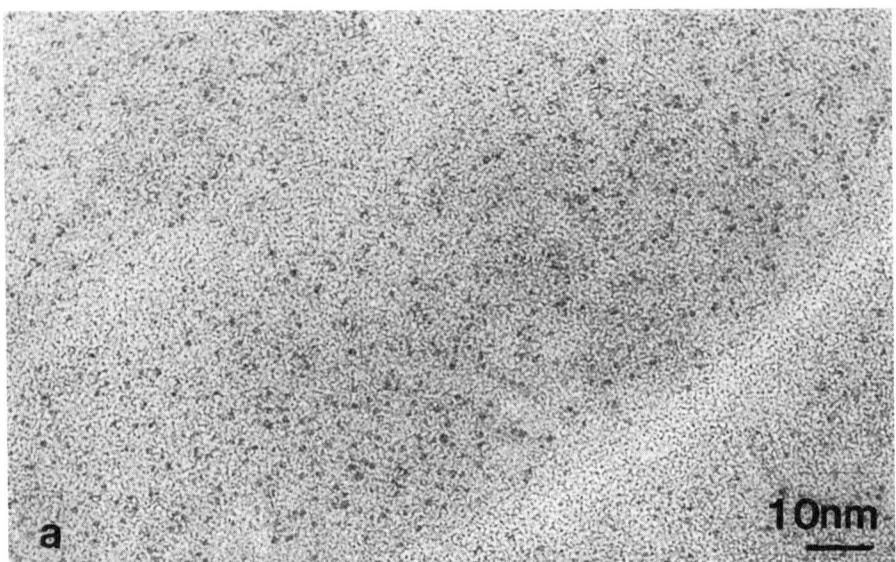

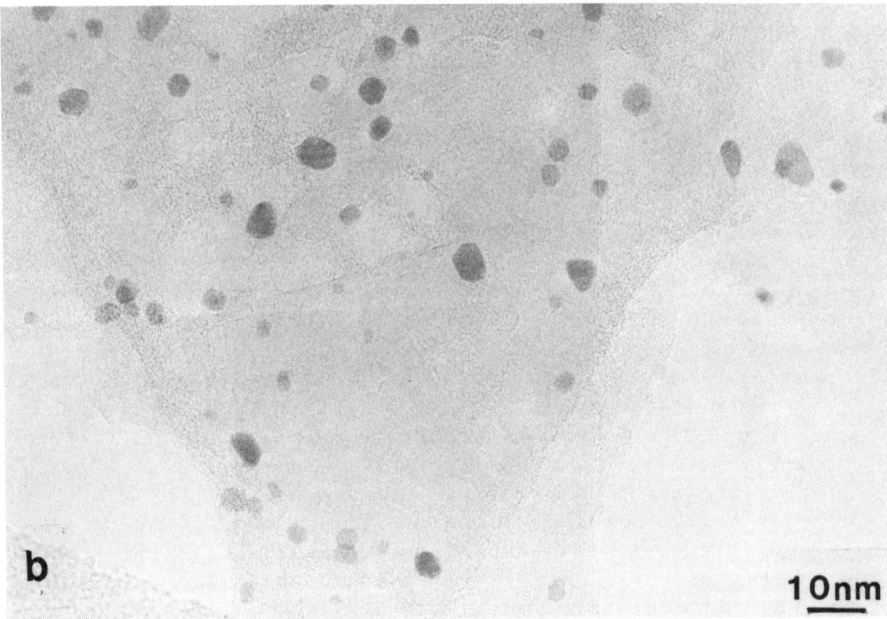

Figure 1.5   TEM views (a) through a section of Pt/Y zeolite catalyst and (b) through Pt/G_{HT} catalyst.

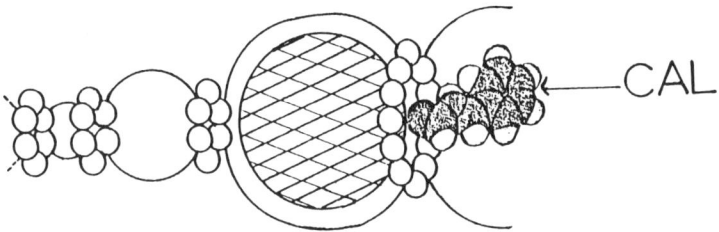

Figure 1.6 The CAL molecule can only adsorb end-on, that is, via the C=O group.

The higher selectivities of platinum and rhodium encaged in zeolite can be attributed to molecular constraint effects (shape selectivity) experienced by the CAL molecule in the zeolite micropores. As illustrated in Figure 1.6 the CAL molecule can only adsorb end-on (i.e., via the C=O group) on a metal cluster located in a contiguous supercage.

The hydrogenation of the C=O bond is highly favored with respect to the C=C bond, which cannot approach the metal. Ideally, the selectivity should be 100%; it is smaller because (1) a small fraction of the metal is always present on the external surface of the zeolite and (2) metal clusters can be much smaller than the cage, which means that there is enough space for the CAL molecule to enter the cage and to adsorb laterally on the metal (i.e., via the C=C bond). Indeed in another platinum zeolite prepared with larger particles but still encapsulated in the zeolite, the selectivity is even higher (sample Pt/Y*, Table 1.1). Preliminary results on the selectivity of metal zeolite catalysts have been published [12].

C. Effect of the Morphology of Metal Particles

The catalyst Pt/G_{HT} was prepared by heating Pt/G_{ex} (described in Section A) at 773 K under hydrogen, then at 1173 K under vacuum. During this treatment, the small particles present in Pt/G_{ex} sinter into large particles (3-6 nm) that have a faceted outline on TEM photomicrographs (Fig. 1.5b). Similarly, the Rh/G_{HT} catalyst obtained by thermal sintering of a rhodium-exchanged graphite Rh/G_{ex} exhibits 3-9 nm faceted particles. A Pt-Adams catalyst obtained from a commercial source was reduced in the liquid phase under H_2 pressure either at 333 K ($PtAd_1$) or at 373 K ($PtAd_2$) prior to catalytic hydrogenation. TEM photomicrographs show that in $PtAd_1$ platinum is in the form of agglomerates of 2-5 nm crystallites (Fig. 1.7a), whereas 20-200 nm faceted particles are present in $PtAd_2$ (Fig. 1.7b).

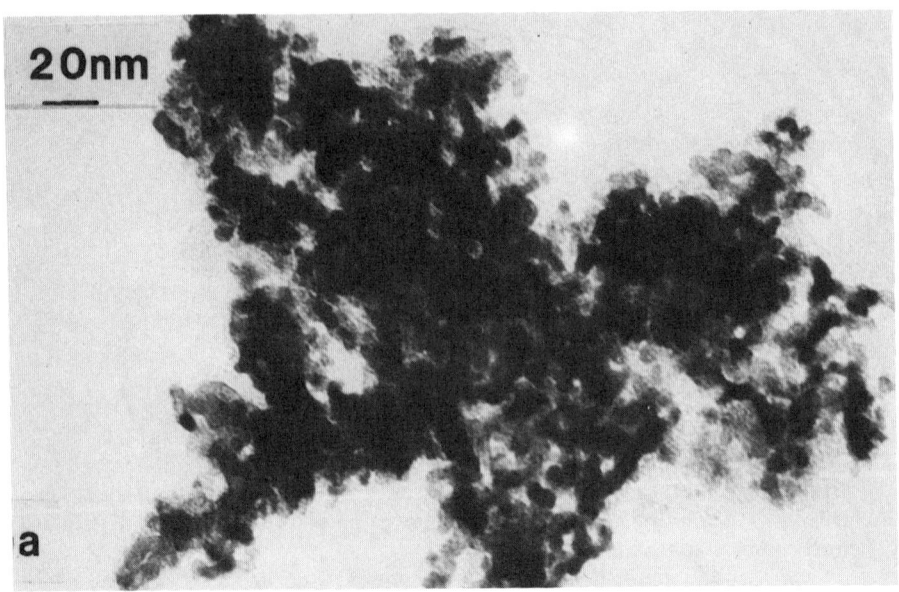

Figure 1.7   TEM views of (a) PtAd$_1$ catalyst (333 K reduction) and (b) PtAd$_2$ catalyst (373 K reduction).

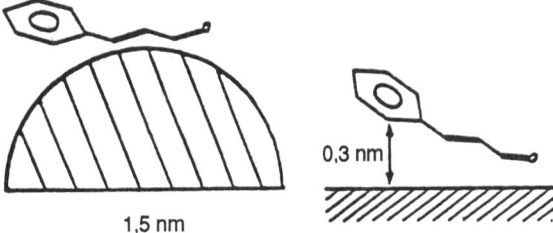

Figure 1.8 The phenyl group lies beside the metal surface.

The selectivities at different conversions are given in Table 1.1. For the same graphite support, the selectivities are much higher on the faceted particles of Rh/G_{HT} and Pt/G_{HT} than on the small particles of Rh/G_{ex} and Pt/G_{ex}. The selectivities are also much higher on $PtAd_2$ than on $PtAd_1$. Clearly the different selectivities can be correlated to differences in particle morphology, more specifically to the presence in the selective catalysts of metal particles with extended crystal faces (at least compared to the dimension of the CAL molecule). This can be attributed to the different steric constraints experienced by the CAL molecule when it adsorbs on a flat or on a curved (stepped) metal surface. As the planar molecule approaches a flat surface, the phenyl group hampers a parallel landing because there is an energy barrier that exerts a repulsive effect [10]. Therefore, the adsorbed CAL molecule should be tilted as illustrated in Figure 1.8.

Simple molecular models show that there is no steric hindrance for the adsorption of the carbonyl group, which can be activated and hydrogenated more easily than the C=C bond, thus accounting for the high selectivities of Pt/G_{HT} and $PdAd_2$ catalysts. In contrast, there is no steric constraint for the approach and adsorption of both the C=C and C=O bonds on particles smaller than 2-3 nm because the phenyl group lies beside the metal surface, as suggested in Figure 1.8. Accordingly, the rate of HCAL formation is 10 times faster on Rh/G_{ex} than on Rh/G_{HT}.

D. Synergetic Effects Induced by a Second Metal

Addition of a second metal to a monometallic catalyst suspended in the reaction medium was the usual way to improve the selectivity to unsaturated alcohol before the recent advent of bimetallic catalysts [13-15]. The purpose of this study, reported previously [16], was to reproduce well-known methods of metal salt addition and to analyze the composition of the catalyst after catalytic reaction in order to learn more about the mechanism favoring the selectivity. Catalyst Pt/C described in Section A was suspended in isopropanol at 373 K in a well-stirred autoclave pressurized under 40 bars of hydrogen. Required amounts of $FeCl_2$ in water solution were introduced under H_2 pressure, and the mixture was stirred for 2 hours. The temperature was lowered to 333 K and 0.1 mol of cinnamaldehyde was transferred under H_2 pressure from a reservoir to the autoclave; the reaction was then conducted at 333 K under 40 bars of hydrogen. Figure 1.9a is a TEM view taken through an ultramicrotome section of the Pt-Fe/C catalyst (Fe/Pt = 0.2:0.8) after CAL hydrogenation. The metal particles are still homogeneously distributed throughout the charcoal grain, their sizes are slightly larger than in the starting Pt/C catalyst (see Section A). The composition of individual metal particles was measured by EDX analysis with a FEG-STEM. Figure 1.9b gives the EDX spectrum of a 2 nm particle. Quantitative analysis shows that the Fe/Pt ratio is 0.15. In a given charcoal grain, the Fe/Pt ratios vary in a small range; that is, the composition is homogeneous, but in different grains the ratios are found in the range 0.1-0.2. The average Fe/Pt ratio is close to 0.12 (i.e., smaller than the chemical composition), which means that part of the iron is not associated with platinum. Accordingly a few, large iron particles were detected in some catalyst grains. Still, this study shows that by adding $FeCl_2$ to the reaction medium, bimetallic particles in which iron is associated with platinum are formed. Thus, in the present treatment conditions, the Fe^{2+} ions in solution are reduced by hydrogen atoms adsorbed on platinum and the iron atoms deposit on the platinum surface.

The initial rates of hydrogenation and the initial selectivities to COL are given in Figure 1.10a and 1.10b, respectively, as a function of the chemical composition. Both the rate and selectivity curves are volcano-shaped, with a maximum close to Fe/Pt = 0.2. Similar curves were obtained previously [13] on Pt-Fe/active charcoal catalysts prepared by coimpregnation and high temperature H_2 reduction.

Clearly, the presence of iron atoms on the surface of platinum enhances selectivity to an unsaturated alcohol. Because iron is more electropositive than platinum, the iron adatoms can act in two ways to improve the selectivity:

1. The electron transfer from iron to platinum increases the electron density on the latter, which should result in a lower probability for the activation of the C=C, bond as discussed in Section A. However this effect does not account for the enhancement of the rate of C=O bond hydrogenation.

2. The electron-deficient iron atoms act as Lewis adsorption sites for the CAL molecule. As suggested by Figure 1.11, the C=O

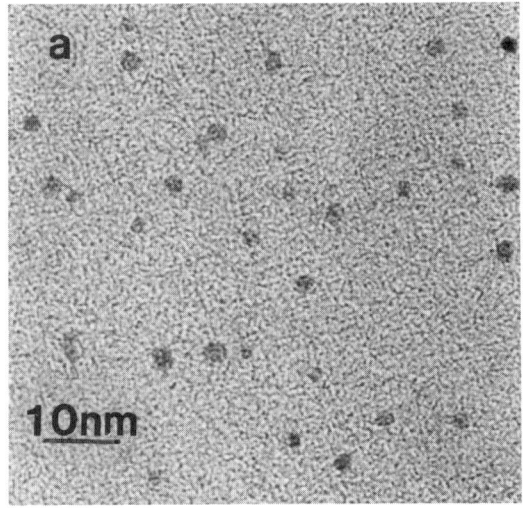

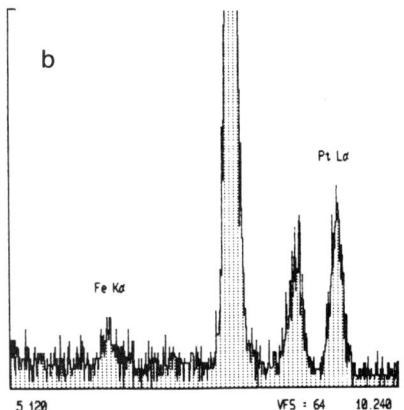

Figure 1.9 (a) TEM view through the Pt-Fe/C catalyst. (b) EDX
spectrum of a 2 nm large particle.

group adsorbs via donation of a lone pair of electrons from
the oxygen atom.

The polarization of the C=O bond favors a nucleophilic attack
on the carbon atom by hydrogen associated on neighboring Pt atoms.
This dual-site mechanism accounts for the volcano-shaped activity
curve. Indeed, the activity for C=O bond hydrogenation (and thus
the selectivity to COL) increases as more and more iron adatoms
cover the platinum surface until an optimal surface composition is

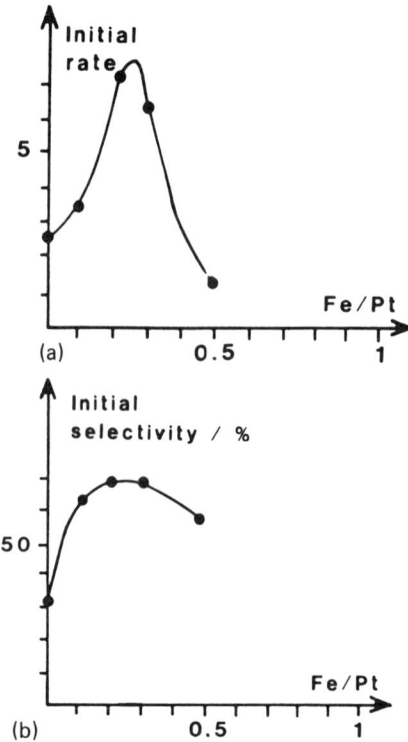

Figure 1.10 (a) Initial rates of CAL hydrogenation versus the composition of bimetallic particles. (b) Initial selectivities to COL.

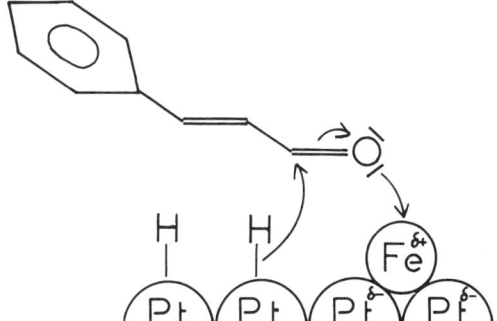

Figure 1.11   The C=O group adsorbs via donation of a lone pair
of electrons from the oxygen atom.

attained. As iron atoms cover too many platinum atoms, the activity
decreases, and the catalyst eventually becomes inactive for a monolayer coverage.

IV.   CONCLUSION

This study shows that in cinnamaldehyde hydrogenation, the selectivity to unsaturated alcohol can be controlled by choosing an appropriate composition and structure of the catalysts. Four different
factors have been highlighted.

1.   Electronic or ligand effects improving the selectivity by increas-
     ing the electronic density on platinum or other Group VIII
     metals, thereby decreasing the rate of C=C bond hydrogenation.
     Thus graphite acting as an electron-donating macroligand toward
     very small metal particles produces an effect that could be
     similar to the addition of Lewis bases.
2.   Geometric effects involving both the morphology of the metal
     surface and the configuration of the cinnamaldehyde molecule.
     Thus, it has been shown that on a flat metal surface, the
     presence of phenyl groups imposes a tilt of the molecule favor-
     ing adsorption via the carbonyl group.
3.   Geometric effects combining the structure of the support, the
     location of metal particle in the support, and the configuration
     of the organic substrate. Thus the cinnamaldehyde molecules
     in the micropores of Y-type zeolites containing encapsulated
     metal particles undergo steric constraints that impose a tip-on
     adsorption on the metal favoring the hydrogenation of the C=O

bond. This is a very interesting inorganic supramolecular catalytic system, akin to enzyme catalysts.

4. Synergetic effects due to a dual-site mechanism on bimetallic
   catalysts, where noble metal and electropositive metal atoms
   are associated in the same particles. The latter act as adsorption
   sites for the molecule via the oxygen atom of the carbonyl group,
   and the former provide dissociated hydrogen. For an optimal
   composition of the metal surface, the rate enhancement, and
   therefore the selectivity to COL, reach a maximum.

There are probably other effects, not considered in this investigation, which could modify the selectivity to unsaturated alcohol.
Also other factors, such as the nature of the solvent and the reaction
conditions, are obviously very important for selectivity control. The
best results will be obtained with a tailored catalyst used in suitable
conditions.

REFERENCES

1. Rylander, P. N., *Catalytic Hydrogenation in Organic Syntheses*,
   Academic Press, New York, 1979.
2. Tuley, W. F., and Adams, R., *J. Am. Chem. Soc.*, 47, 3061
   (1925).
3. U.K. Patent 1,123,837 (1966).
4. Bakhanova, E. N., Astakhova, A. S., and Khiedekel, M. L.,
   USSR Patent 264,352 (1972).
5. Pascoe, W. E., and Stenberg, J. F., in *Catalysis in Organic
   Syntheses* (W. H. Jones, ed.), Academic Press, New York,
   1979.
6. Richard, D., and Gallezot, P., in *Preparation of Catalysts*,
   Vol. IV (B. Delmon, P. Grange, P. A. Jacobs, and G. Poncelet,
   eds.), Elsevier, Amsterdam, 1987, p. 71.
7. Richard, D., Fouilloux, P., and Gallezot, P., *Proceedings of
   the 8th International Congress on Catalysis*, Verlag Chemie,
   Weinheim, 1984, p. 659.
8. Gallezot, P., Richard, D., and Bergeret, G., in *New Catalytic
   Materials* (E. T. K. Baker, ed.), ACS Symposium Series,
   American Chemical Society, Washington, DC, 437, 1990, p. 150.
9. Richard, D., Gallezot, P., Neibecker, D., and Tkatchenko, I.,
   Catal. Today, 6, 171 (1989).
10. Minot, C., and Gallezot, P., *J. Catal.*, 123, 341 (1990).
11. Gallezot, P., in *Proceedings of the 6th International Zeolite
    Conference* (D. Olson and A. Bisio, eds.), Butterworths, Guild-
    ford, 1984, p. 352, and references therein.

12. Gallezot, P., Giroir-Fendler, A., and Richard, D., *Catal. Lett.*, accepted for publication.
13. Goupil, D., Fouilloux, P., and Maurel, R., *React. Kinet. Catal. Lett.*, *35*, 185 (1987).
14. Poltarzewski, Z., Galvagno, S., Pietropaolo, R., and Staiti, P., *J. Catal.*, *102*, 190 (1986).
15. Galvagno, S., Poltarewski, Z., Donato, A., Neri, G., and Pietropaolo, R., *J. Chem. Soc.*, *Chem. Commun.*, 1729 (1986).
16. Richard, D., Ockelford, J., Giroir-Fendler, A., and Gallezot, P., *Catal. Lett.*, *3*, 53 (1989).

2

n-Butyraldehyde Hydrogenation over Raney Mo-Ni Catalyst in Vapor Phase

You-Jyh Lin

Washington Research Center, W. R. Grace & Company, Columbia, Maryland

I. INTRODUCTION

The production of alcohols by hydrogenation of aldehydes has long been practiced. The common commercial catalysts for this reaction are copper chromite, copper and zinc oxides, and supported nickel and/or cobalt catalysts with one or more promoters [1,2]. Although Raney catalysts have been well known as hydrogenation catalysts since the first patent was issued in 1925 [3], these reagents have not been widely used in aldehyde hydrogenation. A few studies have been performed on the hydrogenation of aldehydes using Raney catalysts in the liquid phase [4-6]. However, only limited information concerning this reaction in the vapor phase is available from the literature [7]. It is, therefore, interesting to investigate the performance of Raney catalysts for the hydrogenation of aldehydes in the vapor phase.

We chose *n*-butyraldehyde as our model compound on the basis of its commercial interest. To simulate poisoning by Raney catalysts, it was assumed that the main impurities found in the *n*-butyraldehyde feed are *n*-butyric acid, water, and triphenylphosphine (TPP). TPP is commonly used in the upstream hydroformylation process to increase catalytic selectivity toward *n*-butyraldehyde in processes using phosphine-modified catalysts [8].

This chapter presents preliminary results of the performance evaluation of Raney Mo-Ni catalyst, including effects of contaminants, the optimal temperature for maximizing the yield, and the minimization of the formation of dibutyl ether, a by-product that is very difficult to remove downstream because of the formation of an azeotrope with *n*-butanol.

II. EXPERIMENTAL

A. Catalyst

The molybdenum-promoted nickel Raney catalyst (Raney Mo-Ni) used here is commercially available and was obtained from the Davison Chemical Division of W. R. Grace & Company—Conn. A 6-8 mesh particle size Raney Mo-Ni catalyst with 28% aluminum extracted was evaluated. The Brunauer-Emmett-Teller (BET) surface area was 33 m^2/g and the density was 3.90 g/cm^3 [9].

B. Chemicals

Hydrogen (99.95% purity) was purchased from Air Products. No gas pretreatment was used on hydrogen for this bench scale experiment. n-Butyraldehyde (BAL) was obtained from Aldrich and was distilled before use. A vacuum-jacketed, 10-plate distillation column was used. The distillation was conducted at atmospheric pressure under nitrogen with a 10:1 reflux ratio. The typical distilled n-butyraldehyde contained 0.4 wt % n-butyric acid, 0.1-0.2 wt % water, and trace amounts of isobutyraldehyde and ethyl hexanal.

The n-butyric acid and triphenylphosphine (TPP) were used as received from Aldrich. It has been found that TPP is not stable in n-butyraldehyde. It is suspected that TPP can be easily oxidized to triphenylphosphine oxide (TPPO) very rapidly. Even when n-butyraldehyde is carefully handled under nitrogen, a trace amount of oxygen may still exist in n-butyraldehyde. For simplicity, therefore, the term "TPP" is used for both TPP and TPPO in this chapter.

Three n-butyraldehyde solutions containing varying levels of contaminants were studied. Table 2.1 shows the compositions of feeds used in these experiments.

Table 2.1   Contaminants in the n-Butyraldehyde Feeds

Code for feed	n-Butyric acid (wt %)	Water (wt %)	Triphenylphosphine (ppm)
WRC-1[a]	0.4	0.1	0
WRC-2	3.8-5.8	3.0	0
WRC-3	2.5-3.1	2.6-3.1	500

[a]Distilled n-butyraldehyde, considered to be pure n-butyraldehyde.

C. Bench Scale Pilot Plant

Figure 2.1 is a schematic diagram of the pilot plant. Hydrogen
was regulated by a Brooks 5850 series mass flow controller at 3700
standard cubic centimeters per minute (SCCM), while the liquid
feed was regulated at 0.56 g/min by a Eldex Laboratory A-30-S
high pressure precision pump. The molar ratio of hydrogen to aldehyde was 20:1. The gas-liquid mixture was vaporized in a 10 ft
long coil vaporizer before entering the reactor. The product flowed
through a gas-liquid separator, and the liquids were collected in

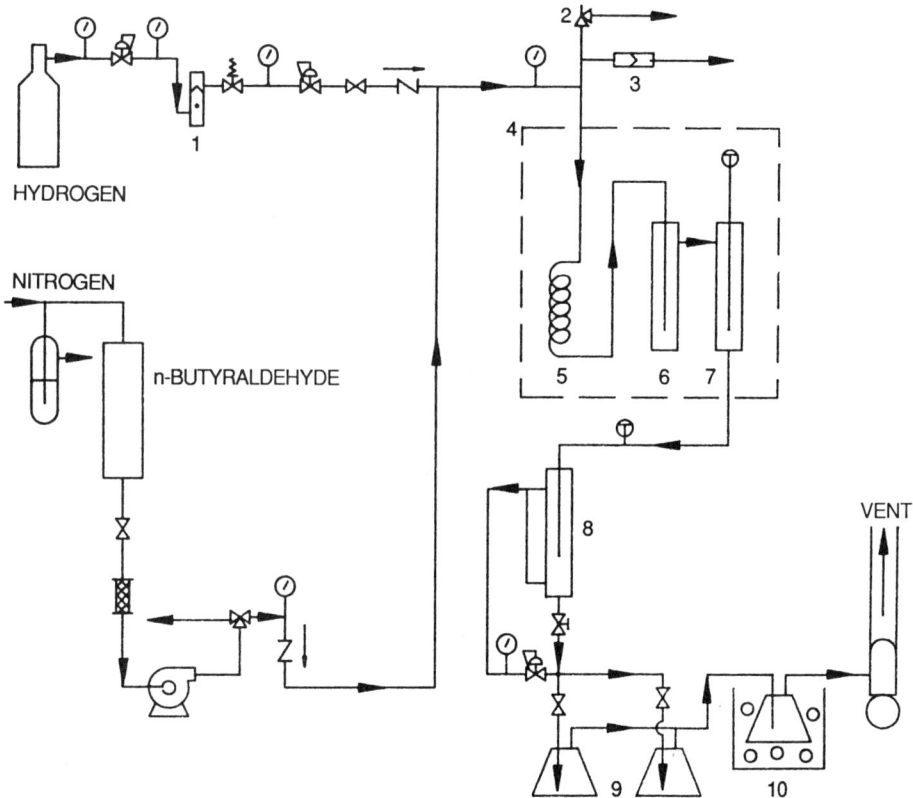

Figure 2.1 Schematic diagram of the system: 1, excess flow valve;
2, pressure relief valve; 3, rupture disk; 4, sandbath; 5, coil
vaporizer; 6, trap; 7, reactor with thermowell; 8, gas-liquid separator with sight glass; 9, liquid collector; and 10, dry ice trap.

a flask. A dry-ice trap was used to collect any residual liquids before the gas was vented.

The reactor was 0.5 in. o.d. (0.45 in. i.d.) and 1 ft long with a thermowell. The temperature was controlled by a Tecam fluidized sandbath, model SBL-2D, equipped with a Tecam TC4D temperature controller. During the reaction, the temperature profile of the reactor was obtained by manually moving a thermocouple inside the thermowell. The reactor was packed with about 35 g of Raney Mo-Ni catalyst. The catalyst was then dried in situ under hydrogen at 120°C overnight before the experiment was started.

The reaction pressure was always 60 psig, while the hot spot temperature (LT/HT) was varied from 90°C (LT) to 185°C (HT) by varying the sandbath temperature from 80 to 163°C. Vapor pressures of all liquid materials were checked to ensure that all liquids would be vaporized in the vaporizer at both sandbath temperatures. Liquid products were analyzed by gas chromatography using J&W Scientific DB-5 capillary columns. Gas effluents were collected in a gas collecting tube with a sampling port, then analyzed by gas chromatography using a Porapak Q column.

III. RESULTS AND DISCUSSION

A. Performance of Raney Mo-Ni with Pure *n*-Butyraldehyde

Using WRC-1 feed, the conversion of *n*-butyraldehyde was 95-96% when the sandbath was maintained at 80°C (LT). The results are shown in Table 2.2. The hot spot temperature was 90°C. Analysis

Table 2.2 Activity Results with Pure Feed

Code for feed	Temperature (°C) of hot spot/sandbath	Conversion (%)	Selectivity (%)	Activity[a] (deactivity)
WRC-1 LT	90/80	95-96	>99.5	Active after 118 hours
WRC-1 HT	185/163	>99.5	95	Active after 138 hours

[a]"Active" indicates that the activity of the catalyst was maintained at the end of the test; that is, the conversion of *n*-butyraldehyde and the hot spot temperature did not decrease, and the hot spot did not move toward the exit of the reactor.

of the product shows that only a trace amount of n-butyl butyrate was formed as a major by-product, and no gaseous by-products formed at this temperature. It is believed that trace amounts of isobutanol and ethyl hexanol found in the products were from the corresponding aldehydes in the feed.

As shown in Table 2.2, the activity of the catalyst was still maintained at the end of the test (118 h); that is, the conversion of n-butyraldehyde and the hot spot temperature did not decrease, and the hot spot did not move toward the exit of the reactor. The conversion was 95-96%, while the selectivity was higher than 99.5%.

The temperature of the hot spot was about 22°C above the sandbath temperature in the HT run shown in Table 2.2. The conversion was higher than 99.5%, while the selectivity was only 95%. The low selectivity was found to be due to the formation of water and paraffins, namely methane (1.5 wt %), propane (4.8 wt %), n-butane (0.1 wt %), and a trace amount of ethane. Similar results have been reported by Wojcik and Adkins: hydrogenolysis of primary alcohols occurred and formed the corresponding hydrocarbons and methane over a nickel catalyst [10]. However, the formation of paraffins is also possible due to the hydrogenolysis of n-butyraldehyde.

Figure 2.2 shows the temperature profile at the center of the reactor for WRC-1 at HT. A similar temperature profile was also

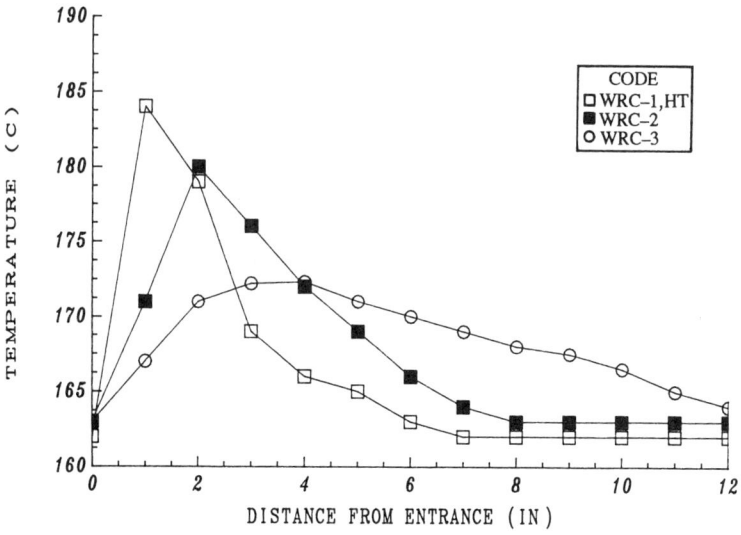

Figure 2.2   Temperature profiles in the reactor.

observed for WRC-1 at LT except the temperature rise was smaller. The sharp peak in the temperature profile, located at the top of the reactor, indicates that only a small portion of the catalyst was used for the reaction. These results suggest that Raney Mo-Ni catalyst is very active for the hydrogenation of n-butyraldehyde. The activity of Raney Mo-Ni was maintained when pure n-butyraldehyde was used, whatever the sandbath temperature was.

B. Optimal Reaction Temperature

As described in Section A, 95% yield of n-butyraldehyde was obtained from both sets of conditions given above. Therefore, a series of optimization experiments was conducted to maximize the yield. The results are shown in Table 2.3. The concentration of dibutyl ether (DBE) in the liquid effluents is also shown in Table 2.3. The concentrations of paraffins in the gas effluents were determined by gas chromatography. The optimal hot spot temperature under these conditions is in the range of 120-145°C. The optimal temperature under plant conditions will depend on the configuration of the reactor, the throughput, etc.

As shown in Table 2.3, the concentration of dibutyl ether was found to increase with temperature up to a maximum, at which point the concentration decreased with temperature. This phenomenon

Table 2.3 Optimization Results[a]

Temperature (°C) for hot spot/sandbath	Conversion (%)	Dibutyl ether (ppm)	Paraffins[b]
90/80	96.0	70	Not found
96/90	97.7	—	Not found
120/110	99.2	106	Not found
135/120	99.6	84	Trace (CH_4 and C_3H_8)
145/130	99.8	78	Trace (CH_4 and C_3H_8)
152/140	99.8	59	0.04% CH_4, 0.05% C_3H_8
180/163	99.9	48	0.39% CH_4, 0.44% C_3H_8, trace of C_2H_6 and C_4H_{10}

[a]The liquid flow rate was 0.5 g/min, while the hydrogen flow rate was 3700 (standard) cm^3/min (SCCM).
[b]"Trace" means that concentration of individual paraffins was less than 1000 ppm.

was also observed when kinetic experiments on the formation of DBE were conducted. The phenomenon of reaction rates passing through a maximum was also observed for benzene hydrogenation on different catalysts [11,12].

Briefly, the coverage (and adsorption equilibrium constants) of DBE precursor on the catalyst surface decreases as the temperature increases, which will then result in a lower reaction rate. Above a certain temperature, this negative effect (decreasing coverages) on the reaction rate outweighs the positive temperature effect on the rate constants and the overall reaction rate decreases.

C. Raney Mo-Ni with Contaminated *n*-Butyraldehyde

Raney Mo-Ni catalyst using feed WRC-2 (doped with *n*-butyric acid and water) showed performance similar to that of catalyst used with feed WRC-1 (pure), except that the observed temperature difference between the hot spot and the sandbath was smaller for WRC-2 (17°C, see Table 2.4). This indicated that the catalyst used with contaminated *n*-butyraldehyde (WRC-2) was not as active as when used with pure feed (WRC-1). This result is also supported by the fact that the temperature profile was broader (Fig. 2.2), indicating that more catalyst was needed to achieve high (> 99.5%) conversion. The activity of the catalyst was, however, maintained after 121 hours of reaction time.

These results suggest that *n*-butyric acid and water can be reversibly adsorbed on the surface of Raney Mo-Ni catalyst. *n*-Butyric acid and water are not poisons; rather, they compete with

Table 2.4 Activity Results with Contaminated Feed

Code for feed	Temperature (°C) of hot spot/sandbath	Conversion (%)	Selectivity (%)	Activity[a] (deactivity)
WRC-2	180/163	>99.5	—	Active after 121 hours
WRC-3	170-175/163	93-100	>95	Deactivating[b]

[a]"Active" indicates that the activity of the catalyst was maintained at the end of the test; that is, the conversion of *n*-butyraldehyde and the hot spot temperature did not decrease, and the hot spot did not move toward the exit of the reactor.
[b]93 mol % conversion after 73.5 hours.

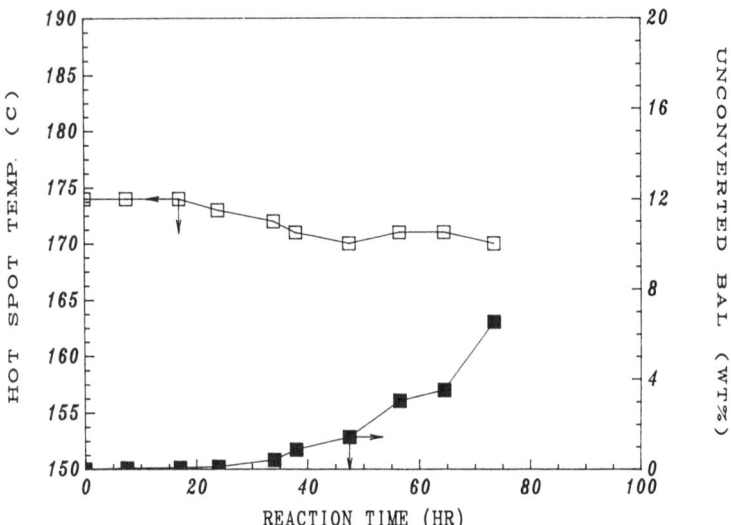

Figure 2.3   Performance of molybdenum-promoted Raney nickel
catalyst at WRC-3 conditions.

hydrogen and *n*-butyraldehyde for the active sites, resulting in
decreased catalyst activity.

When TPP was introduced into the feed (WRC-3), the activity
of Raney Mo-Ni catalyst decreased with reaction time (Table 2.4).
Figure 2.3 shows the relationship between hot spot temperature
and time as well as between unconverted *n*-butyraldehyde (BAL)
and time. The temperature profile taken after a reaction time of
30 hours with WRC-3 is also shown in Figure 2.2. The position
of the hot spot was observed to be slowly moving toward the exit
of the reactor, while the hot spot temperature was decreasing.
These results suggest that Raney Mo-Ni catalyst was deactivating
as a result of the presence of TPP (either TPP alone or from the
synergistic effect of TPP, *n*-butyric acid, and water).

In another experiment, it was found that deactivated catalyst
could be partially regenerated under flowing nitrogen at 163°C.
This suggests that the adsorption of TPP on the surface of Raney
Mo-Ni catalyst is partially reversible. This regeneration implies
that TPP was weakly adsorbed on the surface of the catalyst. This
implication is also supported by the presence in the liquid effluent
of 40-80% of the TPP introduced in WRC-3.

Table 2.5 Minimization of Dibutyl Ether Formation

Flow rates				
BAL Butyraldehyde (g/min)	H_2 (sccm)	Conversion (%)	Dibutyl ether (ppm)	Paraffins[a]
0.5	3700	99	106	Not found
0.8	3700	98	111	Trace
0.8	4900	97	36	Trace
0.25	2200	99.9	44	Trace
0.25	3200	99.6	42	Trace
0.25	4200	99.4	33	Trace
0.25	5200	99.1	29	Trace

[a]"Trace" means that concentrations of methane and propane were less than 1000 ppm.

D. Minimization of Dibutyl Ether Formation

Table 2.5 shows results of the experiments conducted to minimize the formation of dibutyl ether. The reactions were run at 110°C (sandbath temperature) and 60 psig.

As shown in Table 2.5, the formation of dibutyl ether was decreased by increasing the hydrogen flow rate or both hydrogen and *n*-butyraldehyde. These results are confirmed by the kinetic experiments on *n*-butanol dehydration, which suggest that the formation of dibutyl ether from *n*-butanol decreases when the partial pressure of hydrogen increases [13]. The temperature effect due to a change in molar ratio of hydrogen to *n*-butyraldehyde (conversions of *n*-butyraldehyde were also changed) is considered to be small but not negligible.

IV. CONCLUSIONS

Raney Mo-Ni is very active as a catalyst for the hydrogenation of *n*-butyraldehyde. The hydrogenolysis of *n*-butanol (or *n*-butyraldehyde) occurred at higher temperatures and resulted in a lower yield. The yield, however, can be maximized by carefully controlling the reaction temperature.

n-Butyric acid and water were not found to be poisons to the catalyst. Both may, however, compete with *n*-butyraldehyde and hydrogen for the active sites, resulting in lower catalyst activity. The deactivation of the catalyst is due to the presence of triphenyl-phosphine. Also, the catalyst can be partially regenerated under nitrogen at 163°C.

The formation of dibutyl ether is dependent on both temperature and partial pressure of hydrogen. With Raney Mo-Ni catalyst, the formation of dibutyl ether can be minimized and the throughput increased by increasing flow rates of *n*-butyraldehyde and hydrogen.

ACKNOWLEDGMENT

The author thanks W. R. Grace & Company—Conn. for permission to publish these results. Special thanks to D. J. Schuler and T. P. Bresnan for performing the experimental program.

REFERENCES

1. Oxo Alcohols, in *Process Economics Program Report No. 21C*, SRI International, Stanford, CA, 1986, pp. 214-217.
2. Logsdon, J. E., Loke, R. A., Merriam, J. S., and Voight, R. W., U.S. patent 4,762,817 (Aug. 9, 1988).
3. Raney, M., U.S. patent 1,563,587 (Dec. 1, 1925).
4. Fewlass, M. W., and Wilson, T. M. B., U.K. patent 1,182,797, to B.P. Chemicals (U.K.) Ltd. (March 4, 1970).
5. Anikeev, I. K., and Valitov, A. Kh., *Kinet. Katal.*, *15*, 520 (1974).
6. Anikeev, I. K., Valitov, N. Kh., and Panchenkov, G. M., *Kinet. Katal.*, *16*, 544 (1975).
7. Kubicka, R., Huml, M., Bilek, O., and Kozuch, J., Czecho-slovak patent 207,926 (Aug. 2, 1983).
8. Oxo Alcohols, in *Process Economics Program Report No. 21C*, SRI International, Stanford, CA, 1986, p. 46.
9. Data were provided by the Davison Chemical Division of W. R. Grace & Co.
10. Wojcik, B., and Adkins, H., *J. Am. Chem. Soc.*, *55*, 1293 (1933).
11. Yoon, K. J., and Vannice, M. A., *J. Catal.*, *82*, 457 (1983).
12. Lin, Y.-J., Resasco, D. E., and Haller, G. L., *J. Chem. Soc., Faraday Trans. I*, *83*, 2091 (1987).
13. Lin, Y.-J., in preparation.

3

Hydride-Mediated Homogeneous Catalysis: Chemoselective Catalytic Hydride Reductions via Heterolytic Hydrogen Activation

Jeffrey M. Stryker, Wayne S. Mahoney, John F. Daeuble, and Donna M. Brestensky

Department of Chemistry, Indiana University, Bloomington, Indiana

I. INTRODUCTION

Catalytic hydrogenation is essential methodology for the reduction of unsaturated organic molecules. The use of molecular hydrogen as a reductant, in either heterogeneous or homogeneous catalytic reactions, offers reliable and relatively inexpensive procedures equally amenable to laboratory synthesis and large-scale processes. By virtue of mechanistic features inherent to most catalytic hydrogenations, nonpolar unsaturation is typically reduced preferentially over polar functionality [1], thus enabling, for example, selective reduction of carbon-carbon double bonds in the presence of carbonyl groups. While a few homogeneous catalysts have been reported that show reversed chemoselectivity for a limited class of substrates [2], no general catalytic methodology rigorously selective for carbonyl over olefin reduction has been reported.

By contrast, hydride reduction methodology is generally chemoselective for reduction of polar unsaturation [3]. Unfortunately, such reductions are also stoichiometric in hydride, suffering from the practical and economic limitations inherent in such reactions. Our research objective was thus to engineer, by rational design, a class of hydride reductions *catalytic in hydride and rigorously chemoselective for the reduction of polar unsaturated functionality*, combining a catalyst of true hydridic character with the use of hydrogen as the ultimate reductant. Conceptually, our approach to this problem was relatively straightforward: select or create a metal hydride complex sufficiently hydridic to transfer the hydride to the organic substrate and *require* that the resultant metal alkoxide complex react with hydrogen exclusively via a *heterolytic* mechanism, precluding formation of an intermediate metal dihydride complex (eq. 3.1, illustrated for ketone reduction). In this scheme, the hydrogen functions essentially as an acid, simultaneously "protonating" the alkoxide and regenerating the initial hydride complex [4]. This

$$(3.1)$$

mechanistic imposition both reinforces the inherently polar character of the desired transformation and, by restricting the metal's potential for $M^n \rightarrow M^{n+2}$ oxidative addition common to typical hydrogenation mechanisms, minimizes the probability that the complex will also function as a hydrogenation catalyst for olefins.

Heterolytic hydrogen activation is energetically accessible and has been established in a variety of contexts [5,6]. Most pertinent, however, is the facile hydrogenolysis of $[CuOt-Bu]_4$ in the presence of phosphines reported by Caulton [7], and illustrated for the case of Ph_3P (eq. 3.2) [7a]. This apparent heterolytic hydrogen

$$1/4 \ [CuO^tBu]_4 \ + \ xs \ Ph_3P \ \xrightarrow[C_6H_6, \ RT]{1 \ atm \ H_2} \ 1/6 \ [(Ph_3P)CuH]_6 \ + \ HO^tBu$$

$$(3.2)$$

activation of the copper-oxygen bond [8] proceeds at atmospheric pressure and room temperature, producing the known thermally stable copper(I) hydride complex $[(Ph_3P)CuH]_6$ and *tert*-butanol [9]. Moreover, while Cu(0) is a reasonable hydrogenation catalyst [2c], Cu(I) is not, an observation readily rationalized by the general inaccessibility of the Cu(III) oxidation state. Our own investigation of the organic chemistry of this mildly hydridic complex has revealed a rich and synthetically valuable body of reductive transformations [10], in particular, highly regioselective conjugate reduction of α,β-unsaturated carbonyl compounds [10a,d]. Based on the assumption that conjugate reduction proceeds via hydride addition, giving an intermediate copper(I) enolate complex [11] analogous to the alkoxide complex in equation (3.2), a catalytic cycle for conjugate reduction was envisioned (Fig. 3.1, $N = 6$). Based on the inertness of $[(Ph_3P)CuH]_6$ toward ketone functionality, only catalytic conjugate reduction was anticipated.

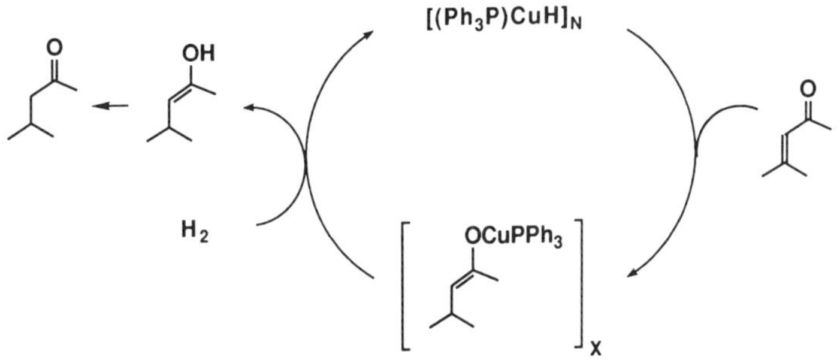

Figure 3.1

II. RESULTS AND DISCUSSION

A. Catalytic Hydride-Mediated Conjugate Reduction

1. Initial Observations [12]

Reaction of 2-cyclohexenone with a catalytic amount of $[(Ph_3P)CuH]_6$ (or $[CuO^tBu]_4/Ph_3P$) under hydrogen ($\geq$ 80 psi) in benzene at room temperature gives slow turnover to cyclohexanone as the exclusive organic product (Fig. 3.2). Faster conversion is obtained under higher hydrogen pressure (200-1000 psi), but complete reduction to cyclohexanol is observed at longer reaction times. No reduction of cyclohexanone by $[(Ph_3P)CuH]_6$ is observed, even at high hydrogen pressure and prolonged reaction time. Under identical conditions, however, the reaction of cyclohexanone containing a small amount of cyclohexenone proceeds with complete consumption of both substrates, leading to the conclusion that although the hydride hexamer is used to initiate the catalytic reduction, more reactive, coordinatively unsaturated, copper hydride fragments are produced by the hydrogenolysis of the intermediate copper enolate complex (Fig. 3.1, $N \neq 6$). Reaggregation of these smaller oligomers is apparently slow relative to reduction of substrate or decomposition: no $[(Ph_3P)CuH]_6$ is recovered upon depletion of the substrate, and the crude reaction mixture is heterogeneous, with the bulk of copper-containing residues recovered as a black precipitate. In addition, under these conditions, the catalytic reduction of unsaturated ketones could not be generalized (Table 3.1), a result attributed to competitive decomposition of the postulated copper(I) enolate and alkoxide intermediates.

2. Catalysis with Added Phosphine [12]

The instability of the highly electron rich copper(I) oxygenates can be attributed primarily to the kinetic lability of the donor

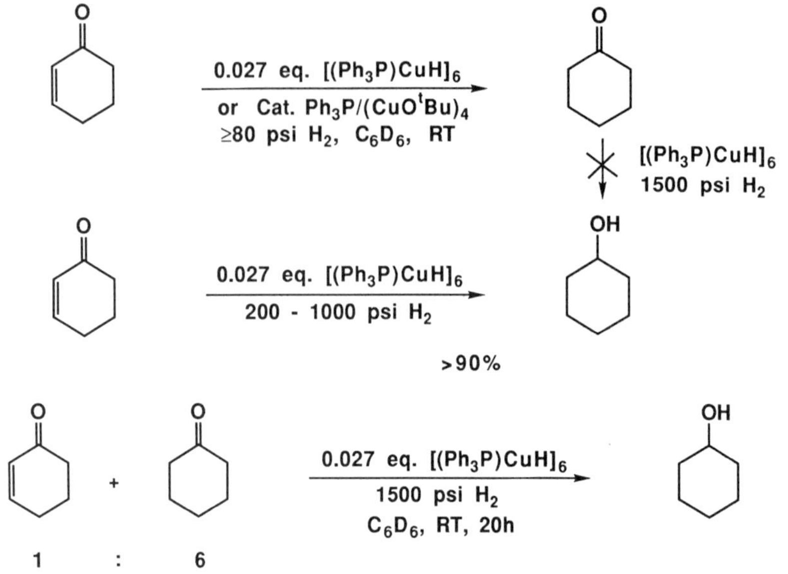

Figure 3.2

phosphine ligand. To stabilize these intermediate complexes, excess phosphine was added to the reaction conditions in an attempt to inhibit destructive dissociation. Several significant effects were observed (Table 3.1). Catalytic reaction mixtures now maintained homogeneity [13], and after complete conversion of the substrate, 80-95% of the copper was recovered as $[(Ph_3P)CuH]_6$. The most dramatic effect of the added phosphine, however, is on the generality of the catalytic process. Both substituted cyclic and acyclic substrates undergo catalytic reduction, although some acyclic substrates require greater phosphine concentration to completely inhibit catalyst decomposition. For cyclic substrates, no direct 1,2-reduction of the carbonyl is observed; for acyclic substrates, the allylic alcohol is a minor (< 10%) by-product. Reactions can be performed in benzene, toluene, or tetrahydrofuran; reductions in the latter solvent proceed significantly more slowly (e.g., Table 3.1, entry 9).

Calculation of turnover numbers and rates for the catalytic reduction is complicated by the unknown nuclearity of the reactive catalyst(s) and the possibility of multiple hydride, alkoxide, and mixed hydridoalkoxide complexes. Although a relatively large amount of catalyst has been used in our routine investigations, "lower limit" turnover numbers (assuming the maximum possible six hydride equivalents per hexamer and two turnovers per reduction of the unsaturated ketone to alcohol) in the range of 40-50 have been obtained (e.g., entry 10). Turnover rate is strongly substrate dependent; in general, cyclic substrates are reduced more rapidly than acyclic substrates [14], and substituents on either the α- or β-position of the double bond significantly slow the reduction. For most substrates, good selectivity for conjugate reduction over complete 1,4- plus 1,2-reduction can be obtained (Table 3.1). For cyclic substrates, the conjugate reduction proceeds with high stereoselectivity, although under these conditions, the stereochemistry of the subsequent ketone reduction is not well controlled.

3. Catalytic Hydride Reduction or Catalytic Hydrogenation [12]?

Under homogeneous conditions, $[(Ph_3P)CuH]_6$ and hydrogen will not reduce isolated double bonds, even under forcing conditions. In addition to the reduction of carvone, a substrate incorporating an isolated isopropenyl substituent (Table 3.1, entries 7-10), competition experiments were conducted by adding several potentially reducible alkenes to active catalytic reductions of 2-cyclohexenone. This procedure subjects the alkene to *all* catalytically active copper species present during the reduction, controlling for the relative inertness of $[(Ph_3P)CuH]_6$ itself. Using excess cyclohexene, 1-hexene, or 4-phenyl-3-buten-2-ol (a styrene derivative and allylic alcohol significantly activated toward hydrogenation), no alkene reduction was detected, even at 1500-2000 psi of hydrogen and prolonged reaction time. Interestingly, the net conversion of the cyclohexenone is inhibited by the presence of the alkene, suggesting that competitive coordination of the unactivated alkene may be mechanistically relevant, although no productive hydride transfer occurs. Under nonhomogeneous conditions, in the absence of excess phosphine, some reduction of 1-hexene and complete reduction of the butenol was observed after 48 hours at 1500 psi. Taken together, these data reveal the occurrence of accompanying, presumably heterogeneous, Cu(0)-mediated hydrogenation processes in the absence of phosphine, but confirm that the homogeneous process is, as designed, unusually resistant to reduction of nonpolar unsaturated functionality.

B. Protolytic Copper Transfer: Catalytic Hydride Reduction at Atmospheric Hydrogen Pressure [15]

1. Catalytic Conjugate Reduction with Added *tert*-Butanol

Although the catalytic reductions as described proceed under relatively mild conditions, both optimum reaction conditions and turnover are highly substrate dependent, a natural consequence of the

Table 3.1 Catalytic Hydride–Mediated Reduction of α,β-Unsaturated Ketones Using $[(Ph_3P)CuH]_6$ and H_2

Entry/substrate	Reduction conditions[a]	Product(s)[b,c]	Yield (%)
1	1000 psi H₂, 5 h	34 : 66	
2	1700 psi H₂, 13 h	0 : 100	
3	1000 psi H₂, 5 h, 6 equiv Ph₃P/Cu	70 : 30	90[d]
4	200 psi H₂, 1 h, 6 equiv Ph₃P/Cu	91 : 9	
5	1000 psi H₂, 20 h	40 : 60	17[e]
6	500 psi H₂, 48 h, 6 equiv Ph₃P/Cu	0 : 100 (38:62 eq:ax)[f]	95[d]
7	1000 psi H₂, 25 h	Stoichiometric conversion only 78 : 22	
8	1000 psi H₂, 25 h 6 equiv Ph₃P/Cu	>95 : <5 (~3:1 eq:ax)	
9	1000 psi H₂, 25 h, 6 equiv Ph₃P/Cu, in THF		
10g	1000 psi H₂, 75 h, 6 equiv Ph₃P/Cu	0 : 100[h]	83[i]

Substrate	Conditions	Products (ketone : alcohol)	Yield
Ph–CH=CH–CO–CH$_3$			
11	1000 psi H$_2$, 10 h	91 : 9	
12	1000 psi H$_2$, 10 h, 12 equiv Ph$_3$P/Cu	20 : 71 (9 allylic alcohol)	
13	1700 psi H$_2$, 24 h, 12 equiv Ph$_3$P/Cu	0 : 92 (8 allylic alcohol)	92[j]
(CH$_3$)$_2$C=CH–CO–CH$_3$			
14	1000 psi H$_2$, 20 h	97 : 3	
15	1000 psi H$_2$, 20 h, 6 equiv Ph$_3$P/Cu	89 : 7 (2 allylic alcohol)	88[d]

[a] All reactions: 2.7 mol % [(Ph$_3$P)CuH]$_6$, C$_6$D$_6$, RT, 0.5 M in substrate, except entry 10.
[b] Product ratios determined by relative NMR integration; zeros denote no material detected.
[c] All products were identified by comparison with authentic materials.
[d] Yield determined by NMR integration against internal standard at long pulse delay.
[e] Remainder starting material.
[f] Stereochemical ratios determined by ^{1}H NMR.
[g] 0.8 mol % [(Ph$_3$P)CuH]$_6$ catalyst used.
[h] Four isomers, two major (85% of total isolated): 2(R)-methyl-5(R)-(2-propenyl)cyclohexan-1(S)-ol and 2(S)-methyl-5(R)-(2-propenyl)cyclohexan-1(R)-ol, 2:1, respectively.
[i] Yield of isolated, purified material.
[j] Total yield of isolated alcohols.

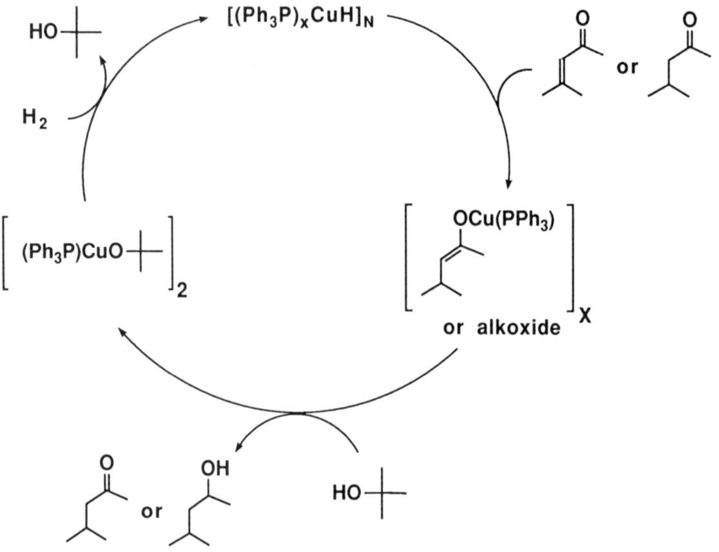

Figure 3.3

incorporation of a substrate molecule into the critical enolate and
alkoxide intermediates. In addition, hydrogen pressures above 1
atmosphere are required for hydrogenolysis to compete effectively
with unimolecular decomposition of the sensitive copper(I) enolate
complexes. To address these issues, a protolytic transfer process
was introduced into the catalytic cycle, to quench the unstable
intermediates with concomitant transfer of the copper to a more
stable alkoxide moiety, where hydrogen activation can proceed under
lower pressure. Provided this transfer is kinetically relevant, the
effect of substrate on the hydrogenolysis should be minimized.
Because it is known that $[(Ph_3P)CuO^tBu]_2$ is both thermally stable
[16] and reactive toward hydrogen at atmospheric pressure [7,10b],
tert-butanol was the obvious initial choice for this function (Fig. 3.3).
      The addition of *tert*-butanol (10-20 equiv/Cu) to the catalytic
reduction conditions enables catalyst turnover at atmospheric pressure (Table 3.2). While essentially no change in the regioselectivity
or stereoselectivity of the reaction is observed, beneficial effects
are noted even at elevated pressure: increased turnover rate (compare Table 3.1, entry 6, and Table 3.2, entry 1), smoother conversion, and greater reproducibility. Under these conditions, a more
detailed investigation of the effects of added triphenylphosphine
was conducted, leading to a greater mechanistic understanding
of the catalytic reduction process. In a series of reductions of

Table 3.2 Catalytic Hydride-Mediated Reduction of α,β-Unsaturated Ketones Using [(Ph$_3$P)CuH]$_6$, H$_2$, and t-BuOH

Entry/substrate	Reduction conditions[a]	Product(s)[b,c]	Yield (%)
1 [3,5-dimethylcyclohex-2-enone]	500 psi H$_2$, 6 equiv Ph$_3$P/Cu, 24 h 10 equiv t-BuOH, 24 h	[3,5-dimethylcyclohexanone] 2 : [3,5-dimethylcyclohexanol] 98 (40:60 eq:ax[d])	94g
2 [isopropenyl cyclohexenone]	1 atm H$_2$, 6 equiv Ph$_3$P/Cu, 10 equiv t-BuOH, 24 h	12 : 59[e]	
3	1 atm H$_2$, 4 equiv Ph$_3$P/Cu, 15 equiv t-BuOH, 30 h	[isopropenyl cyclohexanone] 2 : [isopropenyl cyclohexanol] 90[e,f]	73h
3 [isopropenyl cyclohexenone]	1 atm H$_2$, 6 equiv Ph$_3$P/Cu, 10 equiv t-BuOH, 40 h	70 : [isopropenyl cyclohexanol] 30[f]	
4 [mesityl oxide]	1 atm H$_2$, 6 equiv Ph$_3$P/Cu, 15 equiv t-BuOH, 30 h	[4-methylpentan-2-one] 80 : [4-methylpentan-2-ol] 14[f,i]	84g

[a] All reactions: 2.7 mol % [(Ph$_3$P)CuH]$_6$, C$_6$D$_6$, RT, 0.5 M in substrate.
[b] Product ratios determined by relative NMR integration.
[c] All products were identified by comparison with authentic materials.
[d] Stereochemical ratios determined by ^{1}H NMR.
[e] Remainder starting material.
[f] No further conversion observed.
[g] Yield determined by NMR integration against internal standard at long pulse delay, corrected for conversion.
[h] Combined yield of isolated, purified products.
[i] Allylic alcohol (6%) also formed.

3,5-dimethylcyclohexenone using $[(Ph_3P)CuH]_6$ at 500 psi H_2 for 10 hours in the presence of t-BuOH, the phosphine dependency of both the conjugate reduction and the subsequent carbonyl reduction was determined.

Plotting the disappearance of starting material against the number of added phosphine equivalents (Fig. 3.4) revealed that 4 equivalents of phosphine is required for optimal catalyst stabilization and sustained turnover. Lower phosphine concentration results in eventual catalyst decomposition and accompanying loss of homogeneity. Higher concentration of added phosphine strongly inhibits the conjugate reduction, suggesting that precoordination of the substrate is mechanistically relevant, presumably via a $d \rightarrow \pi^*$ donor interaction from the electron-rich metal to the electron-deficient double bond.

To investigate the effect of phosphine concentration on the carbonyl reduction, the extent to which the ketone produced by conjugate reduction is further converted to alcohol was plotted as a function of phosphine equivalents (Fig. 3.5). In contrast to the conjugate reduction, the carbonyl reduction is *accelerated* by phosphine, with turnover increasing linearly with increasing phosphine concentration. This suggests both that a similar type of precoordination is not required for carbonyl reduction and that a more electron-rich copper hydride complex having two coordinated phosphines per copper may be an effective catalyst for ketone reduction.

Catalytic reductions at 1 atmosphere in the presence of *tert*-butanol proceed substantially but not entirely to completion, under optimum conditions reaching greater than 90% conversion. At higher conversion, the increasing concentration of the alcohol product

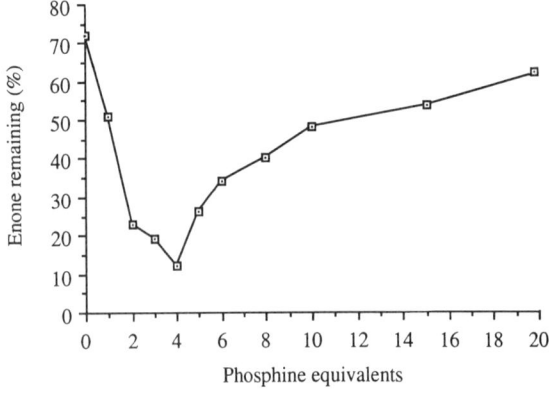

Figure 3.4   3,5-Dimethylcyclohexenone: disappearance of enone.

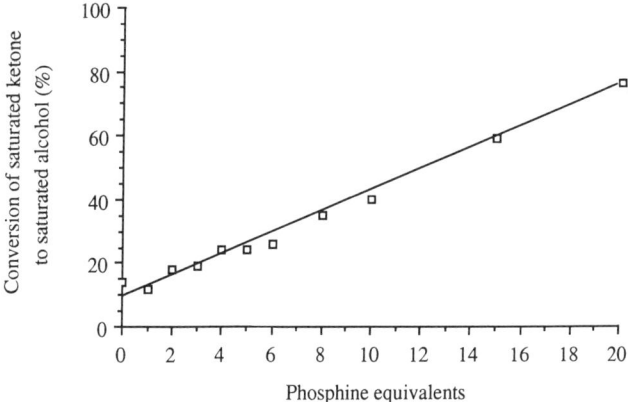

Figure 3.5 3,5-Dimethylcyclohexenone: saturated ketone reduction.

from the reduction apparently interferes with the function of the transfer reagent, relocating the copper among alcohol moieties that do not form thermally stable alkoxide complexes and leading to eventual catalyst decomposition. The use of alcohols other than *tert*-butanol is limited by the requirements of both relative stability and reactivity toward hydrogen: most copper(I) alkoxide complexes are not thermally stable [17] and stabilized alkoxides such as copper(I) phenoxide [18] do not activate hydrogen at atmospheric pressure. Optimization for the nature of the transfer reagent is under investigation.

2. Intramolecular Protolytic Transfer: Phosphine Ligands Incorporating Tertiary Alcohol Functionality

One conceptually simple approach to the problem of sustaining turn-over in low pressure reductions is to attach the protolytic transfer agent directly onto the stabilizing phosphine ligand of the catalyst. This creates an intramolecular transfer reaction that is expected to remain insensitive to exogenous alcohol even at high concentration. Provided that the resultant chelated phosphinoalkoxide inter-mediate remains reactive toward hydrogen (eq. 3.3), giving an active copper(I) hydride complex, substantially improved catalyst properties might be anticipated. Preliminary results using a series of diphenylalkylphosphine derivatives (1a-c) and one triarylphosphine (2) have confirmed the validity of this approach. Although the complex resulting from use of the phosphine 1a, a highly stabilized five-membered chelate, is unreactive toward hydrogen even at 1500 psi, sustained turnover is observed with each of the other phos-phines, some operating efficiently at 1 atmosphere. The diphenyl-

HO PPh$_2$
OCu

$\longrightarrow$

OH

$\left[\begin{array}{c} \text{O} \quad \text{PPh}_2 \\ \text{Cu} \end{array} \right]_X$

$\xrightarrow{\text{H}_2}$

$\left[\begin{array}{c} \text{HO} \\ \text{Ph}_2\text{PCuH} \end{array} \right]_N$

(3.3)

HO (CH$_2$)$_n$PPh$_2$

1a-c (n = 1-3)

PPh$_2$

OH

2

$$\begin{array}{c}
\text{Ph}_2 \quad \text{H} \quad \text{Ph}_2 \quad \text{PPh}_2 \\
\text{P} \quad \quad \text{P} \\
\text{Cu} \quad \text{Cu} \\
\text{P} \quad \text{H} \quad \text{P} \\
\text{Ph}_2\text{P} \quad \text{Ph}_2 \quad \text{Ph}_2
\end{array}$$

3

alkylphosphines, but not the triarylphosphine, show an interesting increase in the formation of allylic alcohol, the product of direct 1,2-addition of hydride.

C. Direct Catalytic Hydride Reduction of Ketones [19]

The thermodynamically stable copper hydride hexamer [(Ph$_3$P)CuH]$_6$ is incapable of initiating the catalytic reduction of isolated carbonyl functionality. Once the coordinatively saturated hexameric aggregate has been disrupted by reduction of the conjugated double bond, however, carbonyl reduction occurs readily and is accelerated by the presence of excess triphenylphosphine. This suggests that copper(I) hydride complexes reactive enough to reduce carbonyl compounds directly are accessible by enhancing the hydridic character of the catalyst and reducing the thermodynamic stability of the aggregate. In principle, this can be achieved by increasing the electron-donating ability of the ancillary phosphine ligand or by increasing the number of phosphines coordinated to the metal, or both. The dimeric copper(I) hydride complex, [(tripod)CuH]$_2$ [3, tripod = 1,1,1-tris(diphenylphosphinomethyl)ethane], reported by Caulton et al. [20], satisfies both criteria: two phosphine moieties coordinated per copper, with each diphenylalkyl arm more electron rich than triphenylphosphine. It is interesting to note that despite the availability of the third phosphine arm of the nominally tridentate ligand, the copper achieves coordinative saturation preferentially via bridging hydrides rather than by coordinating a third donor ligand.

Using this hydride as catalyst, direct catalytic hydride reduction of ketones is observed under hydrogen at room temperature. Sustained turnover is obtained optimally at 50 psi hydrogen pressure and preferentially in tetrahydrofuran solvent. The presence of greater than 1 equivalent of tripod per copper is required, leading to the unusual conclusion that even this potentially tridentate ligand is substitutionally labile in this system. Both cyclic and acyclic ketones are reduced in high chemical yields, although acyclic substrates in general reduce more slowly, requiring longer reaction times or greater catalyst concentration. In contrast to the triphenylphosphine-based catalysts, the tripod catalyst reduces cyclic ketones with promising stereoselectivity.

III. CONCLUSIONS

Chemoselective catalytic hydride reduction has been developed, using hydrogen as the source of active hydride. This new class of homogeneous catalytic reactions is selective for the reduction of polar unsaturated functionality and inert toward isolated double bonds, fully complementary to existing methods of catalytic hydrogenation. The chemoselectivity is mechanism-based, arising from the use of hydridic copper(I) complexes and the imposition of a heterolytic mechanism for hydrogen activation. Catalysts currently under development, both for conjugate reduction and for direct carbonyl reduction, operate efficiently at room temperature under hydrogen pressures as low as 1 atmosphere and give high chemical yields. Extension of these initial investigations into selective hydride reduction of more highly functionalized organic substrates and the eventual development of catalytic asymmetric hydride reductions are anticipated.

ACKNOWLEDGMENTS

The authors thank Professor K. G. Caulton for many helpful discussions. W. S. M. acknowledges partial financial support from an Amoco Fellowship. Financial support from PPG Industries Foundation Grant of the Research Corporation and the National Institutes of Health (GM 38068) is gratefully acknowledged.

NOTES AND REFERENCES

1. See, Rylander, P. N. *Catalytic Hydrogenation in Organic Chemistry*, Academic Press, New York, 1979.
2. Selected examples: (a) Halpern, J., Harrod, J. F., and James, B. R., *J. Am. Chem. Soc.*, *88*, 5150 (1966). (b) Mestroni, G., Spogliarich, R., Camus, A., Martinelli, F., and Zassinovich, G., *J. Organomet. Chem.*, *157*, 345 (1978). (c) Jenck, J., and Germain, J.-E., *J. Catalysis*, *65*, 141 (1980), and references therein. (d) Farnetti, E., Kaspar, J., Spogliarich, R., and Graziani, M., *J. Chem. Soc. Dalton Trans.*, 947 (1988), and references therein; Farnetti, E., Nardin, G., and Graziani, M. *J. Chem. Soc.*, *Chem. Commun.*, 1264 (1989).
3. Hudlicky, M., *Reductions in Organic Chemistry*, Wiley, New York, 1984.
4. Under rather vigorous conditions (130-200°C and 100 atm hydrogen), potassium *tert*-butoxide catalyzes the homogeneous hydrogenation of benzophenone to benzhydrol, a process presumably involving deprotonation of hydrogen by the alkoxide, a heterolytic hydrogen activation closely related to that proposed here. See Walling, C., and Bollyky, L., *J. Am. Chem. Soc.*, *86*, 3750 (1964), and references therein.
5. Reviews: (a) Brothers, P. J., *Prog. Inorg. Chem.*, *28*, 1 (1981). (b) James, B. R., *Adv. Organomet. Chem.*, *17*, 319 (1979). (c) James, B. R., *Homogeneous Hydrogenation*, Wiley, New York, 1973, and references therein.
6. During the course of this work a catalytic reduction of benzaldehyde and cyclohexanone was reported using $AcOM(CO)_5^-$ (M = Cr, W) and acetic acid at 600 psi H_2 and 125°C, a process that also appears to proceed via heterolytic hydrogen activation. The chemoselectivity of this system has not been explored: Tooley, P. A., Ovalles, C., Kao, S. C., Darensbourg, D. J., and Darensbourg, M. Y., *J. Am. Chem. Soc. 108*, 5465 (1986).
7. (a) Goeden, G. V., and Caulton, K. G., *J. Am. Chem. Soc.*, *103*, 7354 (1981). (b) Lemmen, T. H., Folting, K., Huffman, J. C., and Caulton, K. G., *J. Am. Chem. Soc.*, *107*, 7774 (1985).
8. Net heterolytic activation of hydrogen by other copper(I) salts has been extensively documented: Calvin, M., *Trans. Faraday Soc.*, *34*, 1181 (1938); *J. Am. Chem. Soc.*, *61*, 2230 (1939). Weller, S., and Mills, G. A., *J. Am. Chem. Soc.*, *75*, 769 (1953). Wright, L. W., and Weller, S., *J. Am. Chem. Soc.*, *76*, 3345 (1954). Wright, L., Weller, S., and Mills, G. A., *J. Phys. Chem.*, *59*, 1060 (1955). Calvin, M., and Wilmarth, W. K., *J. Am. Chem. Soc.*, *78*, 1301 (1956). Wilmarth, W. K., and Barsh, M. K., *J. Am. Chem. Soc.*, *78*, 1305 (1956). Chalk,

A. J. and Halpern, J., *J. Am. Chem. Soc.*, *81*, 5846 (1959). Chalk, A. J., and Halpern, J., *J. Am. Chem. Soc.*, *81*, 5852 (1959), and references therein.

9. Churchill, M. R., Bezman, S. A., Osborn, J. A., and Wormald, J., *Inorg. Chem.*, *11*, 1818 (1972); Bezman, S. A., Churchill, M. R., Osborn, J. A., and Wormald, J., *J. Am. Chem. Soc.*, *93*, 2063 (1971).

10. (a) Mahoney, W. S., Brestensky, D. M., and Stryker, J. M. *J. Am. Chem. Soc.*, *110*, 291 (1988). (b) Brestensky, D. M., Huseland, D. E., McGettigan, C., and Stryker, J. M., *Tetrahedron Lett.*, *29*, 3749 (1988). (c) Daeuble, J. F., McGettigan, C., and Stryker, J. M., *Tetrahedron Lett.*, *31*, 2397 (1990). (d) Koenig, T. M., Daeuble, J. F., Brestensky, D. M., and Stryker, J. M., *Tetrahedron Lett.*, *31*, 3237 (1990). (e) Brestensky, D. M., and Stryker, J. M., *Tetrahedron Lett.*, *30*, 5677 (1989).

11. A discussion of bonding and stability of copper(I) enolate and alkoxide complexes, and related metal-carbon bonded gold enolate complexes is found in reference 12.

12. Mahoney, W. S., and Stryker, J. M. *J. Am. Chem. Soc.*, *111*, 8819 (1989).

13. Although reaction homogeneity is difficult to determine unambiguously, under conditions of excess phosphine, the presence of elemental mercury had no significant effect on the course of the catalytic reduction for several substrates. In the absence of added phosphine, the presence of mercury resulted in complete suppression of catalytic activity. For discussion on the use of mercury to determine reaction homogeneity, see: Anton, D. R., and Crabtree, R. H., *Organometallics*, *2*, 855 (1983), and references therein.

14. Hydride reductions of cyclic ketones are generally observed to proceed more rapidly than corresponding acyclic substrates: see, e.g., Brown, H. C., and Muzzio, J., *J. Am. Chem. Soc.*, *88*, 2811 (1966); Davis, R. E., and Carter, J. *Tetrahedron*, *22*, 495 (1966).

15. Daeuble, J. F., Mahoney, W. S., and Stryker, J. M., manuscript in preparation.

16. (a) Lemmen, T. H., Goeden, G. V., Huffman, J. C., Geerts, R. L., and Caulton, K. G., *Inorg. Chem.*, *29*, 3680 (1990). (b) [CuO^{t}Bu]$_4$ itself is thermally stable in the absence of additional ancillary ligands: Tsuda, T., Hashimoto, T., and Saegusa, T., *J. Am. Chem. Soc.*, *94*, 658 (1972). See also: Tsuda, T., Habu, H., Horiguchi, S., and Saegusa, T., *J. Am. Chem. Soc.*, *96*, 5930 (1974).

17. Whitesides, G. M., Sadowski, J. S., and Lilburn, J., *J. Am. Chem. Soc.*, *96*, 2829 (1974); Bochmann, M., and Wilkinson, G., Young, G. B., Hursthouse, M. B., and Malik, K. M. A.,

J. Chem. Soc. Dalton Trans., 1863 (1980); Yamamoto, T.,
Kubota, M., and Yamamoto, A. *Bull. Soc. Chem. Jpn.*, *53*,
680 (1980), and references therein.

18. Reichle, W. T., *Inorg. Chim. Acta*, *5*, 325 (1971); Kawaki, T.,
    and Hashimoto, H., *Bull. Soc. Chem. Jpn.*, *45*, 1499 (1972);
    Eller, P. G., and Kubas, G. J., *J. Am. Chem. Soc.*, *99*, 4346
    (1977). Kubota, M., and Yamamoto, A., *Bull. Soc. Chem. Jpn.*,
    51, 2909 (1978), and references therein. Berry, M., Clegg, W.,
    Garner, C. D., and Hillier, I. H., *Inorg. Chem.*, *21*, 1342
    (1982). Fiaschi, P., Floriani, C., Pasquali, M., Chiesi-Villa, A.,
    and Guastini, C., *Inorg. Chem.*, *25*, 462 (1986), and references
    therein.
19. Brestensky, D. M., and Stryker, J. M., manuscript in prepara-
    tion.
20. Goeden, G. V., Huffman, J. C., and Caulton, K. G., *Inorg.
    Chem.*, *25*, 2484 (1986).

4

Catalysis in the Chiral Syntheses of Enalapril, Lisinopril (Semisynthetic Dipeptide ACE Inhibitors), and L-658,758 (a Cephem-Based Elastase Inhibitor)

Thomas J. Blacklock, Richard F. Shuman, John W. Butcher, Paul Sohar, Theresa R. Lamanec, Willard E. Shearin, and Edward J. J. Grabowski

Merck Sharp & Dohme Research Laboratories, Rahway, New Jersey

I. INTRODUCTION

The phenomenon of "catalysis" plays an enormous part in our world, bearing direct responsibility for enabling life as we know it (enzymatic control) and in more mundane concerns, such as establishing the commercial viability of an organic reaction. Not all catalysis is beneficial, however. It can easily work against you, as well as for you, and often steps must be taken to understand and prevent the undesirable reactions. We have found our syntheses of enalapril and lisinopril, the latter being a catalyst "poison" of angiotensin-converting enzymes used for control of hypertension [1,2] and congestive heart failure, to be particularly representative of "catalysis" working in these many ways.

II. DISCUSSION

Enalapril (1a) and lisinopril (2a) are prepared according to the scheme presented in Figure 4.1 [3]. Although the main intent is to present chemistry utilizing catalysis, we shall do so within the context of the larger synthetic aims rather than out of context.

Our tale, thus, begins with the preparation of the requisite dipeptides alanylproline (5a) and N_e-trifluoroacetyl-lysylproline (5b). Development of a two-step practical application of N-carboxyanhydride (NCA) chemistry for their preparation was particularly attractive considering the brevity of such a synthesis over the more classical and costly five-step protecting group approach. We knew from the

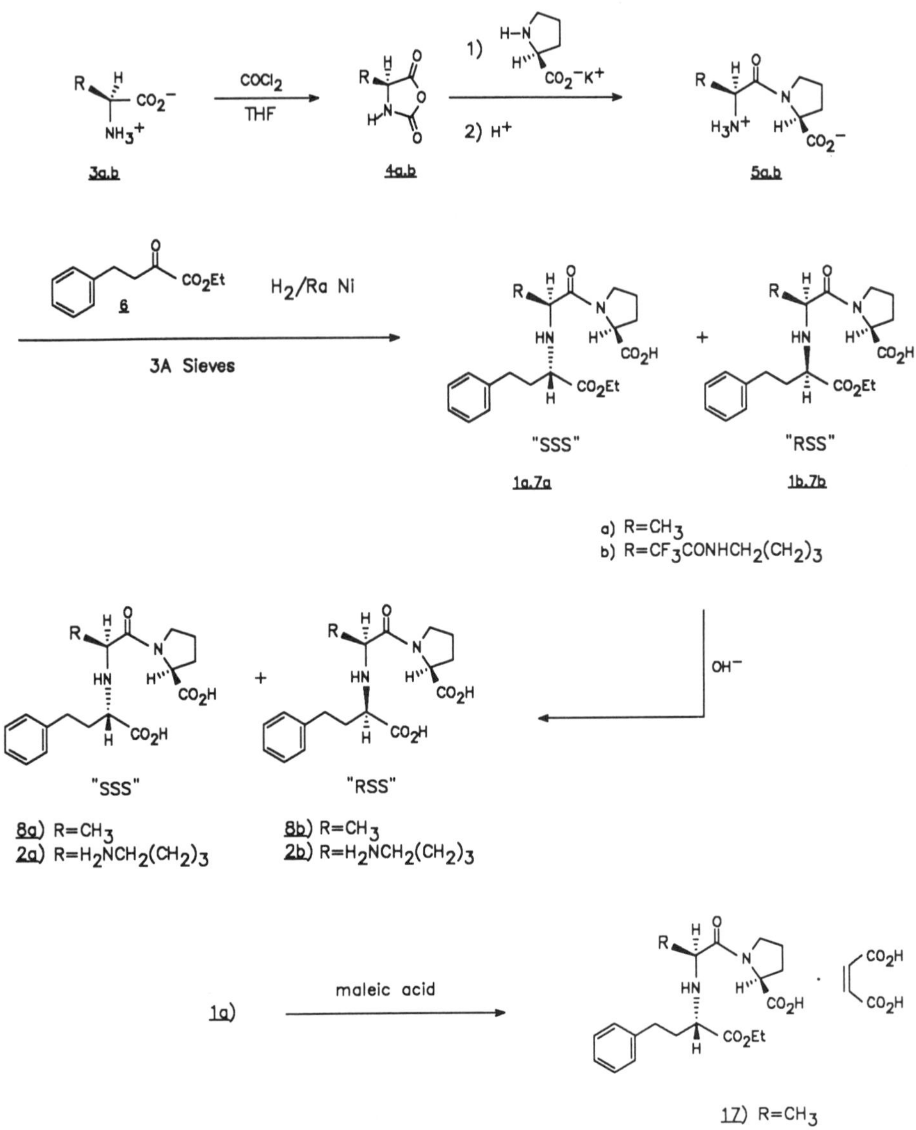

Figure 4.1   Synthesis of enalapril and lisinopril.

start, however, that this task would not be easy; an efficient large-scale synthesis of peptides via NCA chemistry—that is, facile preparation of the NCA and subsequent control of "overreaction" (oligomerization and/or polymerization) during condensation—has long been the goal of many chemists.

A. *N*-Carboxyanhydride Formation

Except for a recent high-yielding but costly and multistep procedure for the preparation of alanineNCA (4a) from Boc-alanine and oxalyl chloride [4], the literature suggests that only moderate yields (60-80%) can be expected for its direct preparation from alanine (3a) and phosgene [5,6]. This is likely due to the near total insolubility of alanine (or alanine hydrochloride) in the reacting medium, which affords a sluggish reaction and thus allows time for side reactions to occur. We have found, however, that alanineNCA (4a) can be prepared in 95% yield (unisolated) directly from alanine via an optimized Fuchs-Farthing procedure [6] similar to that recommended by Goodman et al. [5].

Alanine (3a) or Tfa-lysine (3b) was reacted with a preformed 4-5 molar solution of phosgene in THF; this excess removed phosgene as a rate-limiting reagent, and any large-scale thermal contribution due to the heat of solvation of phosgene with THF was eliminated. A detailed kinetic picture of the various intermediates was obtained (Fig. 4.2).

1. Catalysis Against You . . .

Purity of the tetryhydrofuran used as solvent for NCA synthesis was found to be critical in some instances to success or failure in preparing NCAs. Dimethylformamide (DMF), a common low-level contaminant in many bulk solvents, was found to catalyze a significant reaction loss to the corresponding *N*-chlorocarbonyl amino acid chloride 12 (Fig. 4.3). This product (12) was identified by ^{13}C NMR and presumably arises from a Vilsmeier-Haack-type intermediate. At levels of DMF as low as 1 part per thousand (ppt) a nearly quantitative yield of 12 was obtained. We believe that a side reaction such as this, which can occur during NCA formation, has often led other researchers to conclude that subsequent condensation chemistry was deficient.

With an efficient and well-understood procedure for the large-scale preparation of the requisite NCAs, we turned our attention to the next step, condensation. Heretofore the method had been thought virtually impossible to achieve on a commercially important scale, despite its elegant simplicity. Many of the general arguments presented in the literature attempted to rationalize NCA condensation

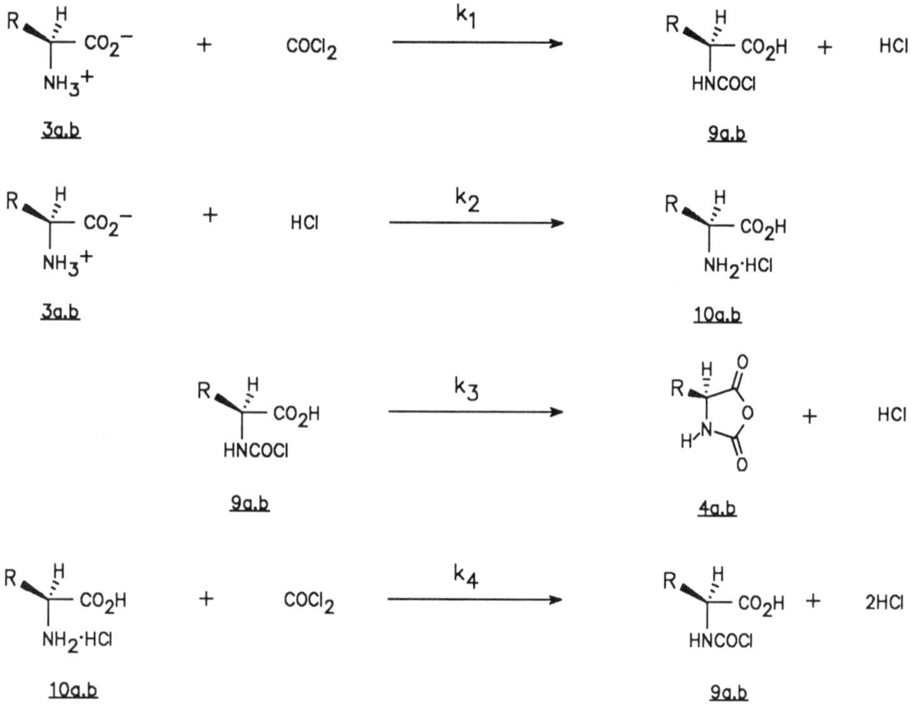

a) R=CH$_3$
b) R=CF$_3$CONHCH$_2$(CH$_2$)$_3$

Figure 4.2 Simplified mechanistic pathway to *N*-carboxyon-hydride formation.

polymerization in terms of pK_a's, thermal stability of carbamate intermediates, reaction medium pH, and mixing [7]. A factor frequently overlooked is the combined substituent effects of both NCA (electrophile) and the nucleophile (free amine, NCA *N*-anion, amino acid, hydroxyl anion, dipeptide anion, etc.). NCAs having very bulky substituents seldom polymerize. In fact, phenylalanineNCA affords a 90% yield of dipeptide, even at 20°C, when condensed with an equimolar amount of glycine benzyl ester [8]. Based on this observation and on studies of *N*- and α-substituent-related effects of NCAs toward homopolymerization and/or dipeptide formation by Akiyama et al. [9] and Oya and Takahashi [8], respectively, one can conclude that the extent of polymerization is roughly inversely proportional to the bulk of the α-substituent of the NCA

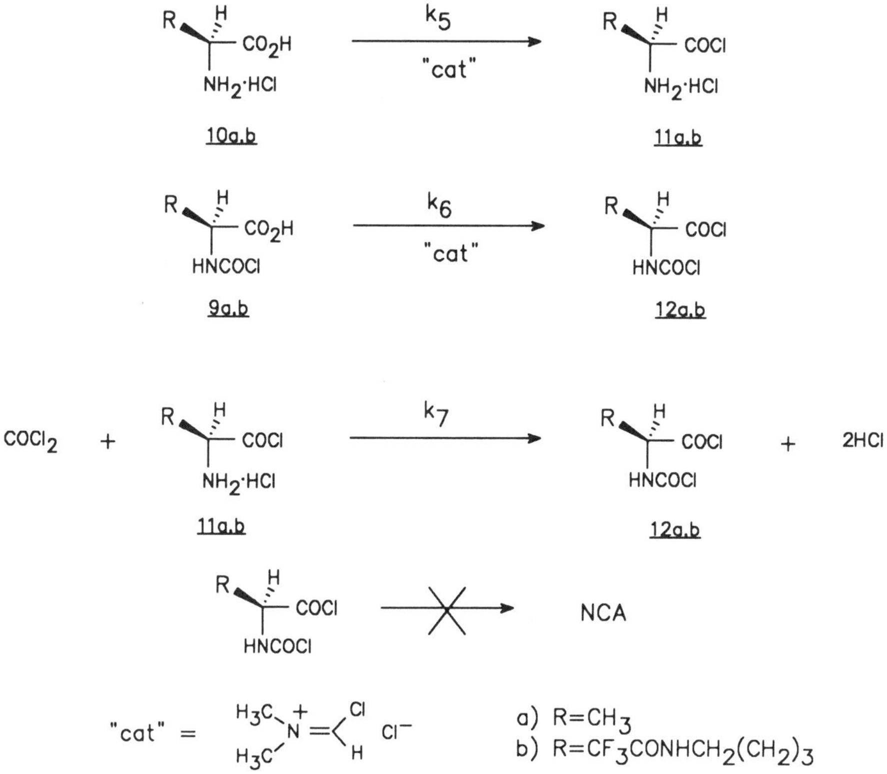

Figure 4.3 Poor NCA formation due to DMF-catalyzed acid chloride formation.

and directly proportional to the bulk of the nucleophile, and also that the nature of the nucleophile may frequently be the controlling factor. Clearly, the factors just named make the condensation of alanineNCA with proline one of the more difficult reactions to achieve efficiently. Consequently, a literature search revealed few examples of NCA condensations with either proline or alanineNCA.

The method for peptide bond formation reported by Iwakura et al [10,11] seemed most promising for the preparation of 5a, but initial attempts using alanineNCA (4a, crystalline and pure) and proline according to their methodology afforded extensive oligomeriza- tion, thus indicating that its application is not without limitation. When alanineNCA was condensed with sodium prolinate in a sodium carbonate-buffered water/acetonitrile medium, the reaction mixtures either froze at the recommended temperature and/or the sodium

Table 4.1   Yield of Ala-Pro Relative to Base/Buffer Counterion
in Proline/AlanineNCA Condensation

MOH	M_2CO_3 buffer	Ratio of NCA to proline to base to buffer	Addition times	Tempera-ture (°C)	Yield of Ala-Pro (%)
Li	Li_2CO_3	1.00:1.05:1.05:1.00	4	0	80.4
Na	Na_2CO_3	1.00:1.05:1.05:1.00	5	-5	78.2
K	K_2CO_3	1.00:1.05:1.05:1.00	5	-5	90.5
Rb	Rb_2CO_3	1.00:1.05:1.05:1.00	5	-5	90.8
Cs	Cs_2CO_3	1.00:1.05:1.05:1.00	5	-5	93.3

prolinate-sodium carbonate mixture crystallized. Thus, once fully
reacted, the product mixture contained as much as 40% oligomers.

Warming the reaction alleviated the freezing problem but did
not reduce the degree of oligomerization. At the recommended 0.1
molar scale, dipeptide yield was shown to be inversely related to
addition time and directly related to mixing efficiency. When the
addition time of the alanineNCA was decreased from several minutes
to less than 5 seconds, the yield of dipeptide increased from about
65% to 90%. This increase was greatly attenuated when the reaction
was carried out at a concentration approaching 1 molar. We discovered, however, that when both the buffer and countercation
were changed from sodium to potassium, the reaction could be run
at higher concentrations (>0.5 molar with respect to the aqueous
phase) and at 0°C without encountering solubility problems. As
a consequence, the yield of L-alanyl-L-proline exceeded 90%. Subsequently, a distinct correlation with the Group I metals versus yield
progressively increased as the Group I counterion increased in
atomic number; that is, the highest yield of 93% was obtained when
the cesium cation was employed (see Table 4.1) [12]. Use of THF
instead of acetonitrile gave comparable results, thus simplifying
use of the alanineNCA solution.

In contrast to the preparation of L-alanyl-L-proline (5a) described above, the buffered biphasic condensation of N_ϵ(Tfa)-L-
lysineNCA (4b) with potassium prolinate proceeded in somewhat
higher yield (92 vs. 89%) than did the similar condensation with
alanine NCA, and it did not require as careful control of reaction
parameters. Addition time of 4b proved to be less critical, and
extensive oligomerization was not observed even when addition time
was extended to 15 minutes. We attribute this phenomenon to the
bulky α-substituent effect as previously discussed.

2. Catalysis for You ...

Reductive Amination. Previously reported syntheses of 1a and 2a suffered from poor yields because of length and/or poor diastereoselectivity in forming the desired asymmetry at the new optical center [13-16]. Our approach centered about a reductive amination procedure whereby Schiff base formation followed by catalytic low pressure hydrogenation in ethanol over Raney nickel affords 1a and 7a in 80-90% yield with high diastereoselectivity.

Catalytic hydrogenation of the presumed imine formed between L-alanyl-L-proline (5a) and ethyl 2-oxo-4-phenylbutyrate (6) [17,18] (EtOH, 3A molecular sieves) affords enalapril free base (1a, *SSS* configuration) and its *RSS* diastereomer (1b). Diastereoselectivity was found to be remarkably dependent on the choice of catalyst; Raney nickel afforded the highest ratio of 87:13 1a (*SSS*): 1b (*RSS*) (see Table 4.2). Reaction workup consisted of filtering the catalyst and molecular sieves, and a solvent exchange from ethanol to ethyl acetate readied crude enalapril for crystallization as its maleate salt.

Diastereoselectivity similar with respect to catalyst choice to that observed with 5a was also observed for the reductive condensation between N_ε(Tfa)-L-lysyl-L-proline (5b) and ethyl 2-oxo-4-phenylbutyrate (6). Raney nickel afforded a 95:5 ratio of the desired *SSS* diastereomer 7a (see Table 4.2) to the corresponding *RSS* diastereomer 7b. The probability of Schiff base formation is implied by appearance of a by-product, ethyl 2-amino-4-phenylbutyrate (16) [17-21], but isolation of the Schiff bases proved elusive because these reagents were unstable (readily hydrolyzed). CHARMm calculations [22] suggest at least a 10 kcal/mol energy difference between imine and reduced product. Phenylbutyrate (15), which

Table 4.2 Screen for Reductive Amination Catalysts

Catalyst	For enalapril (1a) 1a(*SSS*):1b(*RSS*)	For lisinopril (7a) 7a(*SSS*):7a(*RSS*)
5% Pt/C	50:50	50:50
5% Rh/C	50:50	
5% Ru/C	40:60	
P-1 nickel boride [19]	50:50	
P-2 nickel boride [20]	50:50	
IrO$_2$ (Adams catalyst)	60:40	
Raney nickel (Grace #28)	87:13	95:5
Pd/C	60:40	53:47

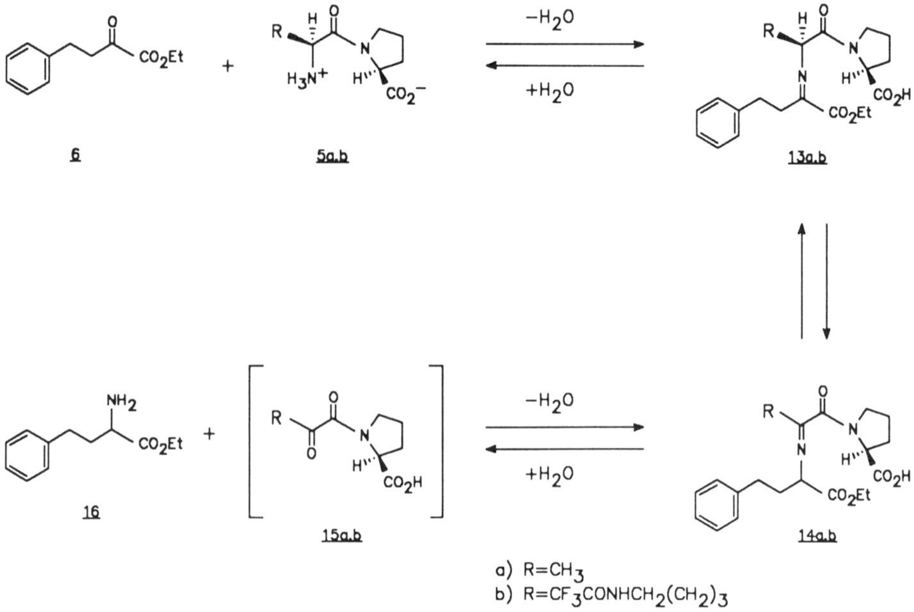

a) R=CH₃
b) R=CF₃CONHCH₂(CH₂)₃

a) $R=CH_3$
b) $R=CF_3CONHCH_2(CH_2)_3$

Figure 4.4   Transamination pathway to ethyl 2-amino-4-phenylbutyrate.

likely forms via the pathway of Figure 4.4, was observed to steadily
increase in prehydrogenation mixtures that had stood for several
hours in the presence of Raney nickel under a nitrogen atmosphere.
Hydrogenation without delay, however, minimized transamination.
No attempt was made to identify the conjugate α-keto derivative 15b.
   Rationalization for such diastereoselectivity with Raney nickel
compared to other catalysts remains elusive. CHARMm calculations
[22] suggest that there is a least-hindered face of a relaxed imine
conformation, which would lead predominantly to the desired SSS
product. Thus, one might conclude "naively" before running any
experiments that the SSS diastereomer would be formed preferentially
from virtually any metal-catalyzed hydrogenation. This is clearly
not the case, as shown by Table 4.2. The association of imine with
nickel might therefore play a very important role prior to hydrogenation. Lower valent nickel does form compounds with π acid ligands
[23]. One might therefore be tempted to invoke complex formation
between imine and Ni⁰ to form a square planar or tetrahedral iminium
species prior to hydrogenation. This is purely speculation at this
time; more work needs to be done before this phenomenon can be
understood. An alternative explanation, however, might involve

enamine formation followed by hydrogenation. Hydrogenation experiments with deuterium are underway.

Saponification/Hydrolysis. Simultaneous deprotection of the N_ε-trifluoroacetyl protecting group (hydrolysis) and saponification of the ethyl ester was accomplished in near quantitative yield in aqueous hydroxide (pH > 11,40°C). The Tfa group, which proved to be remarkably stable through prior reactions, briefly withstanding pH 12.5 at 0°C during the NCA condensation reaction and cold aqueous bicarbonate for extended periods, began to hydrolyze at pH values greater than 10 at 40°C.

Isolation. Large-scale desalination of crude water-soluble lisinopril (2a) from the aqueous salt-laden saponification mixture was accomplished equally well on Dowex ion-exchange resin (acid cycle) and on SP-207 resin, a nonionic brominated polystyrene-divinylbenzene copolymer. These resins gave comparable results. Vacuum concentration of the product-rich cuts and crystallization from 90% ethanol afforded lisinopril containing less than 1% *RSS* diastereomer. A final crystallization from water gave pure 2a (isolated as its dihydrate solvate) in greater than 70% recovery.

Synthesis of Enzyme Inhibitor. Switching to another project, catalysis has played an enormous part in our four-step synthesis of the human leukocyte elastase inhibitor 1,1-dioxo-(*trans*)-7-methoxycephalosporanic acid *t*-butyl ester from 7-aminocephalosporanic acid (7-ACA) in 44% isolated yield (Fig. 4.5) [24].

Neutral proteolytic enzymes, specifically human leukocyte elastase (HLE), released from human polymorphonuclear (PMN) leukocytes, have been implicated in the pathogenesis of adult respiratory distress syndrome (ARDS), emphysema, and rheumatoid arthritis [25]. Inhibition of HLE, therefore, may attenuate the onset of such diseases. Several laboratories, including our own, have instituted a search for compounds that show promise as in vivo inhibitors of HLE. In evaluating modified cephalosporin-based antibiotics as HLE inhibitors, we were presented with the opportunity to devise an adaptable and high-yielding general synthesis for the preparation of 1,1-dioxo-7-substituted cephems from the relatively inexpensive starting material 7-ACA. Our initial target was 1,1-dioxo-(*trans*)-7-methoxycephalosporanic acid *t*-butyl ester (8), which had been established by Doherty et al. as a compound showing strong in vitro inhibition of HLE [26-29]. This compound was subsequently converted to L-658,758 in four steps [29].

Starting with 7-ACA (17) it was obvious that any synthesis of 23 must effect two key chemical transformations: conversion of the *cis*-7-amino group to the *trans*-7-methoxy group, and oxidation of the 1-sulfide to the corresponding 1-sulfone. The original procedure

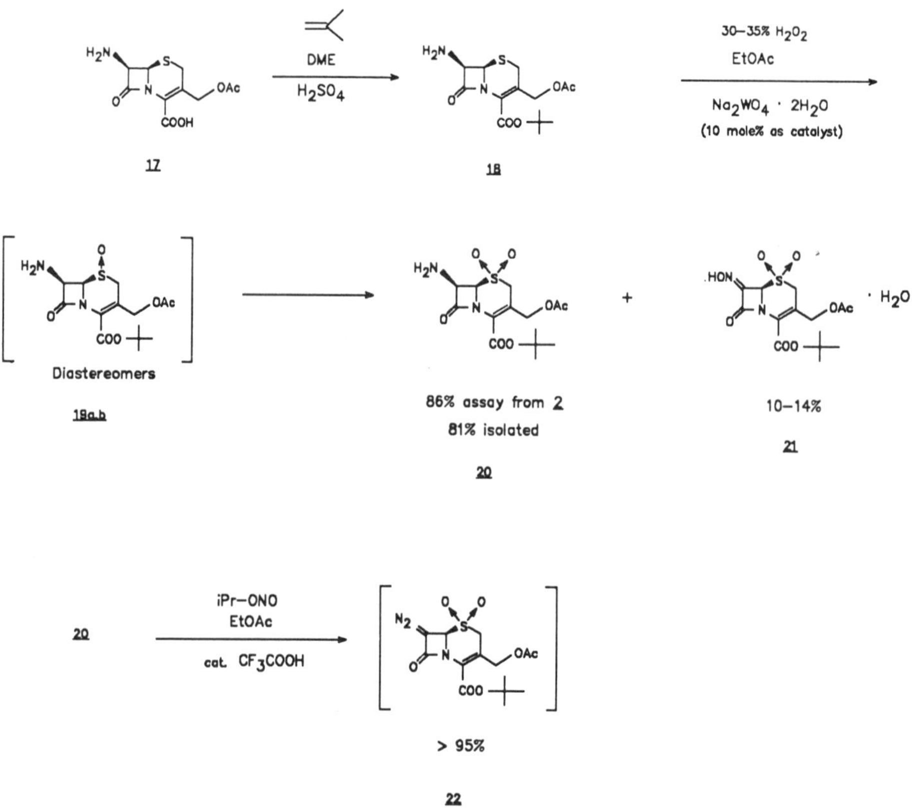

Figure 4.5   Synthesis of 7-methoxy cephalosporin dioxides.

of Doherty accomplished these transformations via diazotization of
7-ACA-O-t-Bu followed by rhodium-catalyzed insertion into methanol
and subsequent oxidation of the 7-methoxysulfide with m-chloroper-
benzoic acid (m-CPBA). Because of the inherent problems associated
with carbene (carbenoid) interaction with sulfide [30], the insertion
reaction on a relatively unstable diazo compound gave low yields
of a key intermediate. During our investigation we discovered that
if we transposed the reaction sequence so that S-oxidation occurred
first, subsequent diazotization/insertion on the resulting sulfone
improved dramatically to better than 90%. Furthermore, the intermediate diazosulfone proved to be particularly useful for the preparation of other 7-substituted sulfone cephems, which heretofore were
either unknown or obtained only with great difficulty.

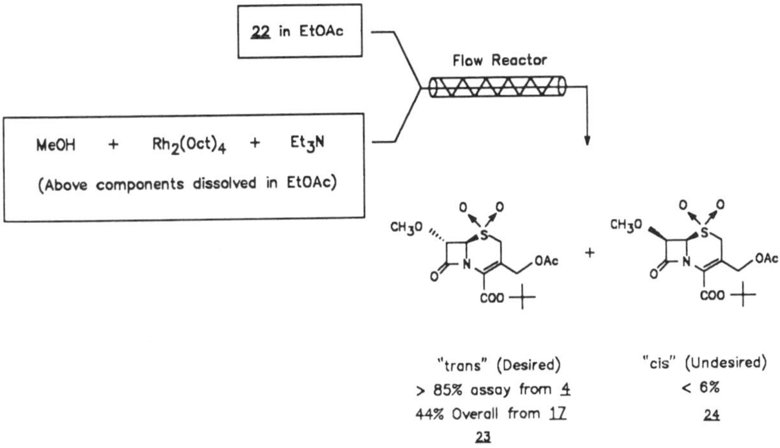
22 in EtOAc
Flow Reactor
MeOH + Rh2(Oct)4 + Et3N
(Above components dissolved in EtOAc)
"trans" (Desired)
> 85% assay from 4
44% Overall from 17
23
"cis" (Undesired)
< 6%
24

Our approach to an adaptable synthesis, which would serve to supply the required 23 and yet be easily modified to supply other 7-substituted cephems from a common intermediate, is described below.

7-Aminocephalosporanic Acid t-Butyl Ester (18). Our preparation of 7-ACA *t*-butyl ester was based on the procedure described by Stedman (52%) [31]. Existing reports in the literature for the preparation of this compound were sketchy, and further optimization of this reaction proved quite instructive. Superficially, this reaction can be viewed as a simple equilibrium between 7-ACA and 7-ACA-O-*t*-Bu mediated by isobutylene. However, the reaction is actually comprised of a complex series of equilibria involving protonated 7-ACA and protonated isobutylene. The role of the acid catalyst is manifold. The acid catalyst must (1) protonate the amino group

of 7-ACA to protect it from electrophilic attack, (2) solubilize the
7-ACA in the solvent medium as an acid/base salt, and (3) be
sufficiently strong to protonate isobutylene and drive the esterification reaction without inducing excessive polyisobutylene formation.
Choice of solvent was critical. Because of safety concerns, p-dioxane,
the classical solvent of choice for the acid-catalyzed esterification
of amino acids with isobutylene, was replaced with 1,2-dimethoxyethane
(DME). Laboratory experiments using DME, however, have consistently
afforded 5-7% lower yields and have taken longer to reach equilibrium.

The ratio of 7-ACA to sulfuric acid to isobutylene was also
critical; a ratio of 1: > 4.8: > 22 was necessary to consistently drive
the equilibrium above an 80:20 ratio of 7-ACA t-butyl ester to 7-
ACA. Larger excesses could drive the reaction further, but we
felt that a point of diminishing returns had been reached. A careful
workup afforded a 71-74% isolated yield of 7-ACA t-butyl ester.

7-Amino-1,1-dioxocephalosporanic Acid t-Butyl Ester (20). A
low yielding four-step preparation of 20 via m-CPBA oxidation of
BOC-7-ACA-O-t-Bu and selective deprotection has been reported
by Durckheimer et al. [32]. The direct oxidation of 7-ACA t-butyl
ester to the corresponding sulfone was, therefore, quite attractive
but posed some unique problems as a result of the potential for
N-oxidation of the 7-amino group. As in step 1 above, we believed
that protection from N-oxidation might come via protonation of the
amine prior to oxidation. Attempts using various combinations of
acetic, formic, and sulfuric acids in conjunction with peracetic acid,
m-chloroperbenzoic acid, and hydrogen peroxide were unsuccessful.
We were partially successful, however, via formation of the p-
toluenesulfonate salt of 7-ACA t-butyl ester. Subsequent oxidation
with m-chloroperbenzoic acid afforded nearly 80% of the desired
sulfone. However, direct oxidation without N-protection was ultimately
effected catalytically with sodium tungstate [33] and 30% aqueous
hydrogen peroxide in a suitable solvent (the oxidizing species likely
being pertungstic acid).

The solvent of choice, ethyl acetate, was key to obtaining a
yield of sulfone exceeding 80%. Methylene chloride, acetic acid,
and methanol afforded much lower yields. The secondary oxidation
of sulfoxide to sulfone failed to complete without by-product formation.
The substitution of vanadyl acetylacetonate under identical conditions
gave similarly low yields. Even in ethyl acetate, the hydrogen
peroxide-sodium tungstate reaction was not without drawbacks.
Some N-oxidation indeed occurred to form 10-14% of a nearly insoluble
and crystalline by-product identified as the 7-oximinosulfone derivative (21) of 7-ACA t-butyl ester. This by-product did not form
as a result of overreaction of sulfone 4, but was shown to have
arisen from primary N-oxidation of starting material and/or subsequent N-oxidation of sulfoxide intermediates.

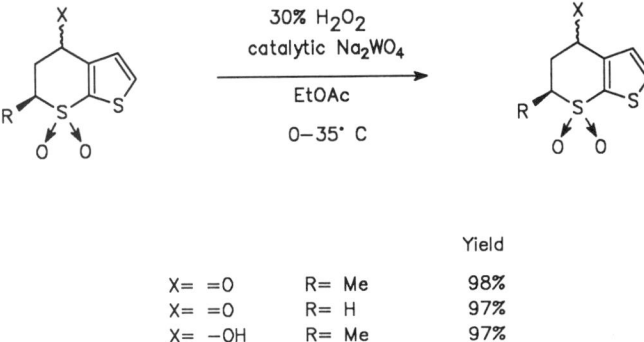

		Yield
X= =O	R= Me	98%
X= =O	R= H	97%
X= −OH	R= Me	97%

Figure 4.6   Examples of hydrogen peroxide/sodium tungstate catalyzed S-oxidations.

Oximesulfone 21 was determined to be potentially detonable in dry solid form. Ethyl acetate-wet oximesulfone, however, could be handled safely. A procedure was devised to avoid its isolation by in situ solubilization and decomposition with aqueous sodium carbonate.

The foregoing oxidation was based on a report by Schultz et al. [34] for the oxidation of water-soluble sulfides to sulfones (less water-soluble sulfides could be run, however, in aqueous alcohols or with dioxane present). Workup, thus, became tedious. As it turned out, our ethyl acetate biphasic modification removed this constraint and made the reaction more manageable. The heat of reaction has been measured at 54 kcal/mol [35] which is moderately exothermic for large-scale work. The reaction can be easily controlled, however, via the rate of addition of peroxide. After oxidation has been completed, excess hydrogen peroxide is decomposed with sodium sulfite, the layers are cut, and the ethyl acetate product layer concentrated to crystallize the sulfone. In our estimation this catalytic procedure is clearly superior to the alternatives mentioned above. Peculiar to oxone oxidations, one need not deal with the enormous bulk of solids that need be efficiently stirred and filtered. This S-oxidation has since been applied to a variety of substrates on the kilomole scale, notably for the oxidation of three thiophene thiopyran derivatives shown in Figure 4.6. In each case, yields of sulfone exceeded 95%.

Before the disclosure by Schultz, the primary use for sodium tungstate-hydrogen peroxide was for the oxidation of primary amines to the corresponding aldoximes [36-38]. To our knowledge no one had placed these oxidations in relative perspective. Thus, as far as 7-aminocephalosporins are concerned, our observations suggest

that under our conditions, the two-step oxidation of sulfide to sulfoxide to sulfone occurs approximately six times faster than the
two-step oxidation of a primary amine to aldoxime (oxime) via
hydroxylamine.

   1,1-Dioxo-(trans)-7-methoxycephalosporanic Acid t-Butyl
Ester (23). During the course of evaluating alternative synthetic
routes to 23, we discovered that diazotization of the sulfone analogue
of 7-ACA *t*-butyl ester (20) afforded the corresponding diazosulfone
22, which was orders of magnitude more stable in solution* and
much less dependent on the purity of the starting material than
the corresponding diazosulfide reported by Wiering and Wynberg
[39]. Furthermore, 22 could be generated under homogeneous reaction
conditions in near quantitative yield from 20 with standard diazotizing
reagents such as *i*-amyl nitrite (acid catalyzed). A switch to *i*-propyl
nitrite [40] was made to facilitate workup, since the reaction by-
product, isopropyl alcohol, could be readily removed from the reaction mixture by vacuum replacement concentration.

   Whereas rhodium-catalyzed insertion reactions of Wynberg's
diazosulfide with methanol afforded poor yields of trans-insertion
adduct (averaging 20-25%), the same insertion reaction using our
diazosulfone 22 showed a dramatic improvement. Under optimum
conditions we have consistently obtained a two-step assay yield
(diazotization and insertion) exceeding 85% of desired 1,1-dioxo-
(*trans*)-7-methoxycephalosporanic acid *t*-butyl ester (23) at trans/cis
isomer ratios greater than 17:1.

   The intermediate carbene (carbenoid) generated from the diazosulfone was much more chemodiscriminate than that derived from
the diazosulfide. Two major products identified as *trans-* and *cis*-
7-methoxy isomers 23 and 24 were formed in high yield.[†] With
a stable diazo precursor such as 22, insertion/displacement reactions
could now be studied. Typical first-choice conditions for the desired
insertion reaction (20 wt % rhodium catalyst in MeOH and CH_2Cl_2)
gave 40-55% yield of a 3:1 mixture of trans/cis isomers. Initial experiments performed at high dilution (0.1 mg/mL) to avoid bimolecular
reactions (dimerization) showed that more polar solvents could raise
the yield dramatically. In methanol, a nearly quantitative, yet non-
stereoselective insertion (1:1, trans/cis) was obtained. The obvious

---

*We note that this diazo compound 7 was determined by our OHEL
to be shock sensitive and potentially detonable when crystalline. It
was also determined, however, to be safely handleable in solutions
of up to 10 wt % in methylene chloride or ethyl acetate.
[†]We believe this to be the first example of a *cis*-alkoxy insertion
product arising from the metal-catalyzed reaction of a diazocephalosporin and with an alcohol.

conclusions were to eliminate the nonpolar cosolvent and minimize the amount of reagent methanol so as to increase stereoselectivity. Among the more polar cosolvents tried, cyclic ethers such as THF and dioxane gave the highest yields at a methanol charge of approximately 5 vol %. Another set of experiments wherein slow addition of catalyst was tested showed that complete conversion (insertion) was obtained after only a small portion of catalyst had been added. Thus, a large excess of catalyst was not necessary, and in fact, a large excess of catalyst was detrimental to optimization of the trans/cis ratio. While a fast reaction rate ensured a higher insertion yield, slowing down the rate of reaction with a reduced methanol and catalyst charge increased steroselectivity. Temperature was also found to influence yield and ratio. Precisely, catalyst and methanol charge directly affected insertion yield and yet inversely affected steroselectivity. Obviously, a balance had to be struck. These conflicting rate requirements were optimized by empirical adjustment of reaction conditions to maintain a 10-30 second half-life relative to diazo reagent. Longer reactions did not reach completion and afforded lower yields.

As scale increased and instantaneous mixing of reactants became nearly impossible, a continuous flow reactor was devised for our preparative efforts. This device allowed for the exact and instant mixing of reagents, critical maintenance of concentration and stoichiometry, an increase in the reagent concentrations to 100 mg/mL (0.27 molar), and controlled evolution of nitrogen gas by-product. Another change necessitated by large-scale preparation was the replacement of ethereal solvents with ethyl acetate, which performed nearly as well when its basicity was increased with a small amount of triethylamine. The present procedure utilizes a Kenics static mixer flow tube. A residence time of 1-3 minutes (six half-lives) was necessary for complete reaction. Shorter residence times resulting in dilution of the reaction mixture in the receiving vessel gave lower yields and poor trans/cis ratios. Two cold (0°C) ethyl acetate solutions, one containing rhodium octanoate,* methanol, and triethylamine, and the other freshly prepared unisolated diazocephalosporin (22), were pumped (metered) together at equal rates so that complete reaction was obtained in the flow reactor and/or appendant tube before discharge into the receiving vessel.

---

*A homogeneous catalyst, rhodium octanoate dimer purchased from Johnson-Matthey, Inc. Chemicals Division (Winslow, NJ 08095), and necessary for flow reactor use, replaced the more traditional and less soluble rhodium acetate dimer. Stationary-bed catalysts were not examined.

B.   Scope

Diazocephalosporin 22 was found to undergo some of the typical
transformations associated with stabilized diazo compounds (Fig. 4.7).
On reaction with HCl, the predominantly *trans*-7-chloro compound
was formed in high yield. As with the insertion reactions described
above, this reaction, too, afforded small amounts of 7-*cis* adduct,
which was readily isolated and identifiable by ^{1}H NMR spectroscopy
on the basis of a characteristic vicinal coupling (J^{cis} = 5.0-6.0 Hz)
in the β-lactam moiety. Even hydrogen fluoride (via HF/pyridine
complex) gave the corresponding *cis*- and *trans*-7-fluoro derivatives,
albeit in poor yield.

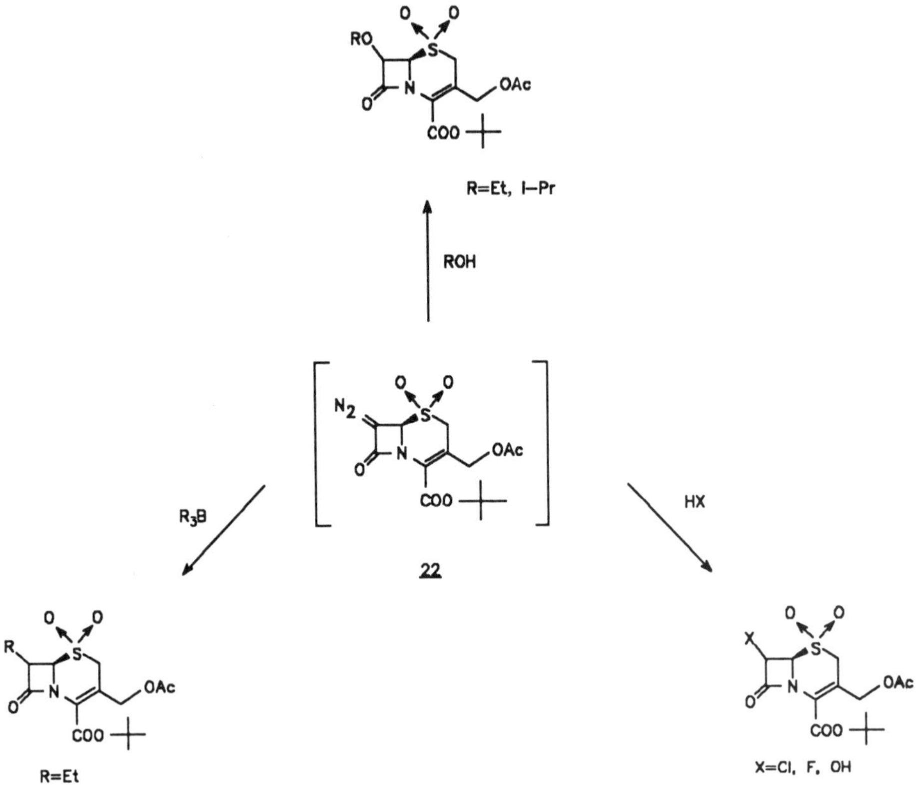

Figure 4.7   Utility of the diazosulfone insertion reaction.

Rhodium-catalyzed insertion reactions worked well on the lower aliphatic alcohols (R = Me, Et, *i*-Pr) to afford predominantly *trans*-alkoxy adducts. Other than the specific case discussed above for methanol, no attempts were made to optimize the reactions with ethanol or isopropanol, either for yield or for isomer ratio. In each case, however, the cis-insertion products were readily isolable by silica gel chromatography and identifiable by their characteristic ^{1}H NMR spectra. In all cases, a small amount of the *trans*-7-hydroxy adduct was observed by high performance liquid chromatography (identified after isolation and characterization from the reaction of 22 with HF in pyridine) and presumed to arise via insertion into water.

Reaction of 22 with triethylborane in similar fashion to the procedure described by Wiering and Wynberg [39] afforded the expected 7-ethyl derivatives in high chemical yield at a cis/trans ratio of approximately 1:1.

The procedures presented above provide easy entry into a variety of 7-substituted sulfone cephems in generally high yields. In particular, use of a flow reactor has safely achieved control of the rapid intermolecular diazo insertion reaction of 7 with methanol while maintaining yield and stereospecificity. The cephems prepared can be further functionalized to provide a vast array of potentially therapeutic agents.

REFERENCES

1. Cleary, J. D., and Taylor, J. W., *Drug Intell. Clin. Pharm.*, *20*, 177-186 (1986).
2. Todd, P. A., and Heel, R. C., *Drugs*, *31*, 198-248 (1986).
3. Blacklock, T. J., Shuman, R. F., Butcher, J. W., Shearin, W. E., Jr., Budavari, J., and Grenda, V. J., *J. Org. Chem.*, *53*, 846-844 (1988).
4. Mobashery, S., and Johnston, M., *J. Org. Chem.*, *50*, 2200-2202 (1985).
5. Fuller, W. D., Verlander, M. S., and Goodman, M., *Biopolymers*, *15*, 1869-1871 (1976).
6. Farthing, A., *J. Chem. Soc.*, 3222-3229 (1950).
7. The preparation and use of *N*-carboxyanhydrides (NCAs) and *N*-thiocarboxyanhydrides (NTAs) for peptide bond formation, Blacklock, T. J., Hirschmann, R., and Veber, D. F., in *The Peptides* (J. H. Meienhofer and S. Udenfriend, eds.), Academic Press, New York, 1987, pp. 39-102.
8. Oya, M., and Takahashi, T., *Bull. Chem. Soc. Jpn.*, *54*, 439-441 (1981).

9. Akiyama, M., Hasegawa, M., Takeuchi, H., Shimizu, K., *Tetrahedron Lett.*, 2599-2600 (1979).

10. Iwakura, I., Uno, K., Oya, M., and Katakai, R., *Biopolymers*, *9*, 1419-1427 (1970).

11. Katakai, R., Oya, M., Uno, K., and Iwakura, Y., *Biopolymers*, *10*, 2199-2208 (1971).

12. Blacklock, T. J., and Shuman, R. F., U.S. Patent 4,510,083 (1985); European Patent 127,108 (1984).

13. Wyvratt, M. J., Tristram, E. W., Ikeler, T. J., Lohr, N., Joshua, H., Springer, J. P., Arison, B., and Patchett, A. A., *J. Org. Chem.*, *49*, 2816-2819 (1984).

14. Kaltenbronn, J. S., DeJohn, D., and Krolls, U., *Org. Prep. Proced. Int.*, 35-40 (1983).

15. Urbach, H., and Henning, R. *Tetrahedron Lett.*, 1143-1146 (1984).

16. Wu, M. T., Douglas, A. W., Ondeyka, D. L., Payne, L. G., Ikeler, T. J., Joshua, H., and Patchett, A. A., *J. Pharm. Sci.*, 352-354 (1985).

17. For a discussion on the asymmetric hydrogenation of Schiff bases derived from α-keto esters and chiral amino acid esters, see Harada, K., and Shiono, S., *Bull. Chem. Soc. Jpn.*, *54*, 1367-1370 (1984).

18. Weinstock, L. M., Currie, R. B., and Lovell, A. V., *Synth. Commun.*, *11*:2, 943-946 (1981).

19. Brown, C. A., and Brown, H. C., *J. Am. Chem. Soc.*, *85*, 1003 (1963).

20. Brown, H. C., and Brown, C. A., *J. Am. Chem. Soc.*, *85*, 1005 (1963).

21. Witkop, B., and Beiler, T. W., *J. Am. Chem. Soc.*, *76*, 5589-5596 (1954).

22. *CHARMm—Chemistry at Harvard Macromolecular Modelling*, ©1987, Polygen Corp., Waltham, MA 02254. Energies were calculated at +1.362, -9.9548, +9.9948, and -1.1094 kcal/mol for 13a, 1a, 13b, and 7a, respectively. Many thanks to Dr. J. Toney for his help.

23. Cotton, F. A., and Wilkinson, G., *Advanced Inorganic Chemistry*, 3rd ed., Wiley-Interscience, New York, 1972, p. 903.

24. Blacklock, T. J., Butcher, J. W., Sohar, P., Rothauser-Lamanec, T., and Grabowski, E. J. J., *J. Org. Chem.*, *54*, 3907-3913 (1989).

25. Bonney, R. J., and Smith, R. J., *Advances in Inflammation Research*, Vol. 11 (I. Otterness et al., eds.), Raven Press, New York, 1986.

26. *Chem. Eng. News*, *65*, 37-38 (1987): preliminary report on the ACS national meeting in Denver, April 6, 1987.

27. Doherty, J. B., Finke, P. E., Firestone, R. A., Shah, S. K., and Thompson, K. R., U.S. Patent 4,637,999 (1987).
28. Doherty, J. B., Ashe, B. M., Argenbright, L. W., Barker, P. L., Bonney, R. J., Chandler, G. O., Dahlgren, M. E., Dorn, C. P., Jr., Finke, P. E., Firestone, R. A., Fletcher, D., Hagmann, W. K., Mumford, R., O'Grady, L., Maycock, A. L., Pisano, J. M., Shah, S. K., Thompson, K. R., and Zimmerman, M., *Nature*, *322*, 192-194 (1986).
29. Doherty, J. B., Shah, S. K., and Finke, P. E., *J. Med. Chem.*, in press.
30. John, D. I., in *Recent Advances in the Chemistry of β-Lactam Antibiotics*, Special Publication of the Royal Society of Chemistry, *52*, 193-208 (1985).
31. Stedman, R. J., *J. Med. Chem.*, *9*, 444 (1966).
32. Durckheimer, W., Klesel, N., Limbert, M., Schrinner, E., Seeger, K., and Seliger, H., in *Recent Advances in Chemistry of β-Lactam Antibiotics*, Special Publication of the Royal Society of Chemistry, *52*, 47-56 (1985).
33. Connon, N. W., *Eastman Org. Chem. Bull.*, *44*, 1-4 (1972).
34. Schultz, H. S., Freyermuth, H. B., and Buc, S. R., *J. Org. Chem.*, *28*, 1140-1142 (1963).
35. In-house determination by L. Williams and R. Cutro.
36. Fieser, L. F., and Fieser, M., *Reagents for Organic Synthesis*, Vol. 1, Wiley, New York, 1967, p. 475.
37. Kahr, K., and Berther, C., *Berichte*, *13*, 132 (1960).
38. Paulissen, R., Reimlinger, H., Hayez, E., Hubert, A. J., and Teyssie, P., *Tetrahedron Lett.*, 2233 (1973).
39. Wiering, J. S., and Wynberg, H., *J. Org. Chem.*, *41*, 1574-1578 (1976).
40. Levin, N., and Hartung, W. H., *Organic Syntheses*, Coll. Vol. III, Wiley, New York, 1955, pp. 191-193.

5

The Reductive Alkylation of Aromatic Amines with Formaldehyde

Anthony P. Bonds and Harold Greenfield

First Chemical Corporation, Pascagoula, Mississippi

I. INTRODUCTION

The purpose of this investigation was to find optimum conditions for the manufacture of N,N-dimethyl aromatic amines by the reductive alkylation of primary aromatic amines with formaldehyde.

$$ArNH_2 + 2CH_2O + 2H_2 \xrightarrow{Pd/C} ArN(CH_3)_2 + 2H_2O$$

The reductive alkylation of a primary aromatic amine with an aliphatic aldehyde forms the secondary N-alkylarylamine, which can further react with additional aldehyde to give the tertiary N,N-dialkylarylamine.

$$ArNH_2 + RCHO + H_2 \xrightarrow{cat} ArNHCH_2R + H_2O$$

$$ArNHCH_2R + RCHO + H_2 \xrightarrow{cat} ArN(CH_2R)_2 + H_2O$$

The initial product of the condensation of the primary amine and aldehyde is a carbinolamine, I,

$$ArNH_2 + RCHO \rightleftharpoons ArNHCHR-OH$$
$$I$$

which can then be dehydrated to the aldimine, II:

$$I + H_2 \rightleftharpoons ArN=CHR + H_2O$$
$$II$$

The secondary amine product can be formed either by hydrogenolysis of *I* or by the hydrogenation of *II*.

$$I + H_2 \xrightarrow{\text{cat}} ArNHCH_2R + H_2O$$

$$II + H_2 \xrightarrow{\text{cat}} ArNHCH_2R$$

The condensation of the secondary amine with a second molecule of aldehyde gives carbinolamine, *III*,

$$ArNHCH_2R + RCHO \rightleftharpoons ArN(CH_2R)-CHR-OH$$
$$III$$

which cannot dehydrate to an imine. Therefore the only route to the tertiary amine product is by hydrogenolysis of *III*:

$$III + H_2 \xrightarrow{\text{cat}} ArN(CH_2R)_2 + H_2O$$

Although *N*-hydroxymethylarylamines (carbinolamine *I*, R = H) from aromatic amines have not been isolated, Abrams and Kallan [1] have shown that *N*-hydroxymethylarylamines, *I*, not the imines, *II*, are formed by the reaction of anilines and formaldehyde in aqueous solution under mild conditions. At higher formaldehyde concentrations, addition of formaldehyde to *I* gives the *N*,*N*-dihydroxymethylarylamine, $ArN(CH_2OH)_2$.

N-Methyleneanilines (*II*, R = H) were not isolated, but were shown to be stable at -60°C [2]. Steric hindrance from alkyl substitution at the ortho positions promotes the stability of *N*-methyleneanilines, presumably by kinetically stabilizing the aldimine against oligomerization [3]. *N*-Methylene-2,6-dimethylaniline was observed, though not isolated, at room temperature in equilibrium with an oligomeric species [3]. *N*-Methylene-2,6-diisopropylaniline [4], *N*-methylene-2-*t*-butyl-6-methylaniline [3], and *N*-methylene-2,6-diethylaniline [5] are all stable

The reaction of arylamines with formaldehyde can be very complex [6]. Depending on conditions, particularly pH, a large number of products, including the cyclic trimer triaryltrimethylenetriamine (hexahydro-s-triazines) and linear oligimers can be obtained.

The nature of the aldehyde species itself is also complex. It may exist to varying degrees as the hydrate in water and as the hemiacetal and acetal in alcohols. The structure of the alcohol has a marked effect on the position of equilibrium. Primary alcohols are more effective than secondary alcohols, which are more effective than tertiary alcohols in hemiacetal and acetal formation [7].

The gradual addition of an aldehyde to an amine in order to maintain low concentrations of the aldehyde is an effective procedure for minimizing side reactions involving the aldehyde during reductive alkylations. This technique has been used for reductive alkylations of aliphatic [8-13], alicyclic [10], and aromatic amines [10,14-17], and also for ammonia [18] and acid hydrazides [19].

Paraformaldehyde, a mixture of polyoxymethylene glycols, can be used as the source of formaldehyde instead of the available aqueous or methanolic solutions of formaldehyde [20]. This permits the addition of all the solid paraformaldehyde reagent to the reaction mixture without the complication of adding a liquid (the formaldehyde solution) to a vessel under pressure. It also permits the use of solvents other than water and/or methanol. The paraformaldehyde gradually depolymerizes, providing a relatively low concentration of formaldehyde. Although the use of paraformaldehyde is convenient on a laboratory scale, it cannot provide the control of formaldehyde concentration possible by addition of a solution of formaldehyde.

Palladium was the catalyst of choice because of its known relatively poor activity for aliphatic carbonyl reduction and aromatic ring hydrogenation [21]. In the reductive alkylation of aniline with a ketone, there is considerable reduction of the ketone to the alcohol with platinum and very little with palladium catalysts [22]. Platinum is severely inhibited for both the desired reductive alkylation and for nuclear hydrogenation [22], presumably because of the strongly adsorbed alicyclic amines formed by initial ring hydrogenation [23]. Indeed, platinum-catalyzed reductive alkylations frequently are conducted with an acid promotor.

II. EXPERIMENTAL

A. Equipment

Laboratory experiments were run in a 1000 mL Parr reactor equipped with a magnetically driven agitator having two turbine-type impellers, one located near the bottom of the vessel and the other located about middle of the reactor, and normally operated at the maximum speed. Other components were a spiral cooling coil, a gas inlet valve, a vent valve, a Bourdon-tube pressure gage, a safety rupture disk, a thermocouple well and thermocouple, a heating mantle, and a time-proportional temperature controller with solenoid valve controlled cooling water.

The autoclave was connected through a pressure regulator to a 2-liter tank fitted with a gage. The tank could be intermittently pressured with hydrogen from a cylinder. Thus, the autoclave could be maintained at a specified pressure while the pressure drop in the tank could be measured.

The pump used to feed the aqueous formaldehyde was an Eldex precision metering pump capable of flows from 0 to 5.0 mL/min at head pressures up to 1000 psig.

B. Materials

The organic feeds used in this study consisted of aniline, o-toluidine (OTOL), m-toluidine (MTOL), and p-toluidine (PTOL). All were produced at and obtained from the First Chemical Corporation. Methanol, 2-propanol, and toluene were reagent grade. Hydrogen was obtained from cylinders and purchased locally.

Water-wet 5% palladium-on-carbon catalysts were received from Degussa, Engelhard, and Johnson Matthey.

Aqueous formaldehyde (37% concentration), containing 10-15% methanol as a stabilizer, and paraformaldehyde (95% purity) were purchased from Aldrich Chemicals.

C. Procedure

A typical procedure is described in detail.

Placed in the 1-liter Parr reactor were 0.8537 g (dry basis) of water-wet 5% Pd/C, 171.4 g (1.60 mol) of o-toluidine, and 55 mL of toluene. The reactor was sealed and purged with hydrogen and then pressured to 120 psig. The vessel was heated, with agitation, to 120°C. Pumping of 277.5 g (3.42 mol) 37% aqueous formaldehyde was then started and continued for 9.0 hours. When the addition of the formaldehyde was complete, the reaction was allowed to continue until gas absorption ceased, which was usually about 1.0 hour. The reaction was then cooled to ambient and the contents filtered through Celite filter-aid to remove the catalyst. The filtrate was then transferred to a 1-liter separatory funnel, where the organic and aqueous layers were allowed to separate for 30 minutes.

D. Analytical

Analyses were done using a Varian 3700 gas-chromatograph equipped with a 30-meter Superox capillary column, flame ionization detector, nitrogen carrier gas flowing at 1.0 mL/min, detector temperature of 270°C, injector temperature of 250°C, and the following column temperature program: 120°C for 5.0 minutes, with an increase to 230°C at 15°C/min, then held at 230°C for 5.0 minutes.

Butylbenzene was used as the internal standard for quantitative analyses. The order of elution was toluene, butylbenzene, the N,N-dimethyl derivative, the N-methyl derivative, and finally the primary aromatic amine.

Qualitative analyses were done using a Hewlett-Packard 59970C gas-chromatograph-mass spectrometer (GCMS) with a column and conditions as described for the GC above, but using helium carrier gas.

III. RESULTS

A. Catalyst Testing

Several commercial 5% palladium-on-carbon catalysts were tested and compared for activity in the reductive alkylation of p-toluidine to N,N-dimethyl-p-toluidine. The results are given in Table 5.1.

B. Effect of the Formaldehyde Addition Rate and Order of Addition

Table 5.2 lists the effects of the formaldehyde addition rate and the order of addition.

C. Effect of Solvent

The effects of various solvents when used with aqueous formaldehyde and paraformaldehyde are given in Tables 5.3 and 5.4, respectively. Table 5.5 shows the effect of varying amounts of a particular solvent, toluene.

Table 5.1 Catalyst Testing[a]

Vendor	Catalyst	Conversion (%)[b]
Engelhard	1	80
Engelhard	2	72
Degussa	3	71
Engelhard	4	66
Degussa	5	64
Degussa	6	64
Johnson Matthey	7	59
Degussa	8	56

[a]Each experiment was run with 180 g (1.68 mol) p-toluidine, 273.8 g (3.38 mol) 37% aqueous formaldehyde, 16 mL toluene, and 0.3428 g (dry basis) of catalyst at 80°C and 120 psig. The formaldehyde was added over 2 hours at a constant rate.
[b]Percentage of theoretical hydrogen consumed.

Table 5.2   Effect of Rate of Formaldehyde Addition and
Addition Order[a]

Addition type	Addition rate (mL/min)	Reaction time (h)	Yield of N,N-dimethyl-o-toluidine (%)[b]
Reverse	0.55	3.8	11
All at once		4.7	22
Fast	1.10	3.7	77
Slow	0.55	9.0	89

[a]Each reaction was run with 171.4 g (1.60 mol) o-toluidine, 55 mL
toluene, 277.5 g (3.42 mol) 37% aqueous formaldehyde, and 0.8537 g
(dry basis) Degussa E196 R/W 5% Pd/C at 120°C and 120 psig.
When the formaldehyde addition was complete, the reactor was
immediately cooled and sampled.
[b]Obtained using GC internal standard methods.

Table 5.3   Effect of Solvent: Aqueous Formaldehyde[a]

Solvent	Catalyst concentration (g/L)	Yield of DMOT (%)[b]
None[c]	5.0	77
Methanol	3.5	64
2-Propanol	3.5	72
Toluene	3.5	77

[a]Each reaction with a solvent used 171.4 g (1.6 mol) o-toluidine,
73.3 mL solvent, 277.5 g (3.42 mol) 37% aqueous formaldehyde,
and 0.8537 g (dry basis) Degussa E196 R/W 5% Pd/C at 120°C
and 120 psig for 3.7-3.8 hours.
[b]Obtained using GC internal standard methods.
[c]This reaction was run with 200.8 g (1.87 mol) o-toluidine,
325 g (4.01 mol) 37% aqueous formaldehyde, and 1.0 g (dry basis)
Degussa E196 R/W 5% Pd/C at 120°C and 120 psig for 3.8 hours.

Table 5.4   Effect of Solvent: 95% Paraformaldehyde[a]

Solvent	Reaction time (h)	Yield of DMOT (%)[b]
None[c]	5.2	54
Methanol	2.3	84
2-Propanol	2.9	84
Toluene	3.5	85

[a]Each reaction with a solvent used 126.8 g (1.19 mol) *o*-toluidine, 129.5 mL solvent, 78.1 g (2.47 mol) 95% paraformaldehyde, and 1.2677 g (dry basis) Degussa E196 R/W 5% Pd/C at 120°C and 250 psig.
[b]Obtained using GC internal standard methods.
[c]This reaction was run with 200.8 g (1.88 mol) *o*-toluidine, 124.5 g (3.93 mol) 95% paraformaldehyde, and 2.0 g (dry basis) Degussa E196 R/W 5% Pd/C at 120°C and 250 psig.

Table 5.5   Effect of Solvent Ratio[a]

o-Toluidine (mL)	Toluene (mL vol %)[b]		CH_2O (mL)	CH_2O addition time (h)	Yield of DMOT (%)[c]	Productivity[d]
200	0	0	300	4.5	70	78.5
192	21	10	287	4.3	78	87.9
171	73	30	256	3.9	76	84.1
143	143	50	214	3.4	62	65.8

[a]Each reaction was run with 0.498 wt % of Degussa E196 R/W 5% Pd/C based on *o*-toluidine at 120°C and 120 psig with a total charge volume of 500 mL.
[b]100 × mL toluene/(mL *o*-toluidine + mL toluene).
[c]Obtained using GC internal standard methods.
[d]100 × g of *N*,*N*-dimethyl-*o*-toluidine/liter-hour.

D. Stepwise Nature of the Reductive Alkylation
of *o*-Toluidine

The relative amounts of primary aromatic amine, *N*-alkylamine, and
N,*N*-dialkylamine are given as a function of the percentage of
theoretical formaldehyde charged in Table 5.6.

Table 5.6   Reductive Alkylation of *o*-Toluidine[a]

Theoretical CH_2O (%)	Yield (%)[b]				Alkylation (%)
	DMOT	NMOT	OTOL	Others	
25	4.3	42.2	51.1	2.5	25.4
50	18.8	61.6	16.7	2.9	49.6
75	49.2	46.4	1.5	2.9	72.5
107	95.9	1.0	0.1	3.1	96.4

[a]Each reaction was run with 171.2 g (1.6 mol) *o*-toluidine, 55 mL
toluene, aqueous formaldehyde added at a flow rate of 0.4-0.5
mL/min, and 0.8518 g (dry basis) Degussa E196 R/W 5% Pd/C at
120°C and 120 psig.
[b]All analyses are GC relative area percent.

Table 5.7   Effect of Structure on Rate: Noncompetitive Reaction[a]

Amine	Yield (%)[b]				Alkylation (%)
	3° Amine	2° Amine	1° Amine	Others[c]	
Aniline	54	30	7	9	69
o-Toluidine	46	43	1	10	68
m-Toluidine	57	29	7	7	72
p-Toluidine	64	25	9	2	77

[a]Each reaction with a toluidine used 171.2 g (1.60 mol) toluidine,
55 mL of toluene, 0.8518 g (dry basis) Degussa E196 R/W 5% Pd/C,
and 195 g (2.4 mol, 75% of theoretical) 37% aqueous formaldehyde.
The aqueous formaldehyde was added over 4.0 hours at a constant
rate at 120°C and 120 psig. The aniline reaction used 148.8 g (1.60
mol) aniline plus an additional 20 mL toluene to keep all volumes
the same. All other charges and conditions were the same for aniline
as for the toluidines.
[b]All analyses were by GC internal standard methods.
[c]By difference.

E. Effect of Structure on Rate

The effects of structure on the rate of reductive alkylation for noncompetitive and competitive reactions are given in Tables 5.7 and 5.8, respectively.

F. Catalyst Reuse

The results of catalyst reuse experiments are given in Table 5.9.

Table 5.8 Effects of Structure: Competitive Reaction[a]

| Amine | Yield (%)[b] | | | Alkylation (%) |
	3° Amine	2° Amine	1° Amine	
Aniline	65	29	6	80
o-Toluidine	5	51	44	31
m-Toluidine	67	27	6	81
p-Toluidine	92	7	1	96

[a]This reaction was run with a mixture of 34.23 g (0.4 mol) of aniline and 42.84 g (0.4 mol) of each toluidine, 55 mL toluene, and 0.8518 g (dry basis) Degussa E196 R/W 5% Pd/C at 120°C and 120 psig; 195 g (2.4 mol) of 37% aqueous formaldehyde was added over 3 hours at a constant rate.
[b]Analyses are GC relative area percent normalized to 100%.

Table 5.9 Catalyst Reuse[a]

State of catalyst	Yield of DMPT (%)[b]
Initial use	99
First use	99
Second use	99

[a]Each reaction used 180 g (1.68 mol) of p-toluidine, 17 mL toluene, and 273.8 g (3.38 mol) of 37% aqueous formaldehyde added over 4 hours at a constant rate at 120°C and 250 psig. The initial reaction was charged with 0.3368 g (dry basis) Degussa E196 R/W 5% Pd/C. After each reaction this catalyst was filtered through Celite and charged (including Celite) to the next reuse experiment
[b]Obtained using GC internal standard methods.

IV. DISCUSSION

The catalyst of choice for the reductive alkylation of primary
aromatic amines is usually a palladium-on-carbon with a metal loading
of about 5% for liquid phase batch processes. Eight such commercial
catalysts were tested (Table 5.1), and a range of activities was
found.

A.   Effect of Rate of Formaldehyde Addition
and Order of Addition

The results in Table 5.2 illustrate the known advantage [8-19]
of maintaining a low concentration of the aldehyde. A gradual addition of the aqueous formaldehyde was far superior to having all
the aldehyde present at the beginning of the reaction, and the
slower rate of addition gave better results than the more rapid
addition. The optimum rate of addition was not determined, and
the addition rate probably should be varied, since the reaction
rate changes during a single experiment. A low concentration of
aldehyde does indeed minimize side reactions and catalyst deactivation, but too low a concentration could produce an undesirable
decrease in the reductive alkylation rate. In the ideal system to
control the addition of the formaldehyde, the rate of addition would
automatically adjust to the rate of formaldehyde consumption.

The reverse addition (addition of the amine to the formaldehyde)
resulted in even more catalyst deactivation than having all of the
amine and aldehyde present at the beginning. The reason for this
is not clear.

B.   Effect of Solvent

The effect of solvents is shown in Tables 5.3 and 5.4.

Toluene results in a faster conversion of o-toluidine (OTOL)
to N,N-dimethyl-o-toluidine (DMOT) than methanol or 2-propanol
with aqueous formaldehyde at 120 psig (Table 5.3). This may be
due, at least in part, to the lower vapor pressure of toluene,
permitting a higher partial pressure of hydrogen. The reaction
rate without solvent was the same as with toluene. It should be
noted that the catalyst level without solvent was the same based
on OTOL, but higher in terms of actual concentration.

With paraformaldehyde as the alkylating agent at 250 psig
(Table 5.4), the three solvents gave comparable rates of conversion
to DMOT, significantly faster than with no solvent. Again, the
reaction without solvent had the same catalyst level based on OTOL,
but a higher actual concentration.

We are dealing with a multistep reaction: some steps are reversible, some are not; some are homogeneous, some are heterogeneous. In addition, some solvents, such as alcohols (and water), interact with formaldehyde. Clearly, a simple understanding of solvent effects is not attainable.

Toluene has some practical advantages over the lower alcohols. Its vapor pressure is lower, it enhances separation of the organic and aqueous phases during workup of the reaction product, it helps to remove water from the organic phase by azeotropic distillation, and it is much less likely to ignite in contact with dry catalyst.

The data in Table 5.5 indicate that the optimum productivity is at about 10 vol % toluene. Although the catalyst level based on OTOL remains the same, the catalyst concentration decreases with increasing amounts of solvent.

C. Stepwise Nature of the Reaction

The stepwise nature of the reductive alkylation of OTOL with aqueous formaldehyde is illustrated in Table 5.6 and Figure 5.1.

The preferential formation of the monoalkylated product, NMOT, during the first half of the formaldehyde addition, suggests that high yields of the N-alkylamine could be achieved by adding one mole of aldehyde per mole of starting primary amine. This has been accomplished quite selectively to produce N-ethyl-m-toluidine by the reductive alkylation of m-toluidine with acetaldehyde [24].

D. Effect of Structure on Rate

The relative reactivities of aniline and the three isomeric toluidines (OTOL, MTOL, and PTOL) have been determined under both non-competitive (Table 5.7) and competitive (Table 5.8) conditions. The results under competitive conditions (i.e., the use of a mixture of the four primary amines) is a reflection of relative strengths of absorption of the hydrogenation intermediates on the catalyst as well as intrinsic reaction rates.

In noncompetitive reactions, based on the total amount of alkylation, PTOL is the most reactive and MTOL slightly more reactive than aniline and OTOL, which are very similar. Although the overall rates of alkylation of the MTOL, aniline, and OTOL are not very different, the ortho-substituted material has particular difficulty in converting the monoalkylated secondary amine to the dialkylated tertiary amine. This suggests that ring substitution at the ortho position would enhance the selectivity to monoalkylation as opposed to dialkylation.

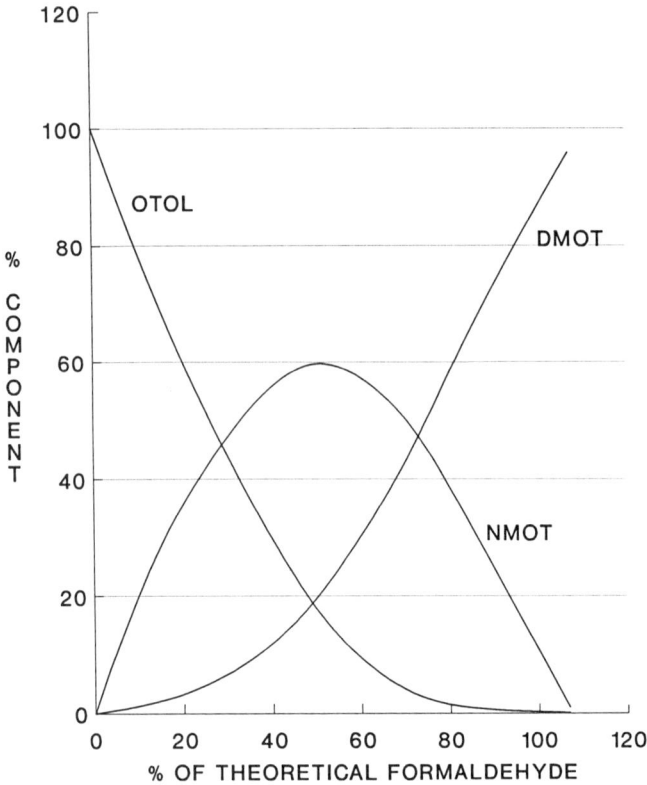

Figure 5.1   Alkylation of *o*-toluidine.

        It should be noted that "Others" in Table 5.7 may include inter-
mediates as well as by-products.
        The differences in reactivity in the competitive reaction
(Table 5.8) are much accentuated, and are in the following order:

        PTOL >> MTOL, aniline >> OTOL

Relative absorption strengths of the hydrogenation intermediates
appear to be playing an important role, particularly with OTOL.
        The activity of the para isomer shows that the reaction rate
is enhanced by the electron-releasing effect of the *p*-methyl substituent. This effect is obviated by the steric requirements of the
ortho isomer. Obviously, the overall rate is favored by a high
electron density at the nitrogen, but it is also very sensitive to

steric hindrance. Applying these facts to a mechanistic interpretation is very difficult, not only because several consecutive reactions and equilibria are involved, but because we lack information as to which step or steps might be rate-determining.

E. Catalyst Reuse

The data in Table 5.9 indicate that at 120°C and 250 psig, at least two reuses can be obtained without any detectable loss of catalyst activity or product yield.

V. CONCLUSION

An efficient process has been developed for the palladium-catalyzed reductive alkylation of aromatic amines with formaldehyde to the corresponding N,N-dimethylaryamines.

REFERENCES

1. Abrams, W. R., and Kallen, R. G., *J. Am. Chem. Soc.*, *98*, 7777 (1976), and references therein.
2. Barluenga, J., Bayon, A. M., and Asenio, G., *J. Chem. Soc.*, *Chem. Commun.*, 1109 (1983).
3. Cortolano, F. P., Pastor, S. D., Ravichandran, R., and Steinberg, D. H., *Tetrahedron Lett.*, *29*, 5875 (1988).
4. Verardo, G., Cauci, S., and Giumanini, A. G., *J. Chem. Soc.*, *Chem. Commun.*, 1787 (1985).
5. Giumanini, A. G., Verardo, G., and Poliana, M. J., *J. Prakt. Chem.*, *303*, 161 (1988).
6. Sprung, M. M., *Chem. Rev.*, *26*, 297 (1940).
7. Ogata, Y., and Kawasaki, A., in *The Chemistry of the Carbonyl Group*, Vol. 2 (J. Zabicky, ed.), Wiley-Interscience, New York, 1970, p. 15.
8. Olin, J. F., and Schwoegler, E. J., U.S. Patent 2,373,705, to Sharples Chemicals (Apr. 17, 1945).
9. Thirion, P., U.S. Patent 3,442,951, to Ugine Kuhlman (May 6, 1969).
10. Brake, L. D., U.S. Patent 3,597,438, to DuPont (Aug. 3, 1971).
11. Hinty, H. J., U.S. Patent 4,163,026, to Monsanto (July 31, 1979).
12. Petree, H. E., and Nabors, J. B., U.S. Patent 4,275,238, to Ciba-Geigy (June 23, 1981).

13. Tahara, S., Nishihira, K., Miyatake, T., Sawada, H., and Kita, J., U.S. Patent 4,373,107, to Ube Industries (Feb. 8, 1983).
14. Martin, J. C. and Hasek, R. H., U.S. Patent 2,947,784, to Eastman Kodak (Aug. 2, 1960).
15. Rosas, C. B., Weinstock, C. M., and Jones, W. H., *Ann. N.Y. Acad. Sci.*, *214*, 94 (1973).
16. Matsui, M., and Misaka, Y., Japan Kokai 77 71,424, to Mitsubishi Chemical Industries (June 14, 1977).
17. Mitsubishi Petrochemical Company, Japan Kokai Tokkyo Koho JP 82 81,444 (May 21, 1982).
18. Thoma, J. A., and Deumens, J. J. M., U.S. Patent 3,658,842, to Stamicarbon N.V. (Apr. 25, 1972).
19. Grim, R. A., Randen, N. A., and Demas, C. L., U.S. Patent 4,071,554, to Ashland Oil (Jan. 31, 1978).
20. Malz, R. E., Jr., and Greenfield, H., in *Catalysis in Organic Synthesis* (W. H. Jones, ed.), Academic Press, New York, 1980, p. 49.
21. Rylander, P. N., *Hydrogenation Methods*, Academic Press, New York, 1985, pp. 66, 117.
22. Dovell, F. S. and Greenfield, H., U.S. Rubber Company, unpublished work, 1962.
23. Greenfield, H., *Ann. N.Y. Acad. Sci.*, *214*, 233 (1973).
24. Bonds, A. P., First Chemical Corporation, unpublished work, 1990.

6

A Comparative Mass Transfer Study in the Reductive N-Alkylation of Aromatic Nitro Compounds

R. J. Malone and H. L. Merten

Herzog-Hart Corporation, Boston, Massachusetts

I. INTRODUCTION

For decades, serious equipment alternatives have been designed and have become commercially available to address the opportunities recognized in three-phased reactions. These opportunities arise from the limitations of the conventional stirred tank in terms of mass and heat transfer requirements, as well as the recognized difficulties associated with the scale-up of such equipment.

Earlier publications [1,2] have already shown that notable rate improvements are achievable with the Buss-Loop system when compared to the customary stirred-tank reactor. The reactions studied for these comparisons were the catalytic reduction of aromatic nitro compounds, the hydrogenation of unsaturated fatty acids, and the carbonylation of several amines and some others. More recent publications [4,5] compare engineering design information. Our brief study is based on a comparison of reaction rate data from plant and laboratory size stirred-tank reactors and from a pilot plant Buss-Loop reactor. These data were obtained on the catalytic N-alkylation reaction of an aromatic amine.

This reaction or reactions differ from above-mentioned chemistry because they encompass several steps that must occur in the proper sequence and require separate operating conditions, which reflect on greatly different heats of reaction. Several means exist to carry out this rather complex sequence, as is shown in the patent literature. The conditions used in the experiments described were essentially those in reference 3. The chemistry involved is shown in Figure 6.1, where 4-nitrodiphenylamine is hydrogenated to form

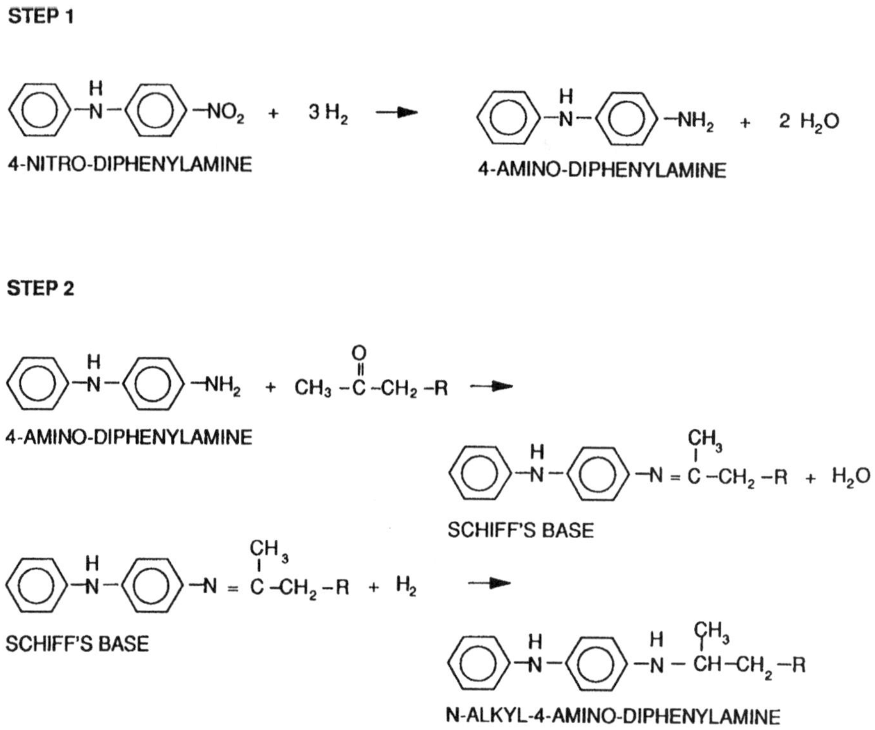

STEP 1

4-NITRO-DIPHENYLAMINE   4-AMINO-DIPHENYLAMINE

STEP 2

4-AMINO-DIPHENYLAMINE

SCHIFF'S BASE

SCHIFF'S BASE

N-ALKYL-4-AMINO-DIPHENYLAMINE

Figure 6.1 Principal chemistry of catalytic reductive alkylation.

4-aminodiphenylamine in the first step. In the second and third steps the Schiff base is generated which, during the hydrogenation, splits off water and converts the intermediate to the mono-N-alkyl-4-aminodiphenylamine.

Those familiar with this chemistry recognize that the reaction temperature and the hydrogen pressure for the first step must stay relatively low, not only to avoid hydrogenation of the aromatic ring, but also to permit the control of a very exothermic reaction, the reduction of the aromatic nitro group. The Buss-Loop reactor facilitates—as has already been recognized—an effective means for the very critical removal of heat with the external heat exchanger, but it also assures the best possible intimate contact of the three phases, which is kinetically needed for the constant regeneration of active hydrogen on the catalyst.

During the second step, an acidic catalyst is needed to form the Schiff base and to encourage the equilibrium, which in turn is brought out of balance during the hydrogenation to the product. Since the second and third steps are far slower and less exothermic when compared with the nitro reduction step, the reaction temperature is raised while observing the limits for the undesirable saturation of the aromatic rings. And, with thorough mixing available, the benefits of a higher hydrogen concentration (higher pressure) are applied.

From a chemist's point of view, one should expect from the use of a Buss-Loop reactor benefits similar to those shown for the other perhaps less complex reactions mentioned; but these assumptions had to be confirmed.

II. EXPERIMENTAL

To reduce the difficulties with other variables (e.g., raw materials), a rather large quantity of commercially manufactured 4-nitrodiphenyl-amine was isolated and blended. Similarly, a whole catalyst shipment

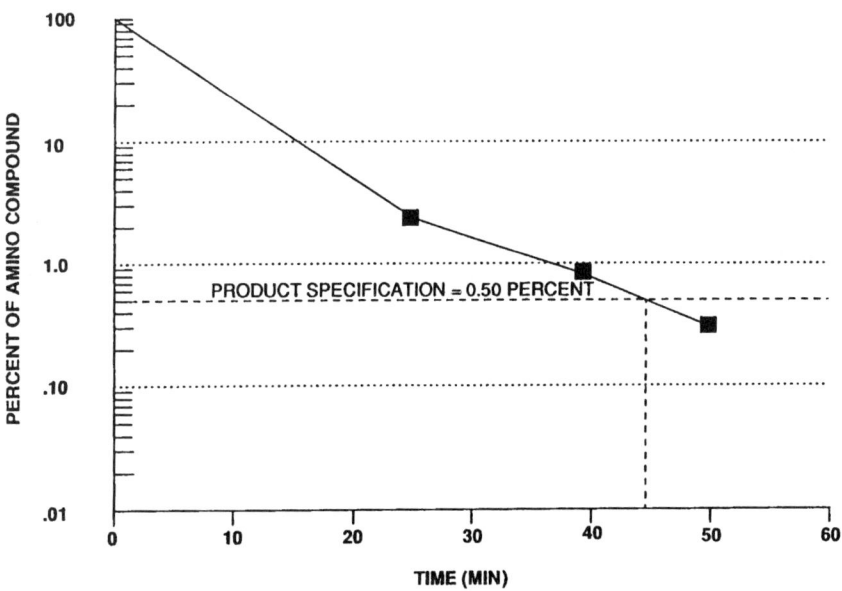

Figure 6.2 Percent of amino compound versus time.

was separated and shipped to the different test sites; the same
was done for the ketone used. Last, only compressed electrolytic
hydrogen was utilized at all locations.

For the initial step, the hydrogenation of the nitro group,
the flow of hydrogen was observed as pressure drop per unit time.
To measure the rates of the reaction during the reductive alkylation
steps, a standardized analytical procedure was agreed upon, samples
were exchanged, and the data confirmed. A semilogarithmic scale
was used on several samples taken throughout the reaction, and
the analytical results were plotted against the time at which the
sample had been taken. This permitted the projection of an end
point of the reaction as long as the rate followed a typical first-order
reaction plot (see Fig. 6.2).

The initial experiments with a stirred reactor were carried out
on plant scale with a charge of catalyst low enough to permit the
observation of a measurable decline in rate after several runs.
The same conditions and catalyst charge were then used in a 1-liter
reactor autoclave. The pilot plant experiments were carried out
in a 50-liter Buss-Loop reactor.

III.   RESULTS AND DISCUSSION

The reference tests were charried out in a plant autoclave, using
a catalyst charge of 0.65 wt %, and gave a total time cycle of
128 ± 12 minutes. The corresponding Buss-Loop reactor cycles
at an identical catalyst charge were 64 ± 6 minutes. This is an
improvement in rate of about 100%. Needless to say, a faster reaction
rate offers an opportunity to better the overall production rate,
provided all other operational steps can be adjusted proportionally.
An alternative consists in lowering the catalyst charge, provided
the minimum charge has not already been reached. Therefore, a
series of experiments in this direction was carried out. All results
are summarized in Table 6.1.

Figure 6.3 is a simplified drawing of the basic elements of a
stirred-tank reactor, including the customary cooling coils and
heating jacket, as well as a cutaway drawing of a typical Buss-Loop
reactor from earlier publications [5]. As pointed out elsewhere [5],
the available heat exchange area in a stirred-tank reactor is limited.
The heat exchange unit in the Buss-Loop reactor is external and
offers opportunities in simplicity and size.

In the loop reactor, the bulk of the turbulent mixing occurs
in the nozzle, which acts as an eductor, making intimate mixing
of the three phases possible. This is where the chemical reaction
occurs, while the body of the reactor functions as a reservoir for

Table 6.1 Experimental Data on Catalytic Reductive Alkylation

Industrial plant			50-Liter Buss loop			1-Liter lab-stirred tank		
Run number	Catalyst (wt %)	Run time (min)	Run number	Catalyst (wt %)	Run time (min)	Run number	Catalyst (wt %)	Run time (min)
48	0.53	130	26	0.53	65		0.65	55
49	0.65	110	27	0.65	57		0.65	68
50	0.65	145	28	0.65	70		0.65	59
51	0.65	115					0.65	70
53	0.65	125			Average 64		0.65	90
		Average 128						Average 68
			13	0.4	86			
			23	0.4	100			
			24	0.4	136			
			25	0.4	141			
					Average 116			

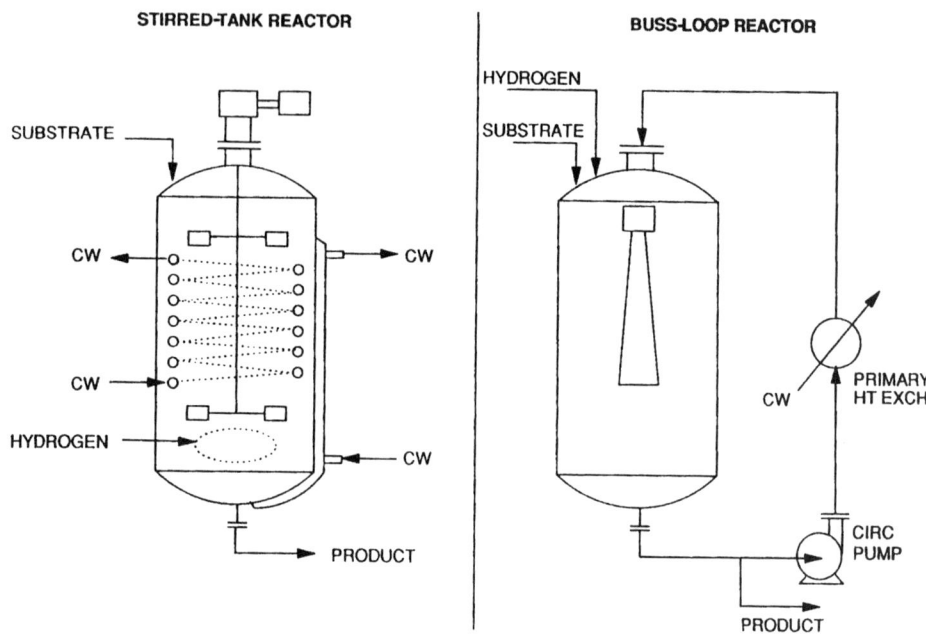

Figure 6.3   Stirred-tank and Buss-loop reactors.

the critical circulating pump. From Figure 6.3 it is understandable
that such a unit tolerates a relative wide margin for the charged
volume without a change in the efficiency of either the mass transfer
or the heat transfer of the system. This observation is supported
by a very satisfactory scale-up history of plant designs of Buss-
Loop reactors by Herzog and Hart in the United States and Buss AG
worldwide.

V.   CONCLUSIONS

1.   The Buss-Loop reactor showed the expected rate improvement
     for the catalytic *N*-alkylation reaction.
2.   Prediction of large-scale Buss-Loop reactor performance based
     on laboratory autoclave tests is possible due to optimized mixing
     in both systems. This has not been possible with stirred-tank
     reactors.
3.   The configuration of the Buss-Loop reactor allows for unlimited
     heat transfer, and the degree of gas/liquid/catalyst mixing is
     not sensitive to charge volume.

REFERENCES

1. Leuteritz, G. M., Reimann, P., and Vergeres, P., *Hydrochem. Process*, June 1976.
2. Malone, R. J., *Chem. Eng. Prog.*, 76:6 (1980).
3. Summers, C. G., U.S. Patent 3,576,767 (April 1971); H. Merten et al., U.S. Patent 4,900,868 (February 1990).
4. Von Dierendonck, L. L., and Leuteritz, G. M., in *6th European Conference on Mixing*, May 1988.
5. Concordia, J. J., *Chem. Eng. Prog.*, 86, 50-54 (1990).

7

Impurity Formation During the Reductive Alkylation of Aniline with Methyl Isoamyl Ketone

Joseph Stieber and Michael Reynolds

Uniroyal Chemical Company, Naugatuck, Connecticut

I. REDUCTIVE ALKYLATION

The reductive alkylation of an aromatic amine will proceed readily under the right conditions of temperature and hydrogen pressure, and use of a selective catalyst. The desired product from the reductive alkylation of aniline and methyl isoamyl ketone (MIAK) can be seen in Figure 7.1.

II. IMPURITIES

Low levels of impurities formed during reductive alkylation arise mainly from two sources. The first source consists of small percentages of undesired ketones present in the feedstock ketone, and side reactions such as (nuclear) ring hydrogenation of the intended product, hydrolysis of the ketimine, and further reductive alkylation of impurities.

A. Impurities from Methyl Isoamyl Ketone

We identified and quantified small percentages of undesired products from the reductive alkylation of aniline with acetone, isobutyraldehyde, and 4-methyl-2-pentanone. These contaminants present in the MIAK led to the formation of 0.05 wt % N-isopropylaniline, 0.01 wt % N-isobutylaniline, and 0.20 wt % N-(1,3-dimethylbutyl)aniline, respectively.

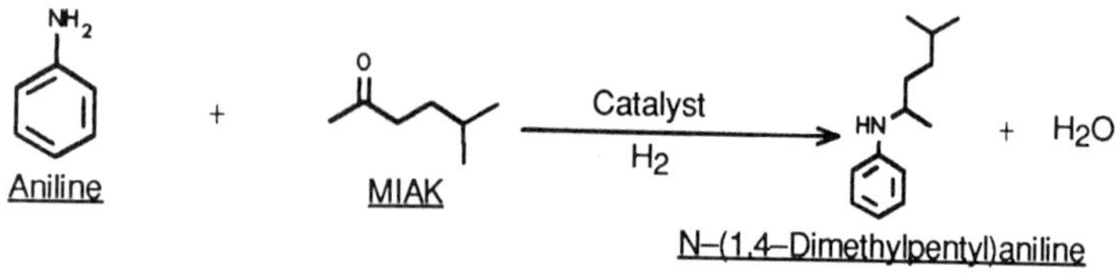

Figure 7.1

B. Impurities from Side Reactions

Upon examination of the reductive alkylation product between aniline and MIAK we identified the following compounds.

Compound	Weight percent
N,N-di(1,4-dimethylpentyl)amine	0.2-2.0
N-Cyclohexyl-N-(1,4-dimethylpentyl)amine	0.2-1.5
N-Cyclohexylaniline	0.5-1.5
5-Methyl-2-hexanol	0.3
N,N-di-(1,4-Dimethylpentyl)aniline	0.2

We postulated two separate mechanisms that would result in the formation of these impurities. The first mechanism can be seen in Figure 7.2, where partial ring hydrogenation of aniline leads to cyclohexeneamine (A), the intermediate for all the impurities we identified. Water formed from the predominant reductive alkylation hydrolyzes A to form cyclohexanone (B), and ammonia (C). Amination of MIAK occurs with the low concentration of ammonia to form E, N-1,4-dimethylpentylamine, which continues to be further alkylated to F, N,N-di-(1,4-dimethylpentyl)amine. The hydrolysis of cyclohexeneamine produces cyclohexanone, B. The reaction of aniline with cyclohexanone, B, produces N-cyclohexylaniline, D.

   The other impurities we identified are produced by the complete ring hydrogenation of the intermediate cyclohexeneamine, A. The hydrogenation of the intermediate leads to the formation of G, N-cyclohexylamine, which undergoes further reductive alkylation with MIAK to create N-cyclohexyl-N-1,4-dimethylpentylamine, H.

   The impurities formed from the common intermediate involve the combination of high concentration compounds, such as MIAK or aniline, with low concentration species. We can eliminate unlikely routes involving species of low concentration, like the alternate route to H, N-cyclohexyl-N-(1,4-dimethylpentyl)amine in Figure 7.3.

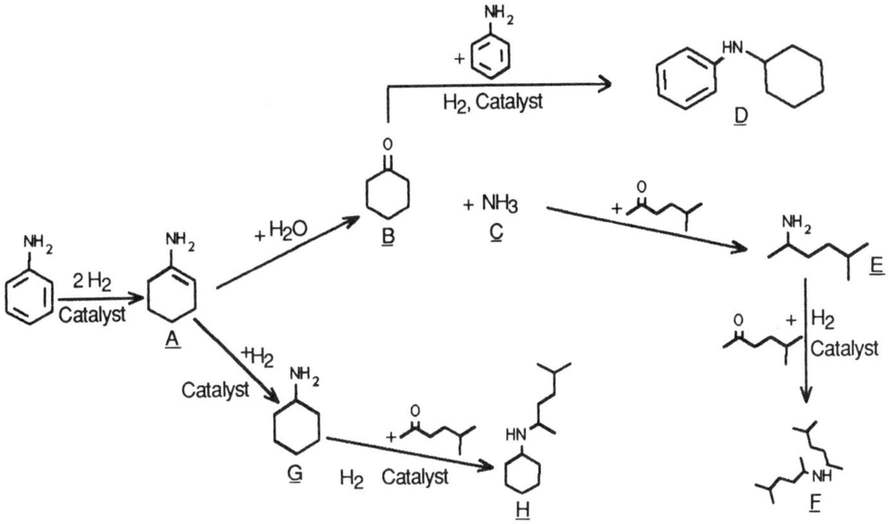

Figure 7.2

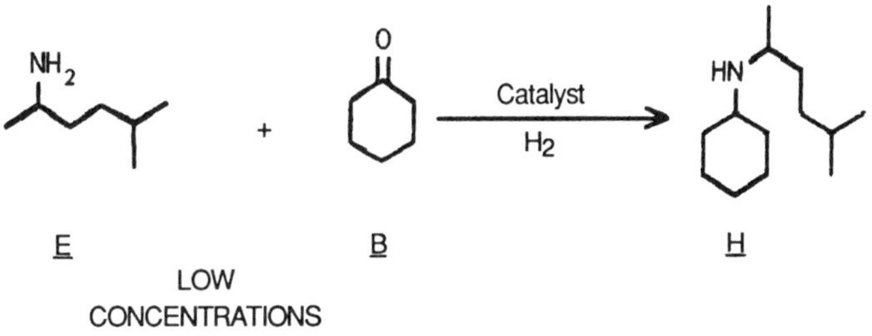

Figure 7.3

The second mechanism that can account for the impurities we identified is similar to the first except the common intermediate results from the partial ring hydrogenation of the desired product, N-1,4-dimethylpentylaniline (Fig. 7.4).

The overhydrogenation of the desired product leads to the eneamine, I, which can be further hydrogenated, or it may undergo hydrolysis. The hydrogenation of the intermediate leads to only one impurity H, N-cyclohexyl-N-(1,4-dimethylpentyl)amine. The impurities resulting from the hydrolysis of the intermediate differ only in the fact that N-1,4-dimethylpentylamine (E) is produced,

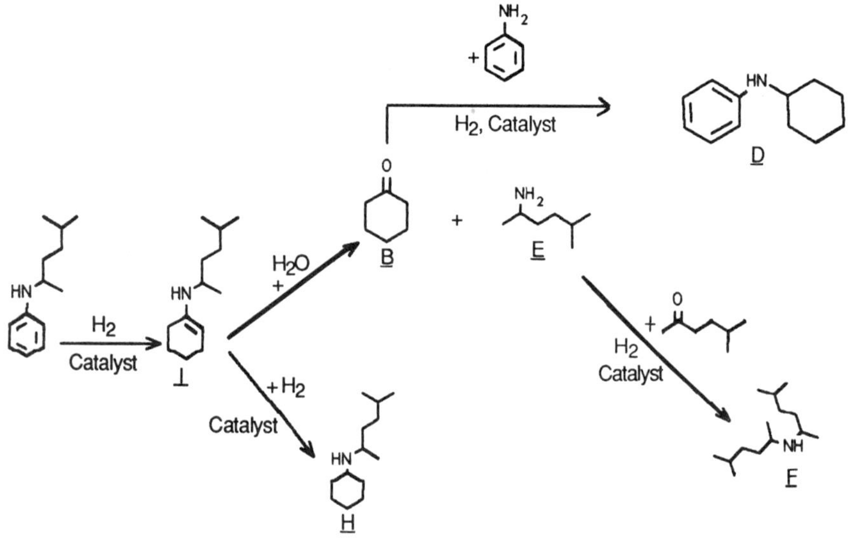

Figure 7.4

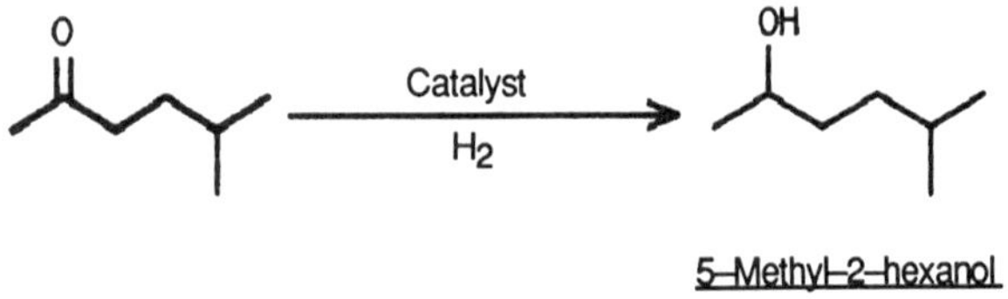

Figure 7.5

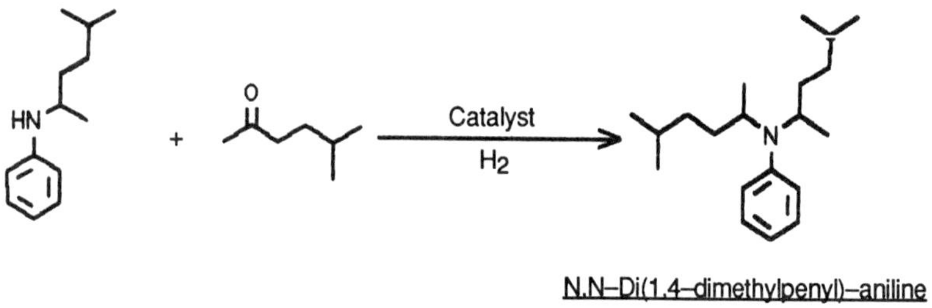

Figure 7.6

not ammonia. The remaining impurities formed from the hydrolysis route are produced by the same mechanisms as in the first postulate.

One of the two final compounds we identified, 5-methyl-2-hexanol, results from the hydrogenation of MIAK, which produces the corresponding alcohol seen in Figure 7.5. The final impurity isolated is from the overalkylation of the desired product with another molecule of MIAK to form N,N-di(1,4-dimethylpentyl)aniline (Fig. 7.6).

III. AVOIDANCE AND REMOVAL OF IMPURITIES

The levels of the various N-alkylanilines can be reduced by use of high quality raw materials—in this case, a grade of methyl isomyl ketone low in acetone, isobutyraldehyde, and 5-methyl-2-pentanone.

The levels of impurities from side reactions can be controlled only by the reaction conditions and by use of a selective catalyst. Once these impurities have been formed, they can be removed from the desired product (N-1,4-dimethylpentylaniline) only through careful distillation.

8

Mechanistic Selectivity in the Hydrogenation of Nitriles to Primary and Secondary Amines

J. L. Dallons and A. Van Gysel
UCB, Drogenbos, Belgium

G. Jannes
CERIA-ISI, Brussels, Belgium

I. INTRODUCTION

The hydrogenation of nitriles, one of the main industrial methods of amine synthesis, leads to a mixture of primary, secondary, and tertiary amines. The selectivity of this catalytic reaction depends on many different factors, such as the nature of the catalyst, the addition of ammonia, the temperature, and the solvent. Results from the literature are generally discussed using a mechanism based on the competition between heterogeneous hydrogenation reactions and homogeneous condensation reactions proposed by Braun et al. as long ago as 1923 [1].

We present some results that cannot be explained by using a mechanism comprising homogeneous intermediate condensations. We therefore propose another mechanism, using only surface reactions, which can explain the literature results as well.

II. EXPERIMENTAL

A. Kinetic Measurements

All hydrogenation experiments were done in a well-stirred autoclave (Labor-Druckruhrwerk Ingenieurburo SFE, Zurich) under a constant pressure of 8 bar. The apparatus allows the introduction of the reactants under pressure and is equipped with sampling and flushing attachments, as well as a thermocouple probe. In a preliminary stage of the work, we studied the experimental conditions, which were found to be free of external diffusion control. The stirring rate was 800 rpm. The introduction of the compound to be hydrogenated was taken as the zero time of the experiment.

B. Analysis

Products analysis was performed by gas chromatography. We used a 2 m glass column filled with Chromosorb 103, 60/80 mesh.

C. Products

All products were laboratory reagents.

III. RESULTS

A. Propionitrile Hydrogenation on Raney Nickel
     and on Rh/C

We hydrogenated propionitrile in methanol (2.5 vol%) on Raney nickel and on rhodium-on-carbon (Rh/C) catalysts at 55°C. The results are represented in Figures 8.1 and 8.2. We used a concentration-concentration graph, which gives the evolution of the mixture composition of the product during propionitrile hydrogenation.

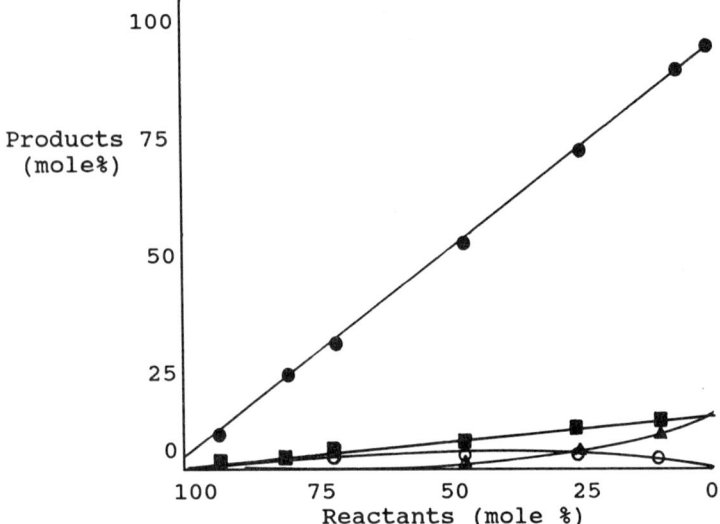

Figure 8.1 Propionitrile hydrogenation on Raney nickel:
(●) propylamine, (▲) dipropylamine, (○) dipropylimine, and (■) dipropylamine and dipropylimine.

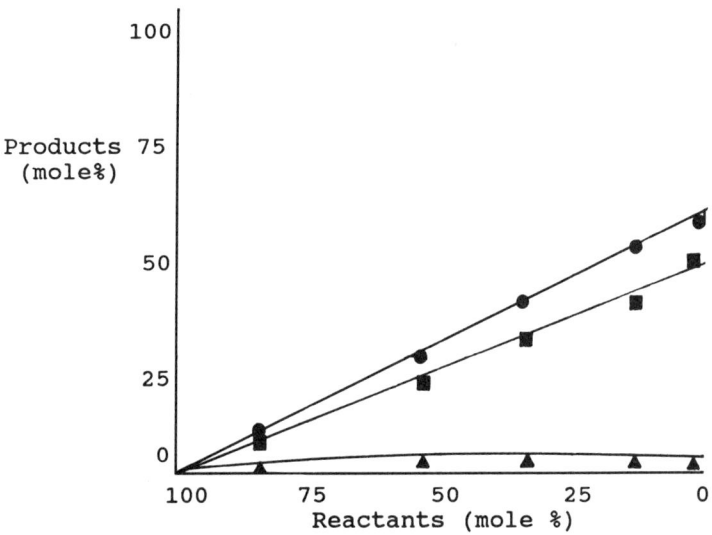

Figure 8.2 Propionitrile hydrogenation on Rh/C: (■) propylamine,
(▲) dipropylimine, and (●) dipropylamine and dipropylimine.

B. Effect of the Catalyst

We used mainly rhodium catalyst because of its selectivity for the
formation of secondary amines, the mechanism of which is not well
established. The evolution of the selectivity for secondary amine
formation is shown in Figure 8.3. The activation energy, using
an Arrhenius plot, was estimated to be 42 kJ/mol, which is an
indication of a chemically controlled reaction.

C. Hydrogenation in the Presence of Primary Amines

The addition of either propylamine or dipropylamine at the beginning
of the reaction (2.5 vol%) does not influence the reaction rate.
We hydrogenated reaction mixtures containing different nitrile-primary
amine couples to check the evolution of the asymmetrical/symmetrical
secondary amine ratios with conversion.

Table 8.1 gives the percentage of nitrile that was transformed
to secondary amines (asymmetrical and symmetrical); the proportion
of both amines at 100% conversion is also given.

Figure 8.4 shows, for example, the evolution of product composition with conversion in the case of the hydrogenation of propionitrile in the presence of butylamine.

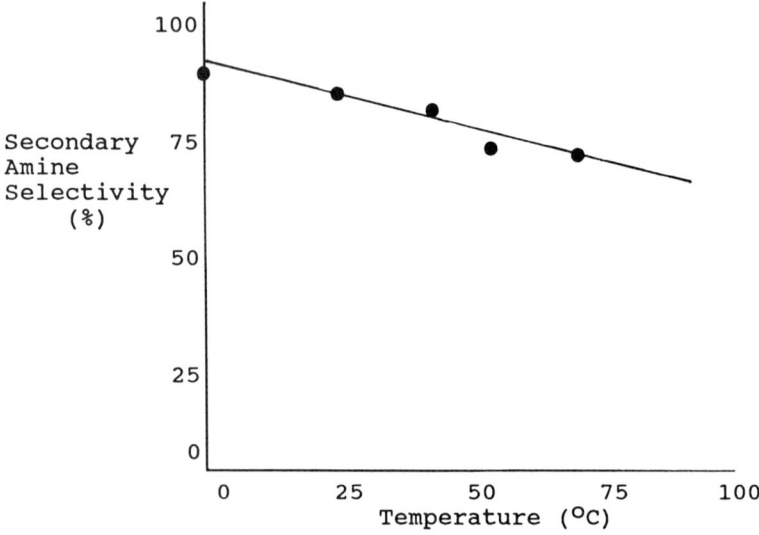

Figure 8.3   Influence of temperature on selectivity.

Table 8.1   Hydrogenation of Nitriles in the Presence of
Primary Amines (55°C)

| Reactants | | | Products (%) | |
| | | | | |
Nitrile	Amine	A_2	A_2 mixed	A_2 symmetrical
Propionitrile		71		
Propionitrile	Butylamine	66.5	67	33
Propionitrile	Pentylamine	66.7	50	50
Butyronitrile		68.9		
Butyronitrile	Propylamine	57	70	30
Butyronitrile	Pentylamine	59.8	64	36

    We hydrogenated propionitrile in methanol on two rhodium cata-
lysts, one supported on carbon and the other on alumina. The
selectivity for the secondary amine formation is totally different,
as indicated in Table 8.2.

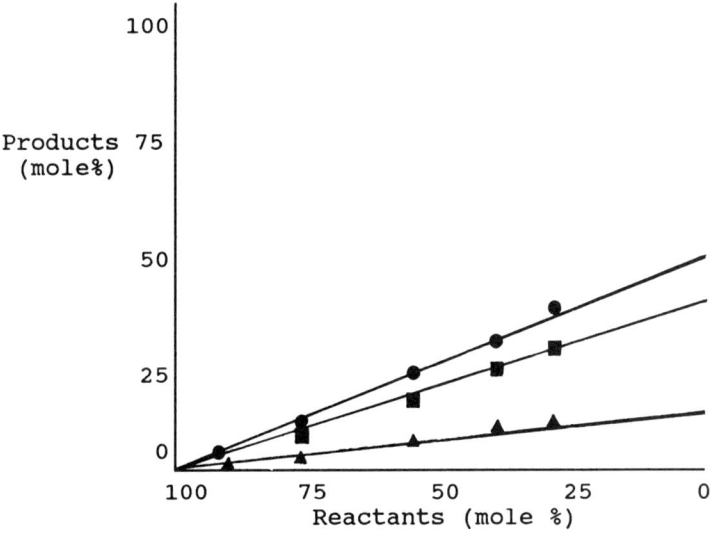

Figure 8.4 Propionitrile hydrogenation in the presence of butylamine: (■) propylamine, (●) propylbutylamine and propylbutylimine, and (▲) dipropylamine and dipropylimine.

Table 8.2 Support Effect on Propionitrile Hydrogenation

Catalyst	Solvent	Temperature ($°C$)	Selectivity, A_2 ($\%$)
Rh/C	MeOH	55	70
Rh/Al_2O_3	MeOH	55	26.8

IV. DISCUSSION

A. Introduction

The assumed mechanism of formation of primary and secondary amines by hydrogenation of nitriles is widely documented in the literature (e.g., refs. 1 and 2). From these works, we can write the general scheme of this hydrogenation (Fig. 8.5), excluding the formation of the tertiary amine, which is not the focus of this chapter and moreover is not formed on the Ni and Rh catalysts used in this work.

The secondary amine would be formed by hydrogenation of the secondary imine, which would be formed through the condensation

R - C ≡ N
R - CH = NH
R - CH2 - NH2
R - CH - NH - CH2 - R
NH2
-NH3
R - CH = N - CH2 - R
+ H2
R - CH2 - NH - CH2 - R

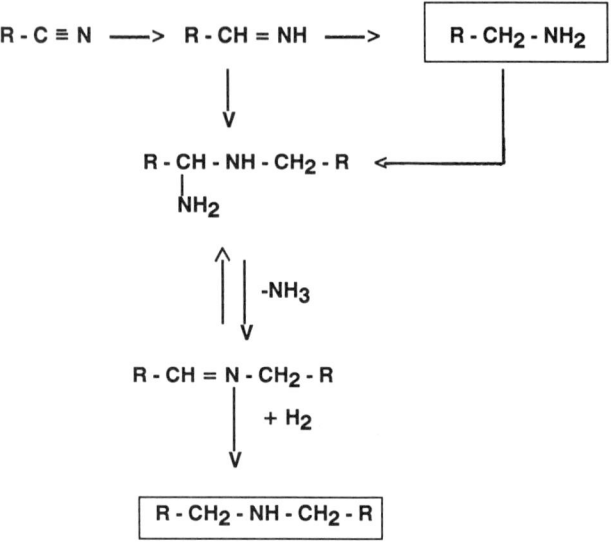

Figure 8.5

of the primary imine and the primary amine with the loss of ammonia. Neither the primary imine nor the 1-aminodialkylamine is detected by analysis of the reaction mixture, whereas the secondary imine is present: in Figures 8.1 and 8.2 it is clear that the formation of dipropylamine results from a consecutive reactions scheme. At this stage of the discussion we can write the chemical reactions pathway shown in Figure 8.6.

The mechanism of formation of the secondary imine is not well evidenced. At first sight, it is tempting to assume that the condensation reaction between the primary imine and the primary amine was proceeding in the liquid phase after desorption of the primary

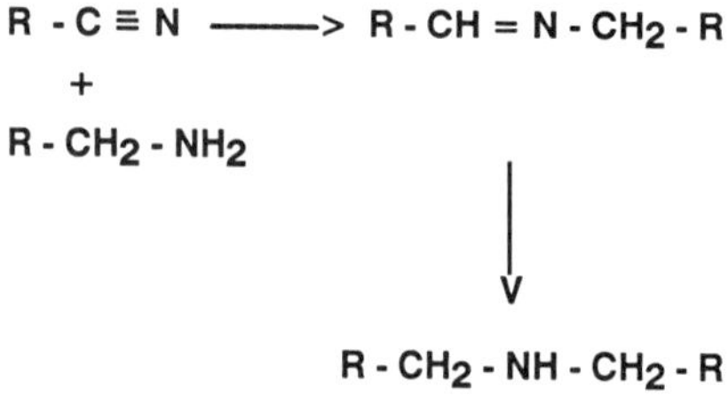

R - C ≡ N
+
R - CH2 - NH2
R - CH = N - CH2 - R
R - CH2 - NH - CH2 - R

Figure 8.6

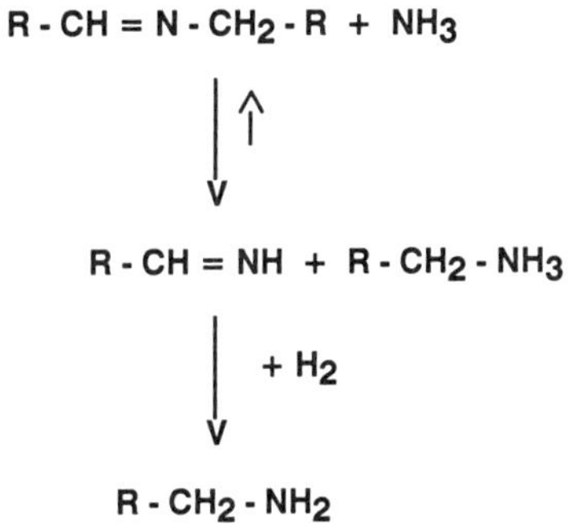

$$R - CH = N - CH_2 - R \ + \ NH_3$$

$$\Big\downarrow \uparrow$$

$$R - CH = NH \ + \ R - CH_2 - NH_3$$

$$\Big\downarrow \ + H_2$$

$$R - CH_2 - NH_2$$

Figure 8.7

imine [3]. The reverse of this homogeneous reaction could explain the well-known influence of ammonia on the selectivity for the primary amine formation. Indeed, addition of ammonia favors the reverse reaction, hence primary amine formation, through readsorption of the primary imine (Fig. 8.7). However, the selectivity toward the formation of the secondary amine must increase with conversion during a nitrile hydrogenation. Indeed, the primary amine concentration in the liquid phase increases with conversion and would favor the formation of the secondary imine. On the contrary, we do not observe any change in the selectivity with conversion (Fig. 8.1 and 8.2).

B. Independent Parallel Reactions

Independent behavior is found in parallel reactions, where the two products do not interfere with each other. The hydrogenation of a nitrile on Ni and Rh catalysts indeed follows a parallel reactions pathway. The nitrile can follow two pathways, one to the primary amine and the other one to the secondary amine (Fig. 8.8).

If we let X be the quantity of primary imine that is hydrogenated to the primary amine and Y the quantity of primary imine that desorbs and reacts in solution with the primary amine to give I_2 in Figure 8.8, the rate ratio of reactions 2 and 3 is:

$$\frac{v_3}{v_2} = \frac{1 - \alpha}{\alpha} = \frac{Y}{X}$$

We can calculate the values of Y and X using the following equations:

$$[A_1] = X - Y$$

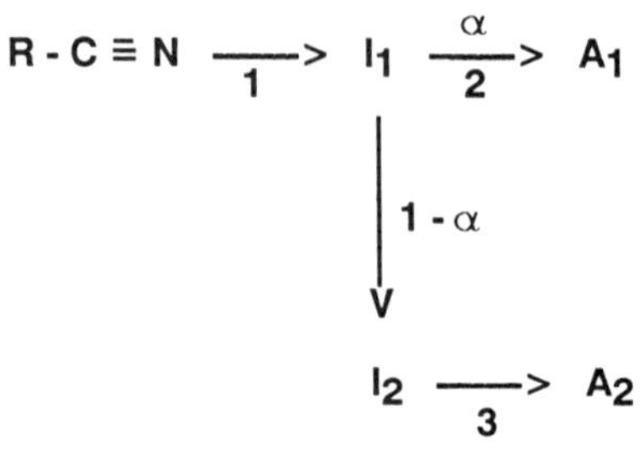

$$R - C \equiv N \xrightarrow{\quad 1 \quad} I_1 \xrightarrow[\;2\;]{\alpha} A_1$$

$$\downarrow 1 - \alpha$$

$$I_2 \xrightarrow[\;3\;]{} A_2$$

Figure 8.8 I_1, primary imine; A_1, primary amine; I_2, secondary imine; A_2, secondary amine; α, fraction of I_1, formed on the catalyst surface which is hydrogenated to the primary amine.

$$[CN] = a - X - Y$$

where a is initial nitrile concentration.

We can plot Y versus X and, in the case of a parallel reactions type of mechanism, obtain straight-line dependence. We indeed obtained such a dependence with nickel and rhodium catalysts (Fig. 8.9). Therefore, nitrile hydrogenation follows a parallel re-actions mechanism in which the two different products do not inter-fere (Figs. 8.1 and 8.2). As a consequence, as soon as the primary imine desorbs, it reacts with the primary amine in solution, regard-less of its concentration. Therefore, the selectivity of the reaction must be controlled by the adsorption-desorption equilibrium of the primary imine, since we do not detect this species in the reaction mixture. The rate-determining step for secondary imine formation in this case would be the desorption of the primary imine. This assumption disagrees with the proposed mechanism for the influence of ammonia on the selectivity. Indeed, ammonia would favor the reverse reaction between the secondary imine and ammonia to give both primary amine and imine. As a consequence, when this reverse reaction is operative, namely at high conversion toward the secondary amine or in the presence of added ammonia, primary imine would be analyzed, since the rate-determining step is assumed to be the adsorption-desorption of the primary imine. On the contrary, we did not detect any primary imine in our reaction mixtures. It is therefore difficult to understand why small quantities of ammonia could so strongly influence the selectivity of the reaction.

C.   Surface Reaction Mechanism

The preceding discussion is based on a homogeneous reaction scheme. If the primary imine reacts with primary amines on the catalyst

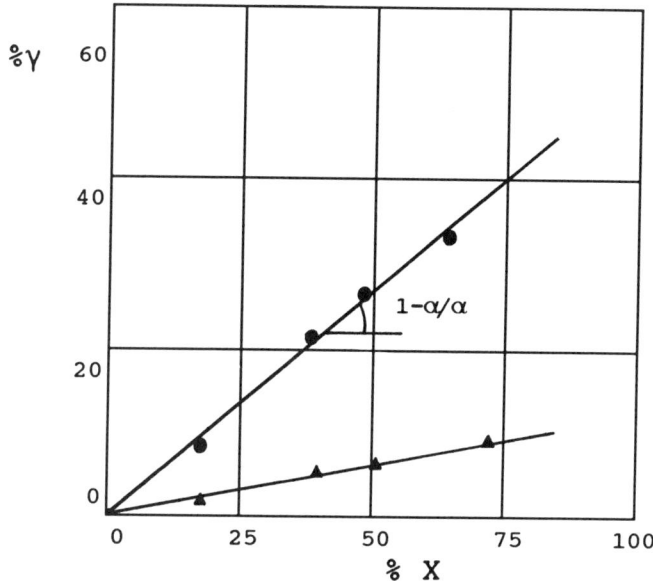

Figure 8.9 Propionitrile hydrogenation, α determination: (●) Rh/C and (▲) Raney nickel.

surface, then the surface concentration of these two products governs the rate of the reaction. If the catalytic surface is completely covered with primary amine, the rate of the reaction could be zero order with respect to the primary amine. But neither primary nor secondary amines influence the reduction rate of propionitrile, either on Raney nickel [4] or on Rh/C. It is therefore obvious that the hydrogenation catalytic sites are not poisoned with primary amines. If our previous assumptions are correct, the amine adsorption sites would be different from the hydrogenation sites. We have made more striking experiments, the results of which deny completely the homogeneous reaction mechanism. The hydrogenation of nitriles in the presence of a primary amine different from the one formed by hydrogenation (Table 8.1, Figure 8.4) produce both asymmetrical and symmetrical secondary amines.

In the case of homogeneous condensation, it is obvious that the selectivity for the asymmetrical secondary amine has to be higher at the beginning of the reaction, where few primary amines are produced by hydrogenation. However, we observe a rather constant selectivity for both secondary amines with increasing conversion (Fig. 8.4). At this stage of the discussion, we must examine another mechanism that could exclude the homogeneous reactions.

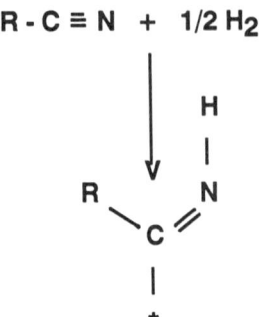

Figure 8.10

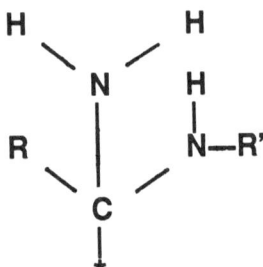

Figure 8.11   R' = H or $C_n H_{2n+1}$.

The first step would be the semihydrogenation of the nitrile function (Fig. 8.10).

We reported evidence of this kind of intermediate in an earlier study of the mechanism of butene nitrile hydrogenation [5]. This intermediate could easily react on the catalytic surface with vicinal NH_2-containing compounds to form a diamino compound (Fig. 8.11). This adsorbed compound could rearrange through further hydrogenation and desorption into ammonia and secondary imine, which could then be hydrogenated into secondary amine. In the particular case of R = H, namely the reaction with ammonia, the diamino compound could rearrange into primary imine and ammonia. That is the reverse reaction on the catalytic surface, which could explain the influence of ammonia on the selectivity. The formation of the diamino compound through the reaction of the primary imine and ammonia has been proposed by Schwoegler and Adkins [7]. This mechanism could explain why we do not detect either primary imine or aminodialkyl-amine.

V. CONCLUSION

We now detail our new proposition for a mechanism of secondary amine formation. Specifically, we suggest that the nitrile may be adsorbed on the catalyst surface, where it may react to form the primary iminelike adsorbed compound described above. This surface intermediate might react with hydrogen, ammonia, or primary amine. Ammonia or primary amine could be physically or chemically adsorbed on nonhydrogenating sites situated in the near vicinity of the catalytic hydrogenating site. This adduct would lose ammonia to form, primary imine or secondary imine in the case of, respectively, ammonia, or primary amine adduct. The secondary imine could desorb, whereas the very reactive primary imine would react on the catalytic surface to form the primary amine or the secondary imine. This mechanism has the advantage of clarifying the striking experiments shown in the present work.

    1. The selectivity toward formation of the secondary amine decreases with increasing temperatures. In our mechanism, the activation energy of reactions between the adsorbed semihydrogenated compound and the NH_2-containing compounds would probably differ from the hydrogenation activation energy. The hydrogenation of a nitrile has a great activation energy (> 50 kJ/mol) and consequently would be superior to the activation energy of the condensation reaction. This could explain the decrease of selectivity toward secondary amine formation with increasing temperatures.

    2. In the hydrogenation of nitriles in the presence of a primary amine different from the one resulting from hydrogenation, we have observed (Table 8.1) that selectivity toward the asymmetrical secondary amine depends on the size of the primary amine added. The shorter the primary amine added, the more important is the asymmetrical amine formation. The nonhydrogenating sites responsible for the adsorption of primary amines would adsorb more short molecules than longer ones, all the more if this adsorption resulted from a weak interaction (e.g., physisorption).

    3. The influence of the support (Table 8.2) on selectivity is huge. This could be another piece of evidence for our mechanism. Indeed, the primary amine could be adsorbed more strongly on the alumina support. This acidic support could attract primary amines far from the hydrogenating sites. As a consequence, secondary amine formation would be inhibited on metal deposited on acidic supports.

We have already shown that oxide species present on the catalyst surface or the weak interaction with a solvent may influence a catalytic process [5,8]. In the hydrogenation of nitriles, a polar solvent such as methanol might better solvate amino groups than an apolar solvent such as cyclohexane. Hence, the adsorption of amine on the catalytic surface must be more important in cyclohexane than in methanol. Consequently, the hydrogenation of propionitrile in cyclohexane on Rh/C must lead to an increased selectivity for secondary amine. We have indeed hydrogenated propionitrile in cyclohexane. The proportion of nitrile that leads to the secondary amine is 96%, versus 71% when methanol is used as solvent.

In the presence of a primary amine at the beginning of the hydrogenation, the number of nitrile molecules that leads to secondary amine formation is weaker than without added amines (Table 8.1). This phenomenon may be also explained by a solvent effect. Indeed, primary amines in solution might help to withdraw from the surface the primary amine formed by catalytic hydrogenation, thereby decreasing slightly the amount of secondary amines that can be formed in a surface process as we propose. This work emphasizes again the importance of second-order interactions in heterogeneous catalysis, which could be more a rule than an exception.

REFERENCES

1. Braun, J., Blessing, G., and Zobel, F., *Chem. Ber.*, *56B*, 1988-2001 (1923).

2. Greenfield, H., *Ind. Eng. Chem.*, *Prod. Res. Dev.*, *6*:2, 142-144 (1967).

3. Volf, J., and Pasek, J., *Stud. Surface Sci. Catal.*, *27*, 105-144 (1986).

4. Turu, M., Étude cinetique de l'hydrogénation des nitriles sur cobalt de Raney, Thèse de doctorat, U.C.L., 1973.

5. Dallons, J. L., Jannes, G., and Delmon, B., *Acta Chim. Hung.*, *119*:2-3, 223-232 (1985).

6. Dallons, J. L., Jannes, G., and Delmon, B., *Catal. Today*, *5*, 257-264 (1989).

7. Schwoegler, E. J., and Adkins, H., *J. Am. Chem. Soc.*, *61*, 3499-3502 (1939).

8. Dallons, J. L., Jannes, G., and Delmon, B., *Stud. Surface Sci. Catal.*, *41*, 115-121 (1988).

9

Chemical Modification of Diene-Based Polymers by Catalytic Hydrogenation

G. L. Rempel and Xiangyao Guo

Department of Chemical Engineering, University of Waterloo, Waterloo, Ontario, Canada

I. INTRODUCTION

Chemical modification of polymers is a field that is achieving increasing importance in macromolecular chemistry. It is considered to be an efficient synthetic route to novel polymers with desirable physical properties and functional groups, which are otherwise inaccessible, or difficult or too expensive to prepare by conventional polymerization techniques. Unsaturated polymers, especially diene polymers, are ideal polymers for chemical modifications because of the technological importance of the parent materials and the reactivities of the double bonds in the polymer chain. A variety of modification reactions of diene polymers have been explored by Schulz, Turner, and Golub [1]. In general, the following are the most common types of chemical reaction on unsaturated polymers:

Hydrogenation [2,3]
Epoxidation [4]
Cyclization [5]
Halogenation and hydrohalogenation [6]
Hydroformylation [7]
Sulfonation and sulfonyl chloride addition [8]
Carbene additions [9]
Metallation [10]
Hydrosilylation [11]

Among these, hydrogenation of diene-based polymers is particularly important due to the great interest in elastomers with higher stability to oxidative degradation [12-15]. For example, styrene-butadiene copolymers have been widely used as thermoplastic engineering elastomers, whereas nitrile-butadiene polymers are mainly used as oil-resistant elastomers in the automobile industry. Because of the presence of C=C unsaturation, however, these polymers undergo degradation or oxidation when exposed to oxygen, ozone, and heat for long periods of time. Therefore, selective hydrogenation of the C=C unsaturation in these diene polymers has become an important topic in rubber chemistry.

Hydrogenation of polybutadiene (PBD) using $RhCl(PPh_3)_3$ as catalyst has been reported by Soga and coworkers [2]; however the reaction required relatively high hydrogen pressure. More recently, Bouchal and his coworkers [13] reported a kinetic study for the hydrogenation of HO-terminated telechelic PBD in the presence of $RhCl(PPh_3)_3$; they found that some of the kinetic phenomena they observed could be interpreted in the sense of Wilkinson's reaction mechanism for the alkene hydrogenation [16]. Another interesting catalyst for diene polymer hydrogenation that was recently reported involves complexes of the type $RuCl(CO)(OCOR)(PPh_3)_2$ [17,18], where R is an alkyl or aryl group. However, no mechanistic information on diene hydrogenation using such a catalyst has been reported.

The main object of the present study was the hydrogenation of a variety of diene-based polymers in the presence of the homogeneous transition metal catalysts: $RhCl(PPh_3)_3$ and $RuCl(CO)(OCOR)(PPh_3)_2$. Plausible reaction mechanisms were suggested for these reaction systems on the basis of the experimental results. For hydrogenation systems catalyzed by $RhCl(PPh_3)_3$, nonlinear least-squares regression methods were used for reaction constant estimation [19].

II. RESULTS AND DISCUSSION

A. Hydrogenation of Polybutadiene

The most simple diene polymer for hydrogenation is polybutadiene. The structure and the properties of the hydrogenated PBD polymers depend on the precursor PBD polymers and the extent of hydrogenation. For example, hydrogenation of PBD of high 1,4 structure converts this elastomeric polymer to a tough semicrystalline polymer:

$$(-CH_2-CH=CH-CH_2-) + H_2 \rightarrow (-CH_2-CH_2-CH_2-CH_2-)$$

$$(9.1)$$

Hydrogenated 1,4-PBD-block-1,2-PBD copolymer has exceptional stress-strain properties for diblock materials, because the hydrogenation produces crystalline polyethylene (PE) blocks that trap the rubbery domains, creating a pseudonetwork:

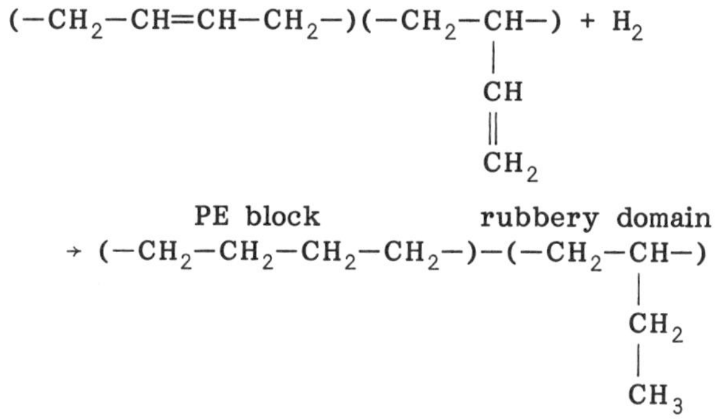

$$(-CH_2-CH=CH-CH_2-)(-CH_2-CH-) + H_2$$
$$\begin{array}{c} | \\ CH \\ \| \\ CH_2 \end{array}$$

$$\text{PE block} \qquad\qquad \text{rubbery domain}$$
$$\rightarrow (-CH_2-CH_2-CH_2-CH_2-)-(-CH_2-CH-)$$
$$\begin{array}{c} | \\ CH_2 \\ | \\ CH_3 \end{array} \qquad\qquad (9.2)$$

1. Hydrogenation of PBD Using $RhCl(PPh_3)_3$

The PBD polymers examined in the hydrogenation study using $RhCl(PPh_3)_3$ as catalyst were as follows:

1,2-PBD: polybutadiene with 90% 1,2-addition units
cis-1,4-PBD: polybutadiene with 98% cis-1,4-addition units
1,4-PBD: polybutadiene with 55% trans-, 36% cis-1,4-addition units
MPBD: monodisperse polybutadiene with $M_w/M_n = 1.06$

The hydrogenation of 1,2-PBD was carried out at mild reaction conditions (50°C, < 101.3 kPa H_2) in toluene solvent. The product polymers were characterized by infrared and NMR spectroscopy and by gel permeation chromatography (GPC). Quantitative hydrogenation of the C=C unsaturation was obtained without any chain scission and crosslinking of the polymers. However the hydrogenation of 1,4-PBD under the same reaction conditions was incomplete due to the precipitation of the partially saturated product polymer [12]. Thus to avoid precipitation of the partially hydrogenated polymer, the hydrogenation of 1,4-PBD was performed in o-dichlorobenzene over a higher temperature range (50-125°C).

Kinetic studies for the hydrogenation of 1,2-PBD and cis-1,4-PBD suggested a similar reaction mechanism for both reaction systems (Fig. 9.1).

Step $A_0 \rightarrow A$ involves the formation of dihydridorhodium complex $H_2RhCl(PPh_3)_3$. This complex (see structure 1) has been synthesized successfully and characterized in detail by Brown and Lucy [20]. The next step in the proposed mechanism—the dissociation of a triphenylphosphine ligand from A to form B—was also followed and characterized by Brown and Lucy [20].

Structure II was proposed as the most likely configuration for species B instead of the traditionally assumed trans-PPh_3 configuration shown as structure III [21,22]. It was also reported that if

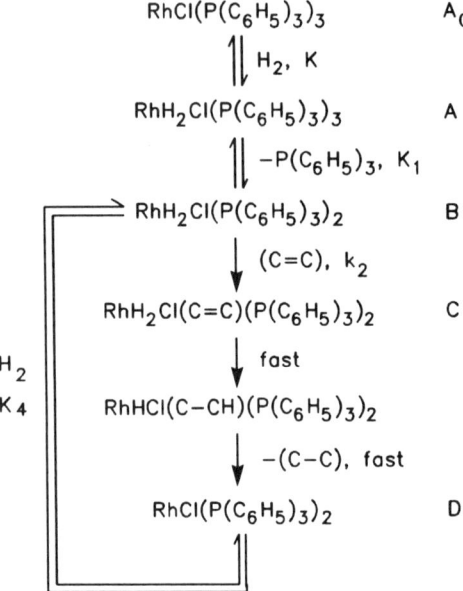

Figure 9.1

cis-phosphine geometry is in fact present as a main intermediate, hydrogen transfer to coordinated olefin is very rapid [23] (i.e., step C → D in Fig. 9.1 would be very rapid). Therefore step B → C, which is the coordination of $H_2RhCl(PPh_3)_2$ with the C=C bonds of the polymer, is assumed as the rate-determining step for the PBD hydrogenation systems above.

$$
\begin{array}{ccc}
\underset{Ph_3P}{\overset{H}{\diagdown}}\hspace{-0.3em}\underset{\underset{PPh_3}{|}}{\overset{\overset{PPh_3}{|}}{Rh}}\hspace{-0.3em}\overset{H}{\diagup}\hspace{-0.3em}Cl
&
H\!-\!\underset{\underset{Cl}{|}}{\overset{\overset{H}{|}}{Rh}}\!\!\diagup^{PPh_3}_{\diagdown PPh_3}
&
\underset{H}{\overset{H}{\diagdown}}\hspace{-0.3em}\underset{\underset{PPh_3}{|}}{\overset{\overset{PPh_3}{|}}{Rh}}\!-\!Cl
\\
I & II & III
\end{array}
$$

Another possible pathway, the "unsaturated route," which involves an initial dissociation of $RhCl(PPh_3)_3$ (A_0) to $RhCl(PPh_3)_2$ (D) with subsequent addition of the carbon-carbon double bond to form a π-alkene complex has been proposed for the hydrogenation of monomeric alkenes [24]. This particular mechanism is considered to be much less significant in the hydrogenation of polybutadiene, mainly because of the observed rapid formation of $RhH_2Cl(PPh_3)_3$ and the increased steric barriers for facile π-alkene complex formation of the polymeric substrate.

A mathematical rate law model for Figure 9.1 was developed as shown in equation (9.3):

$$\frac{-d[C{=}C]}{dt} = \frac{k_2 \underline{K}\underline{K}_1\underline{K}_4\,[Rh]_T\,[H_2][C{=}C]}{\underline{K}_4\,[PPh_3](1 + \underline{K}\,[H_2]) + \underline{K}\underline{K}_1\underline{K}_4\,[H_2] + \underline{K}_1\underline{K}}$$

(9.3)

where $[Rh]_T = [A_0] + [A] + [B] + [D]$.

Relevant reaction parameters were estimated by applying a non-linear regression analysis method [25] to this model. The estimated values of the reaction parameters for the hydrogenation of 1,2-PBD and cis-1,4-PBD systems are listed in Table 9.1. It is important to realize that the 1,2-PBD hydrogenation kinetics were carried out in toluene at 50°C, and the cis-1,4-PBD hydrogenation kinetics were performed in o-dichlorobenzene at 65°C, to accommodate the different solubilities of the product polymers. In the 1,2-PBD hydrogenation system, the equilibrium for the formation of $H_2RhCl(PPh_3)_3$ (eq. 9.4) was found to lie far to the right ($\underline{K}$ estimated > 10 mM^{-1}).

However, the equilibrium constant $\underline{K}$ for this step in the cis-1,4-PBD hydrogenation system was measured as 0.60 (mM^{-1}). The higher reaction temperature in the latter reaction system probably leads to increasing dissociation of the fully formed $H_2RhCl(PPh_3)_3$:

$$RhCl(PPh_3)_3 + H_2 \underset{}{\overset{K}{\rightleftharpoons}} (H_2)RhCl(PPh_3)_3$$

(9.4)

The equilibrium constant for the dissociation of triphenylphosphine:

$$(H_2)RhCl(PPh_3)_3 \underset{}{\overset{K_1}{\rightleftharpoons}} (H_2)RhCl(PPh_3)_2 + PPh_3$$

(9.5)

is estimated as 1.17 mM at 50°C in the presence of 1,2-PBD and 4.45 mM at 65°C in the presence of cis-1,4-PBD. The $\underline{K}_4$ value as defined in Figure 9.1 (step D → B) is estimated as 0.22 mM^{-1} for the 1,2-PBD hydrogenation system and 0.59 mM^{-1} for the cis-1,4-PBD hydrogenation system. Finally, a comparison of the rate-determining steps is important. The k_2 value for the 1,2-PBD hydrogenation system is evaluated as 2.90×10^{-4} (mM·s)$^{-1}$, whereas the $\underline{k}_2$ value of 12.8×10^{-4} (mM·s)$^{-1}$ is obtained for the cis-1,4-PBD hydrogenation system. Although sterically, hydrogenation of

Table 9.1   Hydrogenation of 1,2-PBD and cis-1,4-PBD Catalyzed
by $RhCl(PPh_3)_3$: Comparison of the Estimated Reaction Constants

Polymer	Temperature (°C)	Solvent	$\underline{K}_1$ (mM)	$10^4\ \underline{k}_2$ $(mM.s)^{-1}$	$\underline{K}_4$ (mM^{-1})	$\underline{K}$ (mM^{-1})
1,2-PBD	50	Toluene	1.17	2.90	0.22	>10[a]
cis-1,4-PBD	65	o-Dichloro-benzene	4.45	12.8	0.59	0.60

[a]The equilibrium between H_2 and $RhCl(PPh_3)_3$ lies far to the right-hand side for the 1,2-PBD hydrogenation system.

1,2-PBD is expected to be more favorable than that of cis-1,4-PBD, the higher reaction temperature and better polymer solvation used in the latter system could be responsible for its higher rate of hydrogenation. In general, all the estimated reaction parameters for each step of the reaction are of the same order of magnitude for the two hydrogenation systems above.

An interesting and important objective to be realized in the hydrogenation of PBD concerns the possible synthesis of monodisperse polyethylene or poly(ethylene-co-butylene) as a standard for GPC analysis, since so far there appears to be no report on such a synthetic method for GPC standards. This goal was successfully achieved at a pressure of less than 101.3 kPa H_2 and 65°C using 1.99 mM $RhCl(PPh_3)_3$ for 378 mM MPBD (M_w/M_n = 1.06). GPC results (using universal calibration with polystyrene as standard) showed that the hydrogenated polymer had an M_w/M_n ratio of 1.09. This suggests that the molecular weight distribution of the polymer remained essentially unchanged during hydrogenation. Consequently, this process presents a useful technique for the preparation of monodisperse poly(ethylene-co-butylene).

The selectivity of the catalyst toward the hydrogenation of cis- and trans-1,4 C=C structure was investigated by carrying out hydrogenation reactions of cis-1,4-PBD and 1,4-PBD (55% trans) under the same reaction conditions. The results for the reactions, listed in Table 9.2 show that $RhCl(PPh_3)_3$ exhibits some selectivity for the hydrogenation of cis-1,4 C=C over trans-1,4 C=C structure. This result is consistent with the previously observed selectivity of this catalyst for the hydrogenation of simple alkenes [24].

2.   Hydrogenation Using $RuCl(CO)(OCOR)(PPh_3)_2$

Recent interest in developing more economic hydrogenation processes for polymer hydrogenation in the rubber industry has led to the

Table 9.2 Selectivity of RhCl(PPh$_3$)$_3$ Toward the Hydrogenation of the *cis*-1,4- Versus the *trans*-1,4-C=C Structure[a]

Polymer	Initial [C=C] (mM)	Degree of hydrogenation (%)	Initial rate (mM.s^{-1})
cis-1,4-PBD	200	≈95 (<0.5 h)	0.361
1,4-PBD	200	≈93 (>2 h)	0.226
cis-1,4-PBD[b]	720	≈95 (1 h)	1.030
1,4-PBD[b]	720	≈69 (>1.5 h)	0.700

[a]Reaction conditions: [H$_2$] = 3.9 mM, [Rh]$_T$ = 1.99 mM, *T* = 125°C, [PPh$_3$] = 7.4 mM, *o*-dichlorobenzene = 5.0 mL; *cis*-1,4-PBD: 98% cis structure; 1,4-PBD: 55% trans, 36% cis structure.
[b][Rh]$_T$ = 3.58 mM; no added PPh$_3$.

Table 9.3 Stoichiometric Results for the Hydrogenation of 1,2-PBD Catalyzed by RuCl(CO)(OCOCH$_3$)(PPh$_3$)$_2$[a]

Total pressure (kPa)	Temperature (°C)	Degree of hydrogenation (%)
207	100	≈70
552	100	≈70
4130	150	≈97

[a]Reaction conditions: [C=C] = 115.7 mM, [Ru]$_T$ = 0.836 mM, [PPh$_3$] = 7.2 mM; solvent, toluene.

investigation of the catalytic activities of Ru catalysts [17,18,26-28]. During our first study on the ruthenium systems, RuCl(CO)(OCOCH$_3$) (PPh$_3$)$_2$ was used as catalyst for the hydrogenation of 1,2-PBD. The stoichiometric results for this hydrogenation reaction at different reaction conditions are summarized in Table 9.3. When the hydrogenation was performed at low pressure (207-552 kPa) at 100°C, hydrogen consumption ceased after about 70% hydrogenation. The catalyst, however, still maintained its activity for the hydrogenation of additional polybutadiene added in the form of a toluene solution, and the same degree of reaction completion (70%) was observed again. The isomerization of terminal C=C into internal C=C was believed to be the reason for the incomplete hydrogenation. When the hydrogenation was performed under a higher pressure (4130 kPa H$_2$) at 150°C, 97% hydrogenation was achieved. It is felt

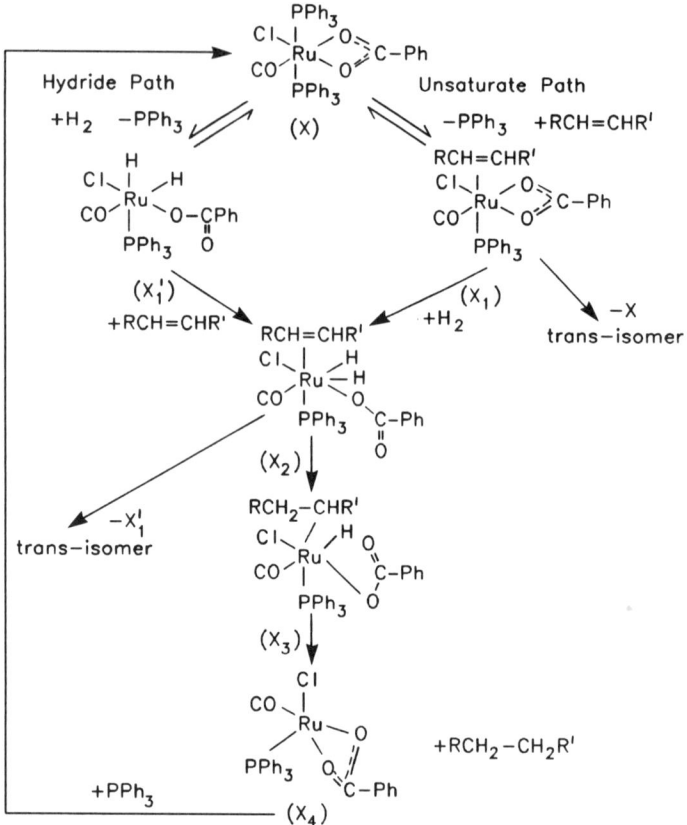

Figure 9.2

that the much higher hydrogen pressure in the latter reaction system increased the rate of hydrogenation and thus decreased the possibility of isomerization. The product polymers of this reaction were analyzed using vapor phase osmometry and dilute solution viscometry. No significant change in large-scale molecular structure was found.

Kinetic studies for the hydrogenation of 1,2-PBD in the presence of $RuCl(CO)(OCOCH_3)(PPh_3)_2$ were performed at mild reaction conditions (207-552 kPa, 100°C). The hydride path in Figure 9.2 is favored as the mechanism at the specified reaction conditions for this hydrogenation system.

The first step of the reaction involves the oxidative addition of hydrogen to the Ru complex. It has been suggested that a bidentate-to-monodentate transformation of the carboxylate ligand is necessary for the creation of a vacant site on the metal (eq. 9.6) [29].

$$(9.6)$$

The hydrogenation of *cis*-1,4-PBD was investigated using RuCl(CO) $(OCOPh)(PPh_3)_2$ as catalyst in *o*-dichlorobenzene. The hydrogenation achieved about 70% completion at 93.4 kPa and 85°C. The mechanism for this reaction possibly involves both a hydride path and an unsaturated path as shown in Figure 9.2. The low hydrogen pressure applied in the study of this reaction system was believed to be responsible for the involvement of the unsaturated path.

At relatively high hydrogen pressure (i.e., 207-552 kPa), the higher hydrogen concentration in the reaction system would be expected to promote the reaction of hydrogen and the Ru catalyst, therefore resulting in a possible preference for the hydride path.

It was also found that cis-trans isomerization was involved during the hydrogenation of *cis*-1,4-PBD catalyzed by RuCl(CO)(OCOPh) $(PPh_3)_2$ (see eq. 9.7). The *trans*-1,4 C=C structure was

$$(9.7)$$

found to be more difficult to hydrogenate than the *cis*-1,4 C=C structure. The formation of the *trans*-1,4 C=C structure was believed to be one of the main reasons for the incomplete hydrogenation. Our experimental results showed that both catalytic and thermal cis-trans isomerizations were involved during the hydrogenation. However, the main cause for cis-trans isomerization was the presence of RuCl(CO)(OCOPh)$(PPh_3)_2$ in the reaction system.

B. Hydrogenation of Styrene-Butadiene Copolymers

The performance range of diene-based copolymers often may be extended by hydrogenation of residual C=C unsaturation. For example, hydrogenation of styrene-butadiene-styrene (St-BD-St) in which the center block contains a moderate amount of 1,2-addition units gives a triblock copolymer with an ethylene-butylene center segment(St-EB-St):

$$(-CH_2-CH-)_{\underline{x}}(-CH_2CH=CHCH_2-)_{\underline{y}}(-CH_2CH-)_{\underline{y}'}(-CH_2CH-)_{\underline{z}} + H_2$$
$$\quad\quad\;\; | \quad\quad\quad\quad\quad\quad\quad\quad\quad\quad\;\;\; |\quad\quad\quad |$$
$$\quad\quad\; C_6H_5 \quad\quad\quad\quad\quad\quad\quad\quad\quad\quad\quad\; CH \quad\quad C_6H_5$$
$$\quad\quad\quad\quad\quad\quad\quad\quad\quad\quad\quad\quad\quad\quad\quad\quad\quad\;\; \|$$
$$\quad\quad\quad\quad\quad\quad\quad\quad\quad\quad\quad\quad\quad\quad\quad\quad\quad\; CH_2$$

St   E   B   St

$$\rightarrow (-CH_2-CH-)_{\underline{x}}(-CH_2CH_2CH_2CH_2-)_{\underline{y}'''}(-CH_2CH-)_{\underline{y}''}(-CH_2CH-)_{\underline{z}}$$
$$\quad\quad\;\; | \quad\quad\quad\quad\quad\quad\quad\quad\quad\quad\quad\quad\; |\quad\quad |$$
$$\quad\quad\; C_6H_5 \quad\quad\quad\quad\quad\quad\quad\quad\quad\quad\quad\quad\;\; CH_2 \quad\; C_6H_5$$
$$\quad\quad\quad\quad\quad\quad\quad\quad\quad\quad\quad\quad\quad\quad\quad\quad\;\; |$$
$$\quad\quad\quad\quad\quad\quad\quad\quad\quad\quad\quad\quad\quad\quad\quad\quad\; CH_3 \quad\quad\quad\quad (9.8)$$

This triblock polymer has a greatly increased thermal and oxidative stability, together with processibility and serviceability at higher temperature over St-BD-St, by virtue of its poly(ethylene-co-butylene) center block [30].

The styrene-butadiene copolymers used for hydrogenation were St-BD-St (styrene-butadiene-styrene triblock copolymer with 28% styrene) and SBR (styrene-butadiene diblock copolymer with 18% styrene). The reactions were carried out in the presence of $RhCl(PPh_3)_3$ at mild reaction conditions (<101.3 kPa, <65°C). Based on gas consumption, IR, and NMR results, it was found that quantitative hydrogenation of C=C was achieved without any hydrogenation of the styrene functionality. The reaction kinetics of the SBR and St-BD-St copolymers also were investigated, to compare them with the corresponding PBD hydrogenation system. The kinetic results can be accommodated by the mechanism postulated for PBD hydrogenation (see Fig. 9.1). Using equation (9.3) as a model, the reaction parameters in Figure 9.1 for these two copolymer hydrogenation systems were estimated. Table 9.4 provides a summary of the values of the estimated reaction constants. Compared with the corresponding PBD hydrogenation systems (see Table 9.1), the same reaction conditions (temperature, solvent), the values of $\underline{K}$, $\underline{K}_1$, and $\underline{K}_4$ for PBD hydrogenation were found to be in very good agreement with the corresponding values for the styrene-butadiene hydrogenation. The differences in $\underline{k}_2$ values probably result from the different solubilities of the polymers and the different steric effects of the polymer chains.

The selectivity of the catalyst in hydrogenating 1,2- versus 1,4- C=C structures for these copolymer hydrogenation systems

Table 9.4 Hydrogenation of St-BD-St and SBR Catalyzed by $RhCl(PPh_3)_3$: Comparison of the Estimated Reaction Constants

Polymer	Tempera-ture (°C)	Solvent	$\underline{K}_1$ (mM)	$10^4\underline{k}_2$ (mM.s)$^{-1}$	$\underline{K}_4$ (mM^{-1})	$\underline{K}$ (mM^{-1})
SBR	50	Toluene	1.25	4.28	0.20	>10[a]
St-BD-St	60	o-Dichloro-benzene	4.70	4.77	0.72	1.23

[a]The equilibrium between H_2 and $RhCl(PPh_3)_3$ lies far to the right-hand side for the SBR hydrogenation system.

Table 9.5 Hydrogenation of SBR by $RhCl(PPh_3)_3$: Selectivity of $RhCl(PPh_3)_3$ on Hydrogenating 1,2- C=C Versus 1,4-C=C

Polymer substrate (%)[a]		Degree of hydrogenation (%)	Product polymer (%)[a]	
1,2 C=C	1,4 C=C		1,2 C=C	1,4 C=C
10	90	≈ 25	5	70
10	90	≈ 38	2	59
10	90	≈ 50	0	50

[a]The percentage of C=C was calculated based on the butadiene content.

was investigated by ^{1}H-NMR analyses of the partially hydrogenated polymer samples. The results listed in Table 9.5 show that $RhCl(PPh_3)_3$ exhibits some degree of selectivity for the hydrogenation of 1,2- C=C over 1,4- C=C structure. This selectivity can be attributed mainly to steric effects of the alkene-Rh intermediates formed during the hydrogenation process.

C. Hydrogenation of Nitrile-Butadiene Copolymer

Highly saturated nitrile-butadiene rubber (HNBR) has greatly improved thermal and oxidative stability compared with the normal nitrile butadiene rubber (NBR). One potentially useful method for obtaining HNBR is via catalytic hydrogenation. One of the major problems encountered in such hydrogenation processes, however, involves the activity and selectivity offered by the catalyst for

removing the C=C unsaturation in the presence of nitrile groups.
Since the copolymer contains both internal and terminal C=C unsaturation, three distinct hydrogenation reactions (eqs. 9.9-9.11)
are required for complete hydrogenation of the C=C unsaturation
in such a copolymer:

$$-H_2C \diagdown \overset{H \diagup \quad \diagdown H}{C=C} \diagdown_{CH_2-CH_2-CH-} \underset{CN}{\quad} \xrightarrow{+\ H_2} -CH_2-CH_2-CH_2-CH_2-CH_2-CH-\underset{CN}{|}$$

$$(9.9)$$

$$-H_2C \diagdown \overset{\diagup H}{C=C} \diagup_{H \quad \diagdown CH_2-CH_2-CH-} \underset{CN}{\quad} \xrightarrow{+\ H_2} -CH_2-CH_2-CH_2-CH_2-CH_2-CH-\underset{CN}{|}$$

$$(9.10)$$

$$\begin{array}{cc} -CH_2-CH-CH_2-CH- \\ \quad | \qquad\qquad | \\ \quad CH \qquad\quad CN \\ \quad || \\ \quad CH_2 \end{array} \xrightarrow{+\ H_2} \begin{array}{cc} -CH_2-CH-CH_2-CH- \\ \quad | \qquad\qquad | \\ \quad CH_2 \qquad\quad CN \\ \quad | \\ \quad CH_3 \end{array}$$

$$(9.11)$$

Our studies on this copolymer revealed that $RhCl(PPh_3)_3$ and
$RuCl(CO)(OCOPh)(PPh_3)_2$ complexes are good catalysts for such
a hydrogenation process. $HRh(PPh_3)_4$ has also been reported as
an efficient catalyst for this process [31]. Selective quantitative
C=C hydrogenation with all these catalyst systems was achieved
without nitrile reduction. Detailed kinetic studies were conducted
for the hydrogenation using $RhCl(PPh_3)_3$ and $RuCl(CO)(OCOPh)$
$(PPh_3)_3$ catalyst systems.

1. Hydrogenation of NBR in the Presence of $RhCl(PPh_3)_3$

Under mild reaction conditions (<101.32 kPa hydrogen pressure,
40-50°C), $RhCl(PPh_3)_3$ functions as an efficient catalyst in 2-
butanone for the hydrogenation of the C=C unsaturation present
in the acrylonitrile-butadiene copolymer [14]. Stoichiometric results
based on gas consumption, as well as IR and NMR analysis, showed
that the hydrogenation achieved completion without reduction of
the nitrile group of the copolymer and the keto group of the solvent.

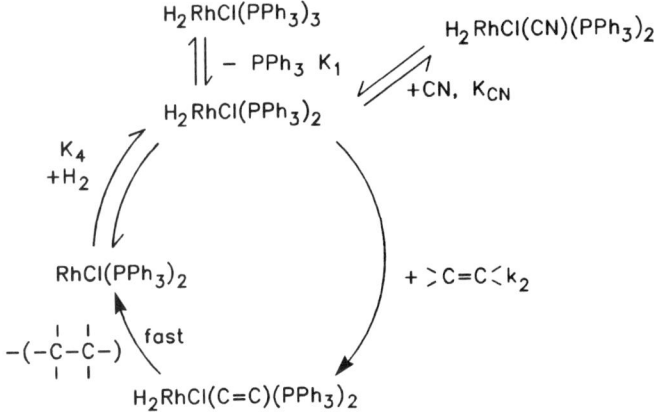

Figure 9.3

An experimental rate law based on the kinetic results showed a first-order dependence on the catalyst concentration, C=C concentration, a first-order to zero-order dependence on hydrogen concentration, and an inverse dependence on the concentrations of nitrile group and added PPh_3. Utilizing the results from kinetic studies and UV and ^{1}H-NMR spectroscopic techniques leads to the postulated mechanism shown in Figure 9.3. As for the cases of PBD and SBR hydrogenation, oxidative addition of hydrogen to the Rh catalyst is proposed as the first step for this reaction (see eq. 9.4).

This step is followed by the dissociation of a PPh_3 group from the Rh catalyst (eq. 9.5). However, the next step could involve either the coordination of the C=C moiety or the complexing of nitrile group of the substrate with the Rh metal center:

$$RhH_2Cl(PPh_3)_2(C=C) \qquad (9.12)$$

$$RhH_2Cl(PPh_3)_2 \quad \begin{array}{c} + C=C \ (k_2) \nearrow \\ \\ + -CN \ (K_{CN}) \rightleftharpoons \end{array}$$

$$RhH_2Cl(PPh_3)_2(CN) \qquad (9.13)$$

The competition between equations (9.12) and (9.13) therefore causes a decrease of observed reaction rate with increasing polymer concentration. This is further substantiated by the decrease in reaction rate that is observed when a saturated nitrile such as acetonitrile or saturated NBR is added to the reaction mixture.

The rate law derived from this postulated mechanism shown
in equation (9.14) is in agreement with the experimental rate law:

$$\frac{-\underline{d}[C{=}C]}{dt} = \frac{\underline{k}_2\underline{K}_1\underline{K}_4\,[Rh]_T\,[H_2][C{=}C]}{\underline{K}_4\,[PPh_3][H_2] + \underline{K}_1 + \underline{K}_1\underline{K}_4\,[H_2] + \underline{K}_1\underline{K}_{CN}\underline{K}_4\,[CN]}$$

$$(9.14)$$

Using this model, the Gauss-Newton method [25] for nonlinear least-
squares regression is used for the computation of the reaction
parameters. Table 9.6 gives the estimates of reaction parameters.
The values for $\underline{K}_1$, $\underline{K}_4$, and $\underline{k}_2$ are in reasonable agreement with
those of the corresponding 1,2-PBD hydrogenation system (Table 9.1).
    The solvent media had an important effect on this hydrogenation
system. When chlorobenzene was employed as the solvent for this
NBR hydrogenation system, it was necessary to increase the reaction
temperature to 80°C to achieve quantitative hydrogenation in a time
comparable to that required for the reaction in 2-butanone at only
40°C. A more interesting effect of the solvent is the selectivity
of the $RhCl(PPh_3)_3$ catalyst toward hydrogenating the 1,2-C=C
over 1,4-C=C structure in chlorobenzene media. It seems that in
2-butanone, $RhCl(PPh_3)_3$ does not exhibit selectivity for the hydrogenation of 1,2-C=C over 1,4-C=C structure. When the hydrogenation
reaction was carried out in ethyl acetate solvent, the catalyst
showed a catalytic behavior similar to that observed in 2-butanone.
These results suggest that the interaction of solvent with the NBR
copolymer is very important in influencing both the selectivity and
the rate of the catalytic process.

2. Hydrogenation of NBR in the Presence of
    $RuCl(CO)(OCOR)(PR'_3)_2$

During the past 2 years, a number of ruthenium complexes have
been reported as promising catalysts for the production of highly
saturated nitrile-butadiene rubber materials [17,18,26-28]. However,
undesirable gel formation during the final stage of quantitative
hydrogenation of NBR was often observed. Consequently, work
has been carried out to further our understanding of how these
newly discovered catalysts interact with the NBR substrate, to
elucidate the possible cause of the gel formation.
    Catalytic hydrogenation of monomeric unsaturated nitrile model
compounds in the presence of Ru complexes revealed that isomerization caused by C=C bond migration was an important side reaction
during the hydrogenation. It was also found that the electronic
and steric effects of the ligands in the Ru complexes had a significant effect on the activities and selectivities of the catalysts. This

Table 9.6   NBR Hydrogenation Catalyzed by $RhCl(PPh_3)_3$ at 40°C: Estimates of Reaction Constant

Polymer	Solvent	$\underline{K}_1$ (mM)	$10^4\underline{k}_2$ $(mM.s)^{-1}$	$\underline{K}_4$ (mM^{-1})	$\underline{K}$ (mM^{-1})	$\underline{K}_{CN}$ (mM^{-1})
NBR[a]	2-Butanone	0.20	4.23	0.28	$>10^b$	0.065

[a]NBR (acrylonitrile-butadiene copolymer): 34% acrylonitrile content).
[b]The equilibrium between H_2 and $RhCl(PPh_3)_3$ lies far to the right-hand side (eq. 9.4) of the NBR hydrogenation system.

work has been extended to NBR hydrogenation. The results obtained suggest that isomerization is one of the main factors inhibiting the accomplishment of quantitative hydrogenation before the onset of side reactions that give rise to gel formation.

Equation (9.15) shows the most likely isomerization caused by C=C migration:

$$-CH_2-CH=CH-CH_2-CH_2-\underset{\underset{CN}{|}}{CH}- \;\leftarrow\; -CH_2-CH_2-CH=CH-CH_2-\underset{\underset{CN}{|}}{CH}-$$

$$\text{(A)} \qquad\qquad\qquad\qquad\qquad \text{(B)}$$

$$\text{(B)} \;\leftarrow\; -CH_2-CH_2-CH_2-CH=CH-\underset{\underset{CN}{|}}{CH}-$$

$$\text{(C)}$$

$$\text{(C)} \;\leftarrow\; -CH_2-CH_2-CH_2-CH_2-CH=\underset{\underset{CN}{|}}{C}- \qquad\qquad (9.15)$$

$$\text{(D)}$$

Among the isomerization products, D is the most stable structure because of the conjugation between the C=C and the −CN (see eq. 9.16). Since this C=C bond is conjugated with the electronic-withdrawing −CN group, it may be expected to have a different reactivity with respect to hydrogenation:

$$-CH_2-CH_2-CH_2-CH_2-CH=\underset{\underset{\underset{N}{\overset{|||}{C}}}{|}}{C}- \;\leftarrow\; -CH_2-CH_2-CH_2-CH_2-CH=\underset{\underset{\underset{N}{\overset{|||}{C}}}{||}}{C}-$$

$$\qquad\qquad D \qquad\qquad\qquad\qquad\qquad\qquad D' \qquad\qquad (9.16)$$

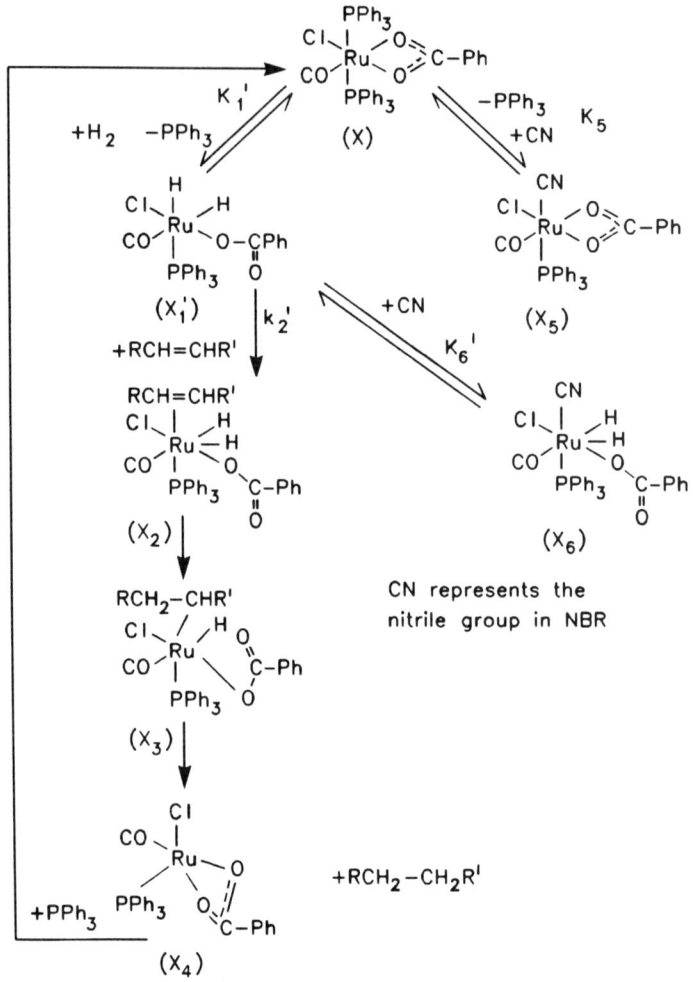

Figure 9.4

Hydrogenation of NBR using RuCl(CO)(OCOPh)(PPh$_3$)$_2$ was performed
under different hydrogen pressures. At 4150 kPa hydrogen pressure,
quantitative hydrogenation was achieved within 3 hours and the
—CN group remained unattacked. However when the reaction was
carried out at 2766 kPa, only about 98.5% hydrogenation was achieved
in a period of more than 10 hours. For the reaction performed at
atmospheric pressure, the maximum degree of hydrogenation observed
was only 40%. It is found that the C=C in isomer D is much more

difficult to hydrogenate than the C=C in the starting NBR polymer at a hydrogen pressure lower than 2766 kPa. This again supports the premise that the formation of isomers during the reaction is one of the reasons for the incomplete hydrogenation.

The results obtained from kinetic studies for this hydrogenation system seem to suggest that the rate of the hydrogenation decreases with the increase of the CN concentration. A possible explanation for this result is that the nitrile group present in the NBR copolymer is potentially a strong coordinating ligand, which may form complexes with the Ru center of the complex [32]. There are three possible types of reaction in the first stage of the NBR hydrogenation: oxidative addition of hydrogen to the Ru catalyst, coordination between C=C and the Ru catalyst, and coordination between CN and the Ru catalyst.

A reaction mechanism as shown in Figure 9.4 is postulated for the hydrogenation of NBR by $RuCl(CO)(OCOPh)(PPh_3)_2$.

The rate law derived from this mechanism shown in equation (9.17) is in agreement with the experimental rate law:

$$\frac{-d[C=C]}{dt} = \frac{k_2'\underline{K}_1'\,[C=C]\,[Ru]_T\,[H_2]}{[PPh_3] + \underline{K}_1'\,[H_2] + \underline{K}_1'\underline{K}_6\,[CN][H_2]} \tag{9.17}$$

III. CONCLUSIONS

Important elastomeric materials with high stability to oxidative degradation have been synthesized in this work by means of homogenous catalytic hydrogenation. $RhCl(PPh_3)_3$ and $RuCl(CO)(OCOR)$ $(PPh_3)_2$ were used as catalysts for the hydrogenation of diene polymers such as PBD, St-BD-St, SBR, and NBR. The emphasis of this work was on gaining scientific understanding of the catalytic processes, to optimize the performance of the catalysts.

Quantitative hydrogenation of the C=C unsaturation of diene polymers was achieved without involving the reaction of functional groups of other types that were present in the polymer substrates. No large-scale changes of chain properties of polymers were found after the hydrogenation reactions. Based on detailed kinetic studies and spectroscopic analysis results, a number of mechanisms were postulated for the different reaction systems. Corresponding rate laws for each mechanism were derived, which were consistent with the observed experimental data. For the hydrogenation systems catalyzed by $RhCl(PPh_3)_3$, reaction parameters were estimated and verified by applying nonlinear least-squares regression analysis to the rate law models. It was found that polymer type and molecular weight have no significant effect on the values of equilibrium constants for reaction steps that did not involve the substrate. Nevertheless the mobility of the polymer chain and the steric environment of the C=C bonds were found to have an important effect on the rate of hydrogenation.

The nature of the reaction solvent was found to be important with respect to the degree of 1,4-PBD hydrogenation. In the case of NBR hydrogenation, the nature of solvent influenced the selectivity.

RuCl(CO)(OCOPh)(PPh$_3$)$_2$ was found to be a catalyst for PBD hydrogenation; it was also found to catalyze the isomerization of *cis*-1,4 C=C to *trans*-1,4 C=C in *o*-dichlorobenzene at 85°C under relatively low hydrogen pressure. In the case of NBR hydrogenation catalyzed by RuCl(CO)(OCOPh)(PPh$_3$)$_2$, isomerization caused by C=C migration and the coordination between Ru and the nitrile group were found to be the main factors resulting in nonquantitative hydrogenation and the onset of side reactions that gave rise to gel formation.

ACKNOWLEDGMENTS

The authors thank Dr. N. A. Mohammadi and Mr. P. Scott for some of the experimental work presented. Financial support of the research by the Ministry of Colleges and Universities of the Province of Ontario, Canada, and the Natural Science and Engineering Research Council of Canada is gratefully acknowledged.

REFERENCES

1. Schulz, D. N., Turner, S. R., and Golub, M. A., *Rubber Chem. Technol.*, *55*, 809 (1982).
2. Doi, Y., Yano, A., Soga, K., and Burfield, D. R., *Macromolecules*, *19*, 2409 (1986).
3. Carella, J. M., Grasessley, W. W., and Fetters, L. J., *Macromolecules*, *17*, 2775 (1984).
4. (a) Udipi, K., *J. Appl. Polym. Sci.*, *23*, 3301 (1979). (b) Udipi, K., *J. Appl. Polym. Sci.*, *23*, 3311 (1979).
5. Cunneen, J. I., and Porter, M., in *Encyclopedia of Polymer Science and Technology*, Vol. 12, Wiley, New York, pp. 318-320.
6. (a) Dreyfuss, P., and Kennedy, J. P., U.S. Patent 4,136,136, University of Akron (1979). (b) Tran, A., and Prud'homme, J., *Macromolecules*, *10*, 149 (1977).
7. Azuma, C., Misubashi, T., Sanui, K., and Ogata, N., *J. Polym. Sci., Polym. Chem. Ed.*, *18*, 781 (1980).

8. Makowski, H. S., Lundberg, R. D., and Bock, J., U.S. Patent 4,184,988, to Exxon Research and Engineering Company (1974).
9. Sang, S., and Tob, M., *J. Rubber Res. Inst. Malays.*, *26*, 48 (1979); *Chem. Abstr.*, *91*, 212357 (1979).
10. Halasa, A. F., and Tate, D. P., U.S. Patent 3,976,628, to Firestone Tire and Rubber Co. (1976).
11. (a) Witte, J., Guenter, L., and Pampos, G., German Offen. 2,344,734, to Bayer A.G. (1975). (b) Tsai, T., U.S. Patent 4,153,765, to Copolymer Rubber and Chemical Corp. (1979).
12. Mohammadi, N. A., and Rempel, G. L., *J. Mol. Catal.*, *50*, 259 (1989).
13. Bouchal, K., Ilavský, M., and Žurková, E., *Angew. Makromol. Chem.*, *165*, 165 (1989).
14. Mohammadi, N. A., and Rempel, G. L., *Macromolecules*, *20*, 2362 (1987).
15. Mohajer, Y., Wilkes, G. L., Wang, I. C., and McGrath, G. E., *Polym. Prepr.*, *22*, 138 (1982).
16. James, B. R., *Adv. Organometal. Chem.*, *17*, 319 (1979).
17. Rempel, G. L., Mohammadi, N. A., and Farwaha, R., U.S. Patent 4,812,528, to University of Waterloo, Waterloo, Ontario, Canada (1989).
18. Rempel, G. L., Mohammadi, N. A., and Farwaha, R., U.S. Patent 4,816,525, to University of Waterloo, Waterloo, Ontario, Canada (1989).
19. Draper, N., and Smith, H., *Applied Regression Analysis*, 2nd edn., Wiley, New York, 1981.
20. Brown, J. M., and Lucy, A. R., *J. Chem. Soc., Chem. Commun.*, 914 (1984).
21. Halpern, J., *Inorg. Chim. Acta*, *50*, 10 (1981).
22. Tolman, C. A., and Faller, J. W., in *Homogenous Catalysis with Metal-Phosphine Complexes* (L. H. Pignolet, ed.), Plenum Press, New York, 1983, p. 13.
23. Landis, C. R., and Halpern, J., *J. Organometal. Chem.*, *250*, 485 (1983), and references therein.
24. Osborn, J. A., Jardine, F. H., Young, J. F., and Wilkinson, G., *J. Chem. Soc. A, Inorg. Phys. Theor.*, 1711 (1966).
25. Neter, J., Wasserman, W., and Kutner, M. H., *Applied Linear Regression Models*, Irwin, Homestead, IL, 1983.
26. Buding, H., Fiedler, P., Koenigshofen, H., and Thormer, J., U.S. Patent 4,631,315, to Bayer AG (1986).
27. Buding, H., Szentivanyi, Z., and Thormer, J., U.S. Patent 4,647,627, to Bayer AG (1987).
28. Fiedler, P., Buding, H., Braden, R., and Thormer, J., U.S. Patent 4,746,707, to Bayer AG (1988).

29. Mohammadi, N. A., Catalytic hydrogenation, hydroformylation and hydroxymethylation of diene polymers, Ph.D thesis, University of Waterloo, 1987.

30. KuDela, V., in *Encyclopedia of Polymer Science and Engineering*, Vol. 7 (J. I. Kraschwitz, ed.), Wiley, New York, 1988, p. 809.

31. Rempel, G. L., and Azizian, H., U.S. Patent 4,464,515, to Polysar Limited (1984).

32. Bryan, S. J., Huggett, P. G., and Wade, K., *Coord. Chem. Rev.*, *44*, 149 (1982).

10

Hydrogenation of Reducible Compounds with High Levels of Sulfur

Victor L. Mylroie and Joseph K. Doles

Chemicals Development Division, Eastman Kodak Company, Rochester, New York

I. INTRODUCTION: A NEED FOR A CATALYST EFFECTIVE IN HYDROGENATIONS IN THE PRESENCE OF SULFUR-CONTAMINATED MATERIALS

This chapter discusses the use of chromium-containing, Raney cobalt catalyst for the hydrogenation of reducible compounds contaminated with high to very high levels of sulfur.

We had a need to reduce nitro compounds to the corresponding amines where the nitro compound was highly contaminated with sulfur. In traditional catalytic reductions using noble metal and Raney catalysts, sulfur is a strong poison, blocking the catalyst sites and rendering the catalyst inactive toward hydrogenation. Problems encountered in reducing compounds with high levels of sulfur include some poisoning of the catalyst (rendering it inactive toward hydrogenation), poisoning or contamination of equipment used, and the difficulties associated with cleanup and disposal.

Materials such as Raney cobalt or Raney nickel can be used as scavengers of sulfur. Although this approach may work as a cleanup technique, it adds handling steps to the chemical preparation and increases the exposure of the catalyst to the air, increasing the potential for fires or fire hazards. In some cases the base alkaline Raney nickel promotes hydrolysis of a labile group, resulting in unwanted side reactions. Palladium and other noble metals are ineffective as sulfur scavengers on the scale involved in this work.

The use of other techniques to eliminate sulfur from a compound has been reported. Harshaw Catalysts, for example, prepared a catalyst especially for desulfurization without hydrogen (Harshaw Catalyst Ni-3266-E and Ni-5124) [1]. Also reported in the literature are two fairly complete studies by Klostermann and Hobert [2] and by Kaps [3].

Another technique is to use catalysts such as supported nickel. We have tried this technique and found that the levels of sulfur we were working with far exceeded the capability of the supported nickels. There are some limited examples of catalytic materials used to reduce reducible organic compounds highly contaminated by sulfur. One such example is incorporated in a European patent [4]. The catalyst described in the patent is formulated as a rare earth oxy-hydride, which could be of a formula containing, among other noble metals, cobalt. There are also a number of patents issued to oil companies in which the feedstock is freed of sulfur-containing mate-rial by use of some catalyst that may contain cobalt [5-14] (for Raney cobalt as a hydrogenation catalyst, see reference 15); how-ever, none of these examples incorporate the use of Raney cobalt promoted with chromium as the catalyst.

The process reported here demonstrates the preparation of primary amines and other reduced materials, which would otherwise be impossible or impractical to prepare because of sulfur contamina-tion and sulfur contaminants, by reduction of nitro groups or other reducible groups attached to organic molecules useful as intermediates in the preparation of photographic materials. Although chemical reductions could be used to reduce sulfur-contaminated materials in the presence of sulfur and sulfur contaminants it is not desirable to do this, first because the disposal of large amounts of metals sludge is not attractive, and second because the presence of metals results in impurities that are photographically active.

II. EXAMPLES OF REDUCTIONS WITH HIGH SULFUR LEVELS

Some examples of organic compounds that were contaminated with sulfur up to a level of 10,000 ppm and were reduced are illustrated by equations (10.1)-(10.3).

A. Pyrazoltriazol-Type Compounds Reduced

$R_1 = C_2$ to C_{14}
$R_2 = H$, CH_3, t-Butyl
$R_3 = H$, Cl
$R_4 = H$, acetyl

(10.1)

B. Pyrazoltriazol (PT) with High Sulfur Contamination: Reduction by Raney Cobalt Catalyst

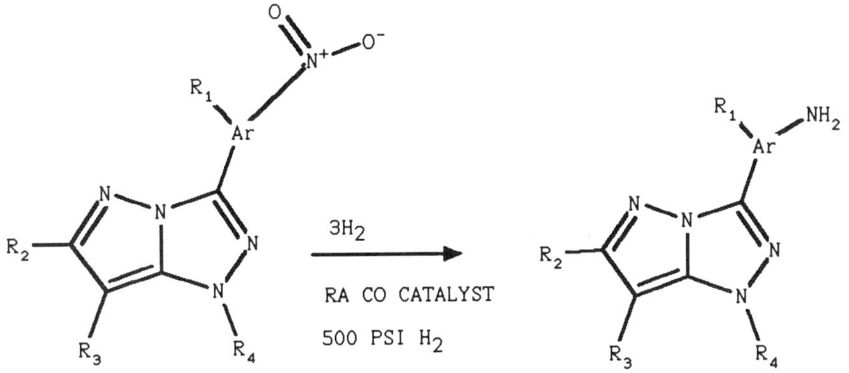

$$\xrightarrow[\text{500 PSI } H_2]{\substack{3H_2 \\ \text{RA CO CATALYST}}}$$

R_1 = H, CH_3 or NO_2
R_2 = H, CH_3 or t-Butyl
R_3 = H, Cl
R_4 = H, acetyl

(10.2)

C.

$$\xrightarrow[\text{THF}]{\substack{H_2 \\ \text{RACO}}} \quad 2$$

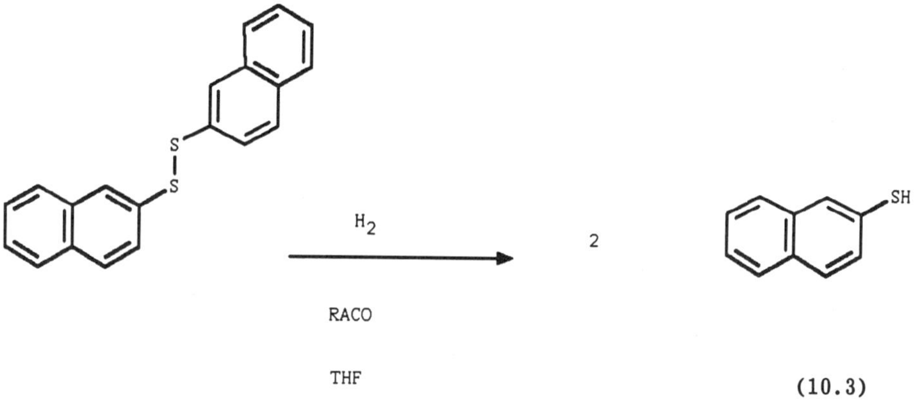

(10.3)

III. CATALYST

The preparation of the Raney cobalt catalyst is covered in a patent by Carl Lentz, issued to Tennessee Eastman Chemicals [1].

IV.   RESULTS AND DISCUSSION

Initially we encountered problems with the reduction of material
shown in equation (10.1) (where $R_1 = C_{12}$, $R_2 = CH_3$, $R_3 = H$,
$R_4 = H$) when we scaled up the reaction from laboratory size (500 mL
autoclave) to a 25-gallon autoclave. We found a gray solid present
upon opening the autoclave to sample for completion of the reaction.
This gray solid was identified as a mixture of hydroxyl amine,
the azo compound, and azoxy compounds. It was impossible to further
hydrogenate this solid material, and consequently the reaction mixture was unfilterable and could not be used.

   We have found that the insoluble materials were formed
because the reduction reaction was "starved" of hydrogen, thus
forming the hydroxylamine and nitroso intermediates, which further
reacted to form the azo and azoxy compounds. Our solution to this
problem was to increase the reaction pressure from several hundred
to 1000 psi and to increase the temperature from room temperature
to 55-60°C. This promoted the immediate formation of the amine

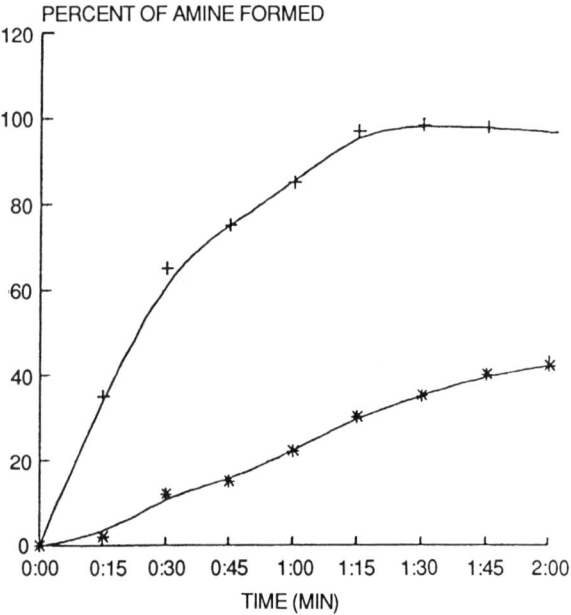

Figure 10.1   Hydrogenation of nitro with Raney cobalt: amine formation versus time. (+) Series 2: 1000 psi H_2, 60°C, and series 3:
100 psi increments, at 60°C.

without building up the amount of hydroxylamine. A combination
of ethyl acetate and tetrahydrofuran (THF) was used as solvent.
The formation of the amine is rapid when the catalyst surface is
kept hydrogen rich (e.g., 1000 psi H_2) and a rapid stirring rate
is used, whereas the formation of the hydroxylamine is promoted
if the catalyst surface is "starved" of hydrogen. This buildup of
hydroxylamine causes an increase of the azo and azoxy compounds,
which cannot be further hydrogenated in these examples, giving
about 45% of the hydroxylamine, azo and azoxy compounds; see
equation (10.4) and Figure 10.1.

Formation of azo and azoxy compounds during reaction:

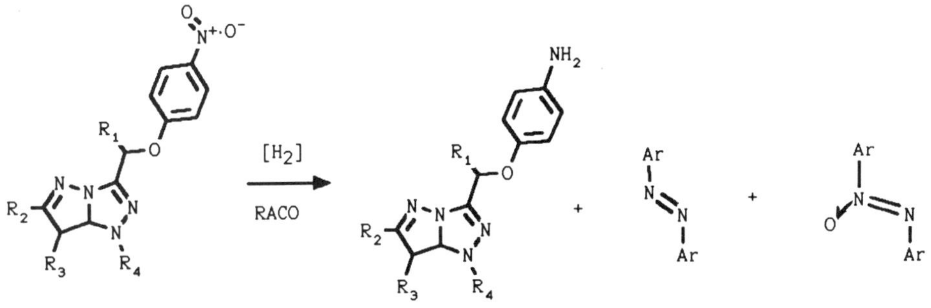

R = C to C   (10.4)
R = C , *t*-butyl, H
R = H, Cl C
R = H, acetyl
Ar = aromatic ring

If the catalyst is not "starved" of hydrogen, the reduction
continues to yield the desired amine and leaves the residual azo
and azoxy compounds in very small amounts, ranging in percentage
from about 1.0 to 2.5%; see Figures 10.1 and 10.2.

If the hydrogenation of the nitro compound is "starved" of hydrogen during the reduction, as is the case by "ramping" the hydrogen
in increments of 100 psi then the intermediates react to form the
azo and azoxy compounds in percentages approaching 45%, causing
the solids to precipitate and coating the catalyst. This stops the
reaction and the reaction is termed a "failure" (Figure 10.3).

The catalyst was originally designed to be used in the reduction
of chloro and bromonitro aromatic compounds to reduce the hydrogenolysis of the halides from the aromatic ring. The examples used in
this chapter also took advantage of this feature, in addition to
being less susceptible to sulfur poisoning [17]. In most cases on
a large scale, the hydrogenolysis was reduced to 0.5-2.0%.

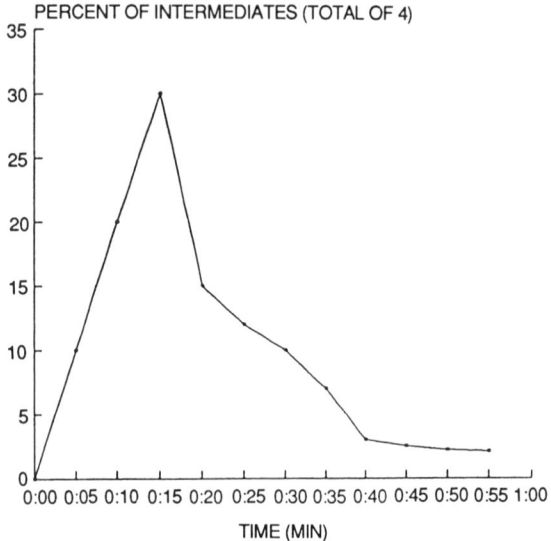

Figure 10.2 Hydrogenation of nitro with Raney cobalt: amine formation of −NO, −NHOH, azo, and azoxy. Products of hydrogen. Data points: series 1% azo/azoxy, 1000 psi at 60°C; prereduced Raney Co.

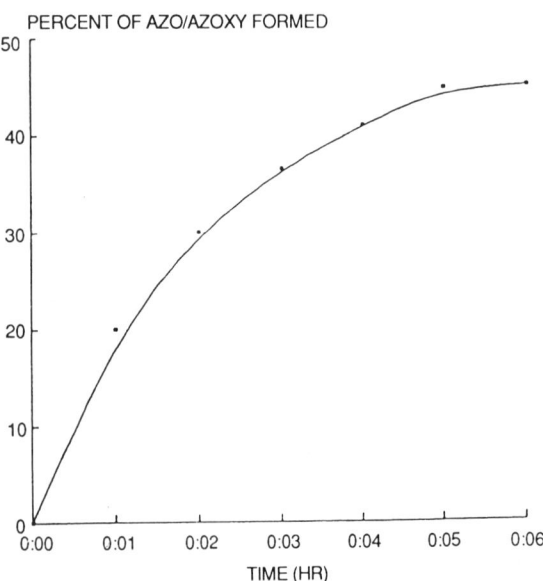

Figure 10.3 Hydrogenation of nitro with Raney cobalt: azo/azoxy formation versus time. Data points: series 1 azo/azoxy, 2 hours with N₂, then 100 psi increments to 900.

V. CONCLUSION

Raney cobalt catalyst, properly prepared and promoted with chromium, can successfully be used to reduce nitro compounds that are highly contaminated with sulfur in the range up to about 10,000 ppm [18]. Vessels that can become contaminated from the sulfur-containing materials must be cleaned periodically to keep the sulfur contamination level on the walls of the autoclave at a minimum.

The catalyst does lose activity upon aging but can be reactivated by a prereduction under conditions of hydrogen at 500 psi and a temperature of at least 50°C. (Tennessee Eastman Organic Chemicals, in conjunction with Kodak Research Laboratories [19] in Rochester, New York, is in the process of developing a test that will predict or establish the relative activity of any particular batch of Raney cobalt catalyst).

ACKNOWLEDGMENTS

We express thanks to the Analytical Technology Division of Eastman Kodak Company for assistance with identification of compounds and with HPLC, particularly Ann Buerschaper and Marsha Peone. Thanks to William Phillips and Kathy McDavid from Tennessee Eastman Chemicals for helpful assistance and dialog. Thanks to Jim Huson from the Chemicals Development Division of Eastman Kodak for assistance throughout the program. Thanks to the Synthetic Chemicals Division of Eastman Kodak for helping meet production schedules, and to William Franklin of the pilot plant. Thanks to Michael Spitulnik, director of the Chemicals Development Division, Organic Products Section, for his direction throughout the work.

REFERENCES

1. Harshaw Catalysts, Desulfurization without hydrogen, Laboratory Technical Report, Beachwood, Ohio.
2. Klostermann, K., and Hubert, H., Reaction of sulfur compounds on silica-supported nickel, *J. Catal.*, *63*, 355-363 (1980).
3. Kaps, U., Veranderung der Aktivität von Nickel-Hydroierkatalysatoren durch Schwefel-und Phosphorverbindungen, Doctoral thesis, Johannes Gutenberg University, Mainz, 1981.
4. European Patent EP 273,575.
5. U.S. Patent 4,727,165.
6. U.S. Patent 3,932,393.
7. German Offen. DE 2,209,020.

 8. French Patent FR 2,127,798.
 9. U.K. Patent GB 1,358,963.
10. U.S. Patent 4,425,312.
11. U.S. Patent 4,083,199.
12. U.S. Patent 4,041,130.
13. U.S. Patent 4,039,621.
14. U.S. Patent 3,974,063.
15. U.S. Patent 4,721,811.
16. Lentz, C., Hydrogenation of halonitroaromatic compounds, Invention Reports, Tennessee Eastman Chemicals, K-209,454 as S.N. 151, 726; U.S. Patent 4,929,737.
17. Mylroie, V. L., Hydrogenation of reducible compounds with high levels of sulfur, Eastman Kodak Company, Rochester, NY, Serial no. 473,006, File date, Jan. 31, 1990.
18. 81st Interplant Technical Conference, Eastman Kodak Research, Rochester, NY, May 14-16, 1990, p. 61 (restricted information).
19. Pressure Reaction, High Pressure Laboratory Operating Procedure SP08361 to SP11300, Eastman Kodak Company, Rochester, NY, February 1987-January 1990.

11

The Synthesis of 1-Deoxynojirimycin Employing the Hydrogenation of a Polymer-Bound Substrate

Mike G. Scaros, James R. Behling, Payman Farid, John R. Medich, and Michael L. Prunier
Department of Chemical Development, Searle, Skokie, Illinois

Richard M. Weier and Ish Khanna
Department of Molecular and Cell Biology, Searle, Skokie, Illinois

I. INTRODUCTION

The potent glucosidase inhibitor 1-deoxynojirimycin (6) has been known for some time [1] and has been synthesized using a variety of methods [2]. Most of these syntheses are encumbered by the requirement for a microbiological transformation or are complex multistep chemical syntheses that are not amenable to large-scale production.

During the course of developing a short, large-scale synthesis of 6 according to the chemistry outlined in Figure 11.1, we encountered difficulties converting 6-deoxy-6-amino-1,2-O-isopropylidene-L-sorbofuranose (4) to its deprotected form 5 using either aqueous trifluoroacetic acid or aqueous acetic acid. Because of the very labile nature of the deprotected amine, hydrogenation at 60 psig of hydrogen at 25-60°C in the presence of palladium on carbon immediately after the deprotection step yielded only small quantities of 1-deoxynojirimycin. Furthermore, when the deprotection was done *in situ* under reductive amination conditions, very little of the desired 1-deoxynojirimycin could be isolated.

Hydrolysis has been reported to proceed well on the corresponding 6-deoxy-6-amino-2,3-O-isopropylidene-L-sorbofuranose using hydrochloric acid. When the deprotected product was isolated as the hydrochloride salt and subjected to the reductive amination conditions above, 1-deoxynojirimycin was isolated in good yield [3]. Unfortunately, the processing of a hydrochloride salt in stainless steel equipment presents as great a problem in large-scale synthesis as the instability of the free base 5 [4].

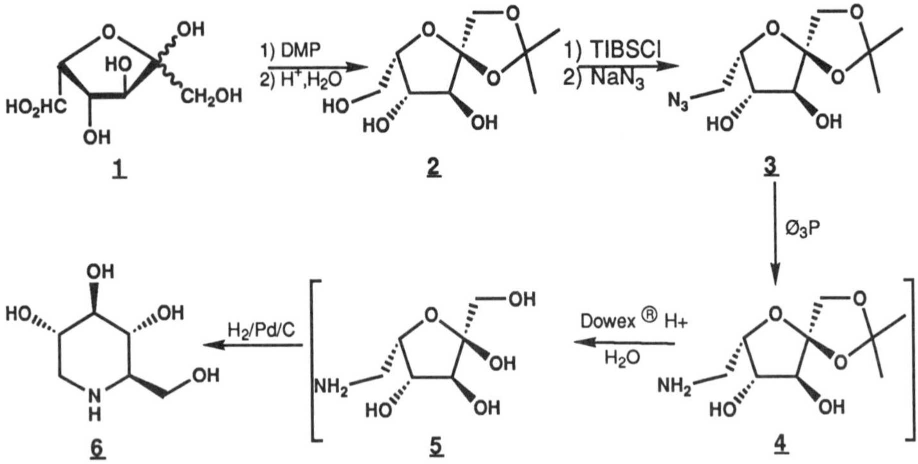

Figure 11.1

To eliminate the need of handling the hydrochloride salt and
the generation of the unstable free amine, we investigated adsorbing
the protected amine 4 on a suitable acidic support followed by performing the required transformations (hydrolysis and reductive
amination) without desorption of unstable 5 from the support.

II. DISCUSSION

6-Deoxy-6-azido-1,2-O-isopropylidene-L-sorbofuranose was prepared
from L-sorbose as shown in Figure 11.1. L-Sorbose was converted
to 1,2-O-isopropylidene-L-sorbofuranose (2) by treating a slurry
of L-sorbose in refluxing tetrahydrofuran with 2.2 equivalents of
2,2-dimethoxypropane and a catalytic amount of stannous chloride
[5] followed by the addition of aqueous sulfuric acid. After 7 hours
at room temperature, the resulting monoacetonide was isolated in
a 40% crystallized yield. Primary selective derivatization was accomplished by treating 2 with 1 equivalent of 2,4,6-triisopropylbenzene-
sulfonyl chloride in a 1:1 mixture of triethylamine and pyridine.
The resulting crude product was subjected to azide displacement
with sodium azide at 100°C for 20 hours. The 6-deoxy-6-azido-1,2-
O-isopropylidene-L-sorbofuranose was isolated and purified by
chromatography (yield, 72% based on 2).
    Conversion of 3 into 6 was carried out without isolation of either
4 or 5, as described below.

Reduction of azide 3 to amine 4 was carried out according to the reduction procedure of Vaultier et al. [6]. 6-Deoxy-6-azido-1,2-O-isopropylidene-L-sorbofuranose (3) in tetrahydrofuran was treated with 1.5 equivalents of triphenylphosphine at ambient temperature for 20 hours. The intermediate iminophosphorane was hydrolyzed by the addition of water, and the tetrahydrofuran was removed by rotary evaporation. The resulting aqueous mixture was filtered to remove triphenylphosphine oxide [Note that catalytic hydrogenation also can be used to reduce the azide to the amine.] The aqueous solution, containing the protected amino sugar 4, was added to an aqueous slurry of sulfonic acid ion-exchange resin (Dowex 50X8-200) and the resultant slurry was stirred well for 1 hour. The slurry was acidic (pH = 5-6). The resin was filtered and the filter cake washed successively with water, methanol, and more water, to remove neutral by-products. The resin was then slurried in water and a 20% loading of 4% palladium on carbon (based on 3) was added. The reaction was subjected to 60 psi of hydrogen for 72 hours. The resin was reisolated by filtration and washed with water followed by methanol. The resin (containing 1-deoxynojirimycin) was slurried in anhydrous methanol-ammonia and refiltered. The filter cake was washed with additional methanol-ammonia. The filtrate and washes were combined and concentrated by rotary evaporation to provide a crystalline solid, which was recrystallized from 10% aqueous methanol to provide pure 1-deoxynojirimycin (6) in 61% yield based on 3.

Success in deprotecting and hydrogenating the unstable intermediate 4 was both intriguing and surprising. We have found no similar literature precedents, and at this time we assume that 5 is in dynamic equilibrium with the adsorbed intermediate during the course of the reductive amination.

With this encouraging result in hand, we sought to further expand the utility of this hydrogenation-reductive amination methodology by defining the stability of 5 in the adsorbed state on the acidic resin by converting it to 1-deoxynojirimycin (6) at various time intervals. The reduction of 6-deoxy-6-azido-1,2-O-isopropylidene-L-sorbofuranose (3) with triphenylphosphine was repeated, and the resulting 6-deoxy-6-amino-1,2-O-isopropylidene-L-sorbofuranose (4) was adsorbed on the acidic ion-exchange resin as described above. The resulting product was divided into portions. One portion was subjected to the reductive amination-desorption protocol immediately, to provide 1-deoxynojirimycin (6) in a similar (60%) isolated yield. The remainder of the resin was stored in a sealed amber bottle at 5°C and was sampled after 5 weeks. The reductive amination was performed in an identical fashion, and the cyclized product 6 was isolated. No significant reduction in product yield was observed over this period of 5 weeks.

III. CONCLUSION

We have demonstrated a synthesis of 1-deoxynojirimycin (6) from L-sorbose that requires the isolation of only two intermediates (2 and 3), as well as the use of only one protecting group. There is no requirement for the isolation of the unstable amine 5, and 1-deoxynojirimycin (6) is conveniently isolated from an anhydrous methanolic solvent mixture as the free base rather than the more commonly used hydrochloride salt [7]. The hydrolysis–reductive amination methodology has been demonstrated on a multikilogram scale in pilot plant equipment. Additionally, we have demonstrated that 6-deoxy-6-amino-L-sorbofuranose (5) is remarkably stable while it is adsorbed on an acidic ion-exchange resin.

REFERENCES

1. Muriao, S., and Miyata, S., *Agric. Biol. Chem.*, *44*, 219 (1980); Scofield, A. M., Fellows, L. E., and Fleet, G. W. J., *Life Sci.*, *39*, 645 (1986).
2. Fleet, G. W. J., Carpenter, N. M., Petursson, S., and Ramsden, N. G., *Tetrahedron Lett.*, *31*, 409 (1990); Fleet, G. W. J., Fellows, L. E., and Smith, D. W., *Tetrahedron*, *43*, 979 (1987); Ziegler, T., Straub, A., and Effenberger, F., *Angew. Chem. Int. Ed. Engl.*, *27*, 716 (1988).
3. Stoltefuss, J., U.S. Patent 4,220,782 (1980).
4. Paulsen, H., Sangster, I., and Heyns, K., *Chem. Ber.*, *100*, 802 (1967).
5. Chen, C., and Whistler, R. L., *Carbohydr. Res.*, *175*, 265 (1988).
6. Vaultier, M., Knioezi, N., and Carrié, R., *Tetrahedron Lett.*, *24*, 763 (1983).
7. Beaupere, D., Stasik, B., Vzan, R., and Demailly, G., *Carbohydr. Res.*, *191*, 163 (1989).

12

Si Modification of Pd and Pt Catalysts

S. Tjandra and D. Ostgard
Department of Chemistry and Biochemistry, Southern Illinois University at Carbondale, Carbondale, Illinois

G. V. Smith
Molecular Science Program and Department of Chemistry and Biochemistry, Southern Illinois University at Carbondale, Carbondale, Illinois

M. Musoiu and T. Wiltowski
Molecular Science Program, Southern Illinois University at Carbondale, Carbondale, Illinois

F. Notheisz and M. Bartók
Department of Organic Chemistry, József Attila University, Szeged, Hungary

J. Stoch
Institute of Catalysis and Surface Chemistry, Polish Academy of Sciences, Kraków, Poland

I. INTRODUCTION

Triethylsilane (Et_3SiH) decomposes in stages by hydrogenolysis of one ethyl group at a time until only a strongly adsorbed Si remains attached to the catalytic surface [1]. This strongly adsorbed Si seems to be difficult to remove and represents a strongly bonded poison. Attempts to reactivate the poisoned catalyst by heat treatment were only partially successful for some of the catalysts. However, when the silicon-poisoned catalyst is treated with oxygen at room temperature, a thin film of SiO_2 or SiO_2-like material forms on the surface [2-4]. The formation of SiO_2 upon exposure of Si to oxygen is commonly observed. Because of its practical importance in semiconductor technology, the growth of SiO_2 on top of Si single crystals has been widely studied [4,5]. For example, Dubois and Nuzzo [4] reported that Ni_2Si exposed to oxygen forms a thin layer of SiO_2.

We have continued our studies of triethylsilane hydrogenolysis
to Si as a poisoning or modifying agent for supported palladium
and platinum catalysts. The results show that not only does triethylsilane poison Pd and Pt catalysts for both the hydrogenolysis of
triethylsilane and the hydrogenation of cyclohexene and cyclopentene,
but also the fully poisoned catalyst can be reactivated by oxidation
and reduction. The poisoning and reactivation are structure sensitive,
with the smaller metal crystallites showing the greatest reactivation.

II.   EXPERIMENTAL

Several types of catalyst were used in this study. The catalysts
were either prepared in this laboratory or provided by others.
Some of the catalysts were also available commercially. The commercial catalysts were Pd black and three 1% Pd/Al_2O_3 (5.5, 22, and
49% dispersion) donated by Engelhard Industries. The 1.48%, 0.825%
Pt/SiO_2, and 1.45% Pd/SiO_2 were obtained from the laboratories
of Burwell and Butt [6]. All the remaining catalysts (0.65, 0.34,
0.083, 0.46, and 3% Pd/SiO_2) were prepared in our laboratory.
Characterization of all these catalysts was also done by hydrogen
chemisorption and electron microscopy.

Both the hydrogenation and the triethylsilane hydrogenolysis
were conducted in a microreactor system consisting of a U-tube
microflow reactor, which was protected from oxygen by a highly
dispersed Mn/SiO_2 oxygen indicator/adsorber. Triethylsilane was
injected directly into the microreactor over the catalyst at 250°C
with flowing hydrogen as a carrier gas. The ethane-ethene mixture
produced from the hydrogenolysis was determined by gas chromatography. Gas phase hydrogenations were carried out in the same
apparatus. Liquid phase hydrogenations were carried out in a separate apparatus by transferring the poisoned catalyst in the liquid
phase hydrogenation reactor with protection from the air. For experiments in which the Si-poisoned catalyst was oxidized, pure oxygen
was passed through the reactor at 24°C for 2 hours. The ESCA
data were recorded on a VG Scientific ESCA-3 photoelectron spectrometer using $AlK\alpha_{1,2}$ radiation (1486.6 eV) from an X-ray source
operating at 13 kV and 10 mA. Working pressure was 2×10^{-8} Torr.
Binding energies were referenced to the C1s peak from carbon
surface deposit at 285.0 eV. Also, the Si poisoning experiments
were performed with the help of a Cahn model 113 microbalance
reactor.

III. RESULTS

Standard experiments show that cyclohexene conversions on the
fresh (untreated) catalysts are the same within experimental error
before and after oxygen treatment. The results also indicate that
the silica gel used for preparation of the catalyst does not have
any catalytic activity for decomposing triethylsilane or hydrogenating
cyclohexene.

During poisoning, addition of triethylsilane to the catalyst results
in the formation of ethane indicating the rupture of C—Si bonds.
The results show that when the amount of ethane generated increases,
the catalytic activity decreases. As an example, Table 12.1 presents
the results obtained for the 0.083% Pd/SiO_2 (D = 83.2%) catalyst.

Figure 12.1 shows the correlation between percentage exposed
(dispersion) and percentage of original hydrogenation activity restored after oxidizing of various Si-poisoned catalysts. The results
indicate that when the poisoned catalysts are exposed to oxygen
and then reduced in hydrogen, the initial activity is partly restored
and, in many cases, restored to even greater than the initial level.
This behavior was observed for the catalysts with dispersions larger
than 49%.

ESCA spectra of the silicon-poisoned Pd-black catalyst reveal
a decrease in the number of Pd surface atoms and confirm that
only part of the Si atoms move to the surface upon oxidation and
that they are mixed with surface metal atoms. Additionally, decomposing silane (SiH_4) on a highly dispersed Pd/SiO_2 catalyst in a
microbalance reactor reveals weight increases equivalent to several
Si atoms for each Pd surface atom. Subsequent oxidations cause
weight increases corresponding to oxidation of only part of the Si.

Table 12.1 Changes of Rate of Hydrogenation with Increasing
Amount of Ethane Formation over 0.083% Pd/SiO_2 (D = 83.2%)

Volume of ethane (μL)	Rate of hydrogenation of cyclohexene ($mL/min \cdot mg$)
	0.066 (fresh catalyst)
7.5	0.065
8.1	0.052
9.6	0.046
12.8	0.019
16.1	0.000

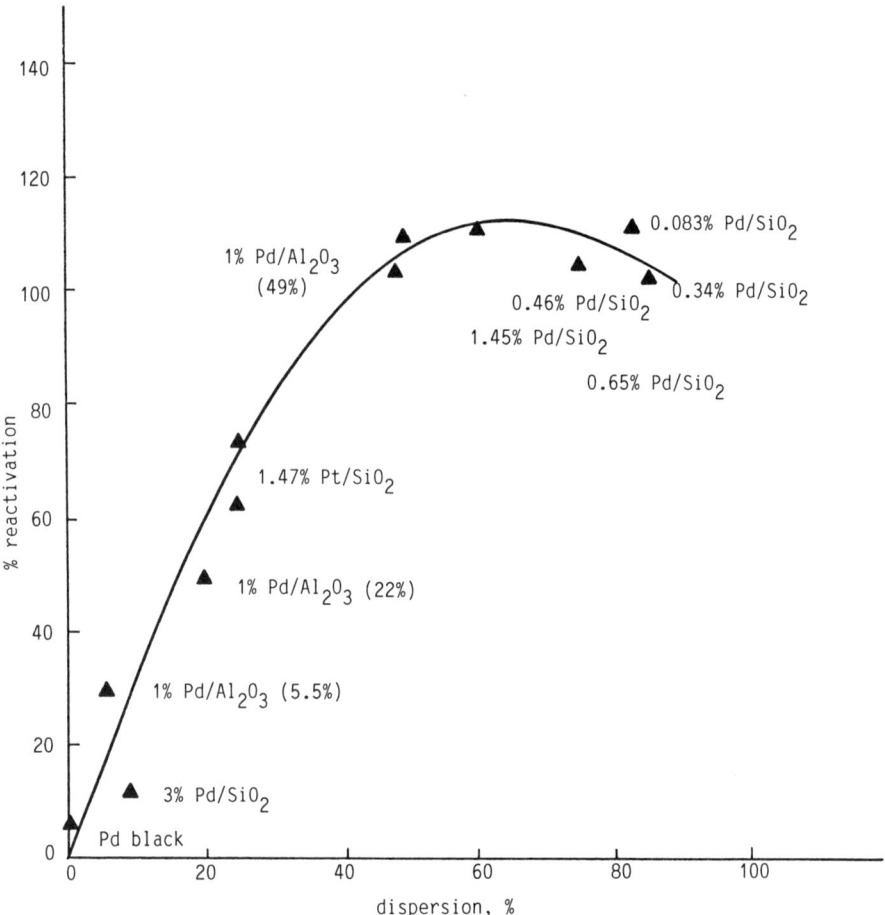

Figure 12.1 The influence of catalyst dispersion on the percentage of reactivation by oxygen for the reaction in cyclohexene hydrogenation.

IV. DISCUSSION

The results presented can be explained by the following process. During poisoning, the silicon atoms dissolve into Pd or Pt crystallites forming a Pd-Si (or Pt-Si) alloy, which gradually loses its activity at higher Si concentrations. Upon oxidation, Si returns to the surface and is oxidized to SiO_2. Metal is also oxidized at the surface, but reduction with H_2 reduces only the metal particles, which now present a new and different surface.

During this study a new method of reactivation of the fully, as well as the partially, poisoned catalysts was found. The poisoned catalysts exposed to oxygen and then reduced in hydrogen regained to a great extent their activities for hydrogenolysis. Furthermore, the initial hydrogenation activities were restored and in some cases restored to activities greater than the initial level, depending on metal dispersion.

REFERENCES

1. Molnár, A., Bucsi, I., Bartók, M., Notheisz, F., and Smith, G. V., *J. Catal.*, *98*, 386, 1986.
2. Abbati, I., Rossi, G., Calliari, L., Braicovich, L., Lindau, I., and Spicer, W. E., *J. Vac. Sci. Technol.*, *21*, 409, 1982.
3. Grunthaner, P. J., Scott, D. M., Nicolet, M. A., and Mayer, J. W., *J. Catal.*, *19*, 641, 1981.
4. Dubois, L. H., and Nuzzo, R. G., *J. Vac. Sci. Technol.*, *2*, 441, 1984.
5. Hiraki, A., Lugujjo, E., and Mayer, J. W., *J. Appl. Phys.*, *43*, 3643, 1972.
6. Uchijima, T., Hermann, J. M., Inoue, Y., Takahashi, G., Burwell, R. L., Jr., Butt, J. B., and Cohen, J. B., *J. Catal.*, *50*, 464, 1977.

13

Thermogravimetric Study of Pd/SiO₂ Poisoning by Carbon Disulfide

T. Wiltowski
Molecular Science Program, Southern Illinois University at Carbondale, Carbondale, Illinois

G. V. Smith
Molecular Science Program and Department of Chemistry and Biochemistry, Southern Illinois University at Carbondale, Carbondale, Illinois

D. Ostgard
Department of Chemistry and Biochemistry, Southern Illinois University at Carbondale, Carbondale, Illinois

I. EXPERIMENTAL

An investigation of CS_2 poisoning was performed on a series of Pd/SiO₂ (1%) catalysts using the thermogravimetric method. Catalysts with dispersion of 36, 63, and 93% were reduced in a Cahn microbalance reactor in flowing hydrogen at three different temperatures (25, 200, and 300°C) until constant weight was attained: 5 hours at 200 and 300°C; 12 hours at 25°C. When reduction was complete, the samples were cooled to 25°C. Then one-microliter pulses of CS_2 were injected into the flowing hydrogen that carried it over the catalyst. Once adsorption of CS_2 had ceased (no further weight increase after additional CS_2 pulses), the temperature was ramped to 300°C to desorb the CS_2.

II. RESULTS

We did not continuously monitor the weight changes during the reduction process at 25 and 200°C; however, at 300°C the weight changes were monitored and the weight losses of the catalysts show two steps (Fig. 13.1). The first step occurs in the temperature range 60-125°C for the 36 and 63% dispersed catalysts, and 120-180°C for the 93% dispersed catalyst. The second step occurs at 150°C for the 36 and 63% catalysts and at 210°C for the 93% dispersed catalyst.

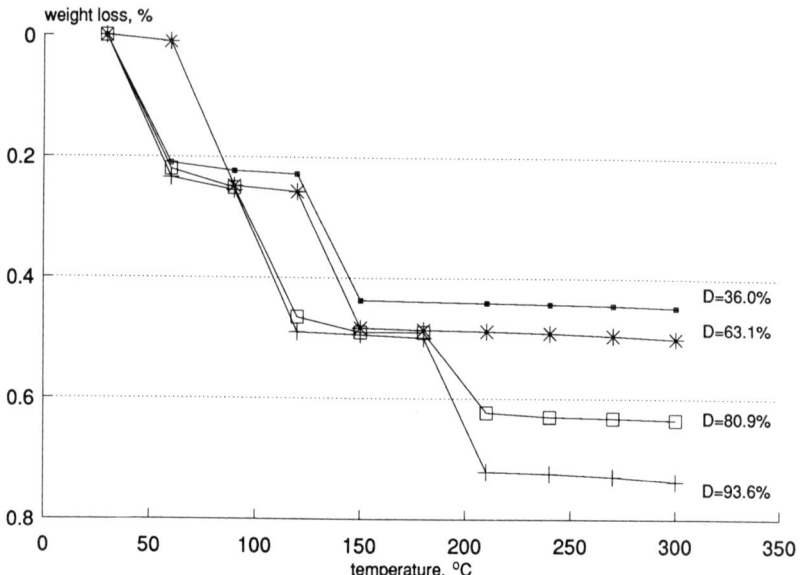

Figure 13.1 One percent Pd/SiO$_2$ reduction in hydrogen: weight loss during temperature ramp.

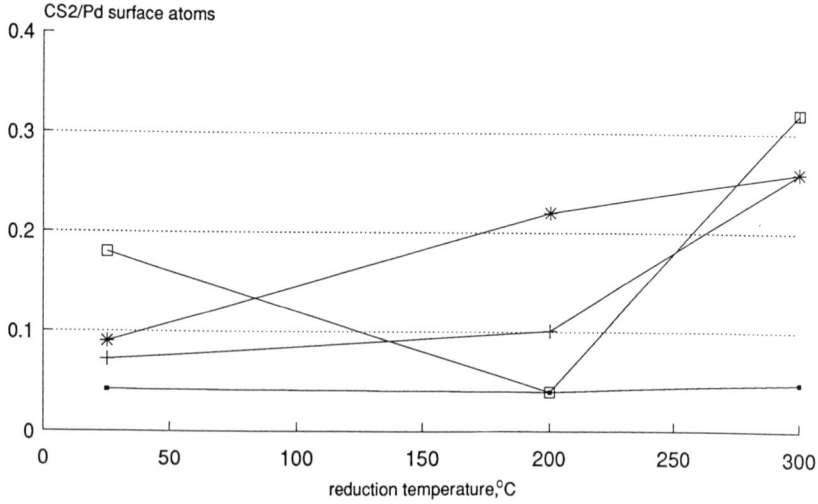

Figure 13.2 One percent Pd/SiO$_2$ microbalance study: (·) D = 36.0%, (+) D = 63.1%, (*) D = 80.9%, and (□) D = 93.6%.

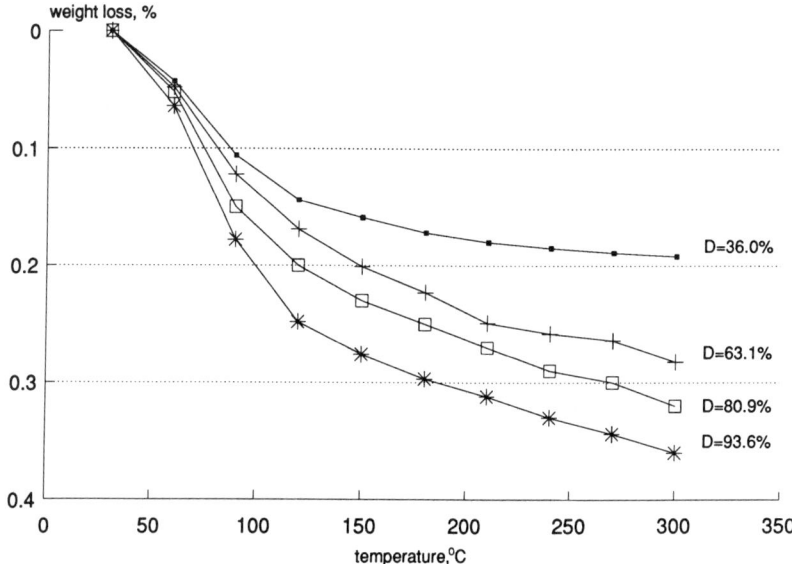

Figure 13.3   CS$_2$ desorption: microbalance study.

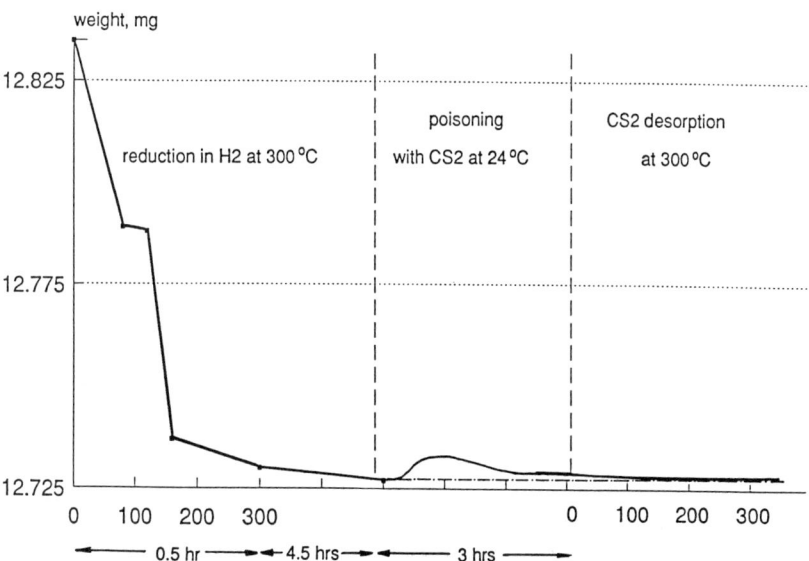

Figure 13.4   One percent Pd/SiO$_2$, D = 63.1%: weight change during the experiment.

During poisoning, the amount of adsorbed CS_2 varies according to the temperature of prereduction of the catalysts (Fig. 13.2). For example, the 36% dispersed catalyst adsorbs CS_2 nearly independently of the prereduction temperature. The 63% dispersed catalyst adsorbs increasing amount of CS_2 with increasing temperature of reduction, and the 93% dispersed catalyst adsorbs CS_2 in a pattern exhibiting a minimum at 200°C as the prereduction temperature increases. During desorption of CS_2, the weights of all catalysts decrease (Fig. 13.3); however, the final weights are always larger than the initial weights of the unpoisoned catalysts, as shown in Figure 13.4 for the 63.1% dispersed catalyst.

These preliminary results show that there is a correlation between the size of the particles and the temperature of the prereduction in the formation of the active sites for CS_2 adsorption. The small particles form more of these active sites when they are reduced at the higher temperature, whereas the large particles form the same number of these active sites during prereduction in the temperature range 25–300°C. The poisoning experiments indicate that CS_2 chemisorbs only on a part of the total Pd surface atoms, in agreement with titration studies using CS_2 to poison liquid phase hydrogenations over these same catalysts [1].

REFERENCE

1.   Smith, G. V., Notheisz, F., Zsigmond, Á. G., Ostgard, D., Nishizawa, T., and Bartók, M., *Proceedings of the 9th International Congress on Catalysis*, Calgary, 1988, Vol. 3, p. 1066.

14

Heterogeneous Catalytic Transfer Hydrogenation: Chemoselective Reduction of Aromatic Nitro Groups in the Presence of Aromatic Cyano Groups

James J. Huson and Louis F. Valente

Eastman Kodak Company, Rochester, New York

I. INTRODUCTION

Heterogeneous catalytic transfer hydrogenation has been used frequently for the removal of protecting groups and reducing nitro groups in peptide chemistry. The phenomenon has been extensively reviewed by Johnstone et al. [1]. This chapter discussed the application of this technology to the manufacture of chemicals on a commercial scale and the benefits of using the method.

Dye intermediates of the following structure are used in photographic systems to produce compounds that form cyan dyes when developed by the proper photographic developers:

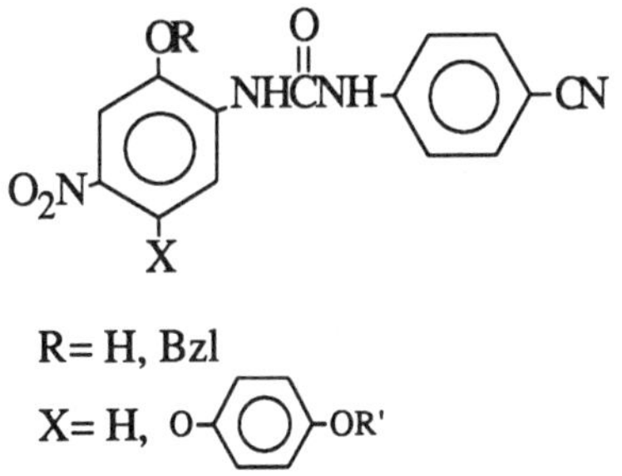

R= H, Bzl

X= H, O-⟨O⟩-OR'

<u>1</u>

Traditionally, heterogeneous high pressure hydrogenation has been used to reduce the nitro group. In some cases, a benzyl protecting group was removed along with the nitro reduction. An intermediate (2) is formed:

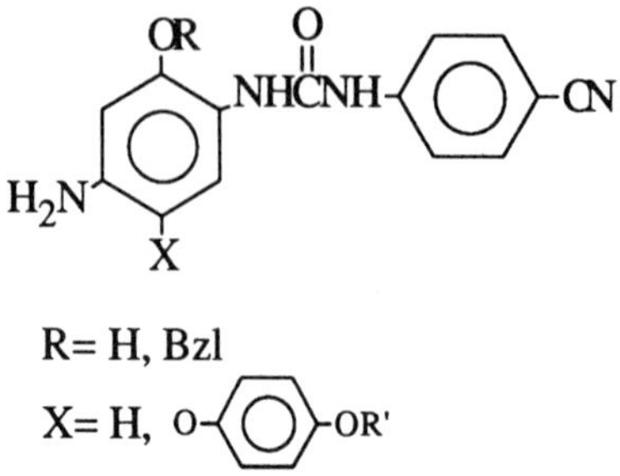

R= H, Bzl

X= H, O—⟨◯⟩—OR'

2

This reduction was hampered by variable results because the cyano group is also susceptible to reduction. The major undesired products are as follows (these impurities will be referred to as overreduced impurities):

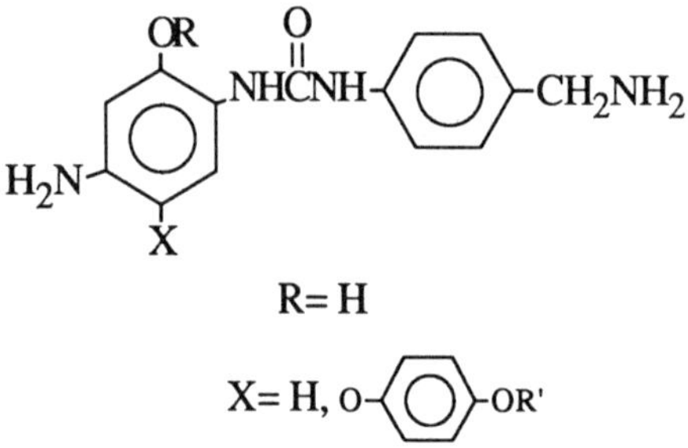

R= H

X= H, O—⟨◯⟩—OR'

3

R= H

X= H, O—⟨◯⟩—OR'

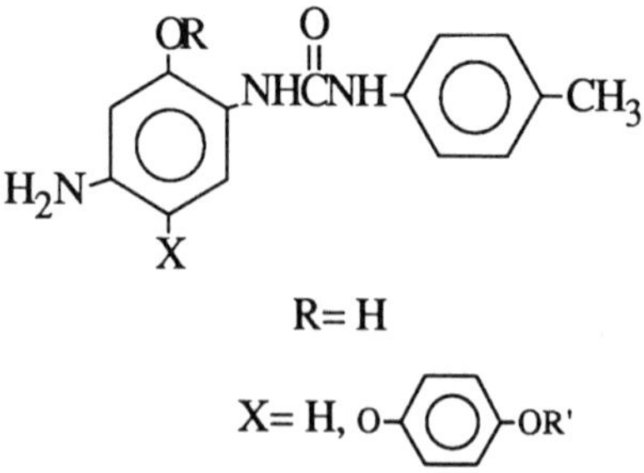

4

Both impurities 3 and 4 cause the dye to absorb at unwanted wavelengths. The removal of the impurities causes yield loss and in some cases makes the product unusable.

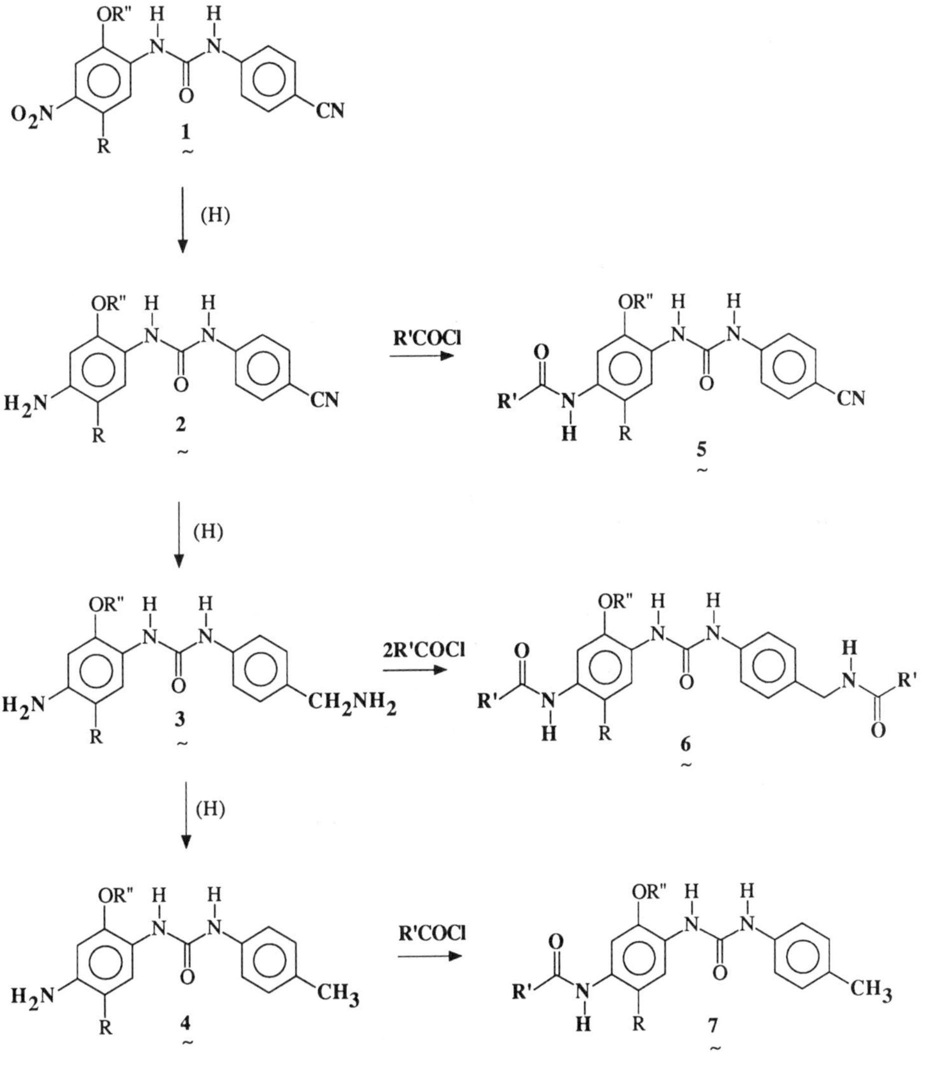

Figure 14.1 Hydrogen transfer versus conventional hydrogenation.

Catalytic transfer hydrogenation is conducted at atmospheric pressure and mild temperatures. It was easily scaled up to production scale by making some simple equipment changes to existing vessels and implementing additional safety measures. Continuous monitoring showed a maximum hydrogen level of 15 ppm throughout the course of the reaction.

II. EXPERIMENTAL

A. Hydrogen Transfer Method

To a round-bottomed flask is added 0.04 mol 4-(benzyloxy)-2-((4-cyanophenyl)ureidyl)-4-(4-methoxyphenoxy)-5-nitrobenzene, 100 mL of tetrahydrofuran, 60 mL of isopropyl alcohol, and 11 g (0.175 m) of ammonium formate, and 1.6 g of Pd/C-S (50% water wet). The reaction is warmed to 50°C and held for 2 hours. The reaction is cooled to 20°C and 0.04 mol of the appropriate acid chloride is added. The product is isolated to yield 76% of the desired product.

Liquid-liquid chromatography (LLC) was used to determine assay of the reaction mixture. The results are listed in the following tables.

B. High Pressure Method

To a 300 mL Parr reactor is added 0.04 mol 4-(benzyloxy)-2-((4-cyanophenyl)ureidyl)-4-(4-methoxyphenoxy)-5-nitrobenzene, 100 mL of tetrahydrofuran, 60 mL isopropyl alcohol, and 1.6 g of Pd/C-S (50% water wet). The vessel is placed on a rocker equipped with a heating mantle and placed under hydrogen pressure (60 psi). The reactor is warmed to 50°C and shaken for 2 hours.

The reaction is cooled, and 0.04 mol of the appropriate acid chloride is added. The product is isolated to yield 69% of the desired product. LLC was used to determine assay of the reaction mixture. The results are listed in Tables 14.1 and 14.2.

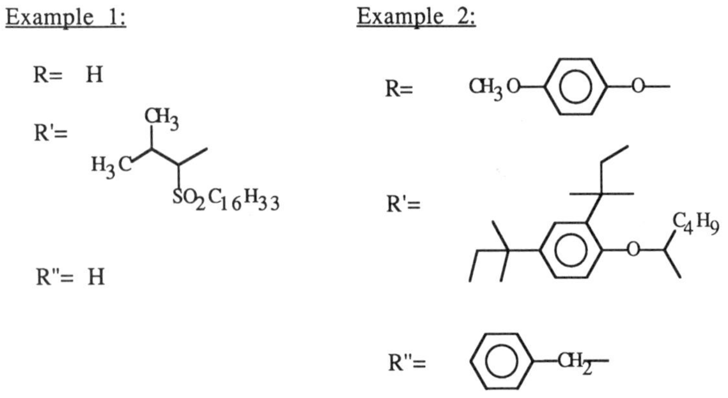

Example 1:   Example 2:

R= H

R'=

R"= H

R=

R'=

R"=

Figure 14.2 Hydrogen transfer versus conventional hydrogenation.

Table 14.1 Results Obtained with Example 1 (Fig. 14.2)

Method[a]	Desired product 5	Over reduced 6	7
A	97.0	1.9	1.1
B	90.7	1.3	8.0

[a]Method A: Hydrogen transfer (see text), 1:30 catalyst load, 5% Pd/C-S (Engelhard Escat 11; 50% water wet); method B: conventional hydrogenation (see text), 1:30 catalyst load, 5% Pd/C-S (Engelhard Escat 11; 50% water wet).

Table 14.2 Results Obtained with Example 2 (Fig. 14.2)

Method[a]	Desired product 5	Over reduced 6	7
A	90.0	8.2	1.8
B	80.0	6.6	13.4

[a]Same as Table 14.1.

15

Dinitrotoluene-to-Toluenediamine Catalyst Test and Reaction Modeling

Geoffrey T. White

Engelhard Corporation, Beachwood, Ohio

I. INTRODUCTION

The hydrogenation of dinitrotoluene (DNT) to toluenediamine (TDA) is facilitated using a heterogeneous catalyst. A laboratory performance evaluation test was developed for this application.

The catalyst evaluation test permits the investigator to obtain reaction rate data for different types of catalyst. These data are used to place the various catalysts in activity rank order. The motivation for performing the evaluations is twofold: to obtain fundamental data on the catalyst performance, and to help predict the relative catalyst performance in a commercial plant reactor.

II. EXPERIMENTAL

The catalyst test is performed by injecting a series of small doses of DNT solution into a TDA-isopropanol solvent heel. The reaction progress during the experiment is monitored by recording the amount of hydrogen consumed from a hydrogen ballast cylinder. The reaction model allows calculation of reaction rates from the ballast cylinder pressure drop.

A mechanism was proposed to explain the observed data. The choice was that the DNT reduction proceeded in a stepwise manner, with the sequential first-order hydrogenation of the two nitro groups. The proposed sequence route is:

2,4-DNT → 4-methyl-3-nitroaniline → 2,4-TDA

This route is a simplified reaction sequence, since each nitro hydrogenation is also a stepwise reaction and various intermediates are possible. Other reaction sequences are also possible, such as:

2,4-DNT → 2-methyl-5-nitroaniline → 2,4-TDA

The two steps for the stepwise reaction sequence are as follows:

1. DNT + $3H_2$ → MNA + H_2O
2. MNA + $3 H_2$ → TDA

The kinetics for this series reaction can be modeled by the following first-order reactions:

$$A \xrightarrow{\; k_1 \;} B \xrightarrow{\; k_2 \;} C$$

The rates of appearance of A (DNT) and B (MNA) can be written in terms of the concentrations of the reactants A and B, where

$$\frac{dC_A}{dt} = -k_1 C_A \quad \text{and} \quad \frac{dC_B}{dt} = k_1 C_a - k_2 C_B \tag{15.1}$$

The full equation for the two-step reaction can be expressed as follows:

$$P = P_0 - (P_0 - P_f)\left\{ 1 - \frac{e^{-k_1 t} + \dfrac{k_2}{k_2 - k_1} e^{-k_1 t} - \dfrac{k_1}{k_2 - k_1} e^{-k_2 t}}{2} \right\} \tag{15.2}$$

where P is pressure at time t, P_0 is pressure at time t = 0, and P_f is pressure at time t = infinity.

Rate constants for the reaction according to the two-step model were determined by the following method. The hydrogen ballast cylinder pressure drop during a dose period was divided empirically into two sections. The rate constant k_1 was determined using beginning-of-dose data, k_2 was determined using end-of-dose data, and the least-squares method was used to fit the straight-line equations:

$$\ln \left[1 - \frac{P_0 - P}{0.5(P_0 - P_f)} \right] = -k_1 t + C \quad \text{and}$$

$$\ln \left[1 - \frac{P - P_f}{0.5(P_0 - P_f)} \right] = -k_2 t + C$$

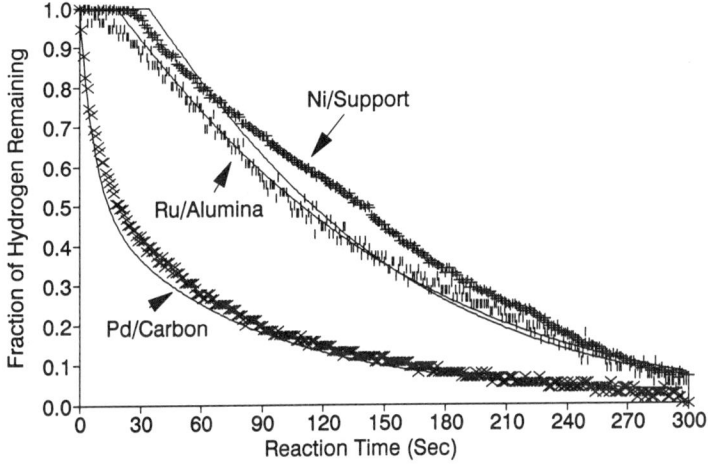

Figure 15.1 DNT → TDA reaction catalyst type comparison. Markers are data points; lines are model fit.

Experimental runs were performed with Pd/C, Ru/Al, and Ni catalysts. For each reactant dose, the ballast cylinder pressure drop was recorded and rate constants were determined. These rate constants were substituted back into the full rate equation and plots were drawn comparing the actual and computed fractional conversions versus time (Fig. 15.1). The model fits the experimental data well. The average deviation of the rate equation fit from the observed data was small (0.019 for Pd, 0.18 for Ru/Al, 0.032 for Ni). The good agreement between the experimental data and reaction model results gives confidence in the use of the DNT → MNA → TDA reaction sequence to model the overall DNT → TDA reaction.

III. RESULTS AND CONCLUSIONS

The performances of the different catalysts during a single dose are compared in Figure 15.1. The behavior of the Pd/C catalyst differs from that of the Ru/Al and Ni catalysts. The initial portion of the reaction with the Pd/C catalyst occurs very rapidly, while the final conversion is relatively slow. The conversion rates for the Ni and Ru catalysts are similar and are more constant throughout the reaction than the Pd rates.

This catalyst test and model offers a useful method for obtaining and interpreting DNT → TDA reaction activity data for different catalysts. The catalyst test is an autoclave test that is reproducible

and allows automated data collection. The model breaks the DNT →
TDA reaction into the steps DNT → MNA and MNA → TDA. First-
order rate constants are determined for each reaction step using
hydrogen consumption data. Predictions made using the experimentally determined rate constants and a two-sequential-step reaction
model agree well with the experimental data.

This test allows distinguishing the relative activity of each
catalyst for each of the sequential nitro group reductions involved
in the reaction. The advantage of this test and use of the model
is that both rate constants may be determined by using hydrogen
consumption data; no other analytical information (e.g., reactant
mixture analysis) is necessary. A disadvantage of this test is that
specific process information is necessary, in some cases, to determine
the relative activity of a catalyst in a given process.

16

Modeling of Aromatic Nitro Reduction in a Three-Phase Semi-Batch Reactor

Makarand G. Joshi and William E. Pascoe

Manufacturing Research and Engineering, Eastman Kodak Company, Rochester, New York

I. INTRODUCTION

Aromatic amines are commercially important chemicals which are key intermediates in the syntheses of a wide variety of products. Catalytic reduction of corresponding nitro compounds, the preferred method of producing these amines, is commonly carried out in mechanically agitated slurry reactors. The reducing agent of choice, gaseous hydrogen, is supplied by bubbling it through the solution of substrate. The reaction takes place on the surface of a solid catalyst, so the reactants have to cross one or two phase boundaries before they can react with each other. Hence the mass transfer processes can have a significant influence on the reactor performance. However, only a few rigorous mathematical models of the reduction that include the mass transfer processes are available [1-3]. This chapter presents a comprehensive mathematical model of the catalytic hydrogenation of nitro-substituted aromatic compounds in a mechanically agitated slurry reactor. The model takes into account various mass transfer resistances and their interaction with the kinetics of the reactions. The motivation for developing the model of the process was to determine the effect of changes in operating parameters, especially the mass transfer parameters, on the reactor performance. It is expected that this model will be useful in design, operation, and troubleshooting of the slurry reactors used for the reductions. It may also be used in safety evaluations.

II. MODEL

The widely accepted reaction mechanism for the reduction of aromatic nitro compounds proposed by Haber [4] is shown for nitrobenzene-

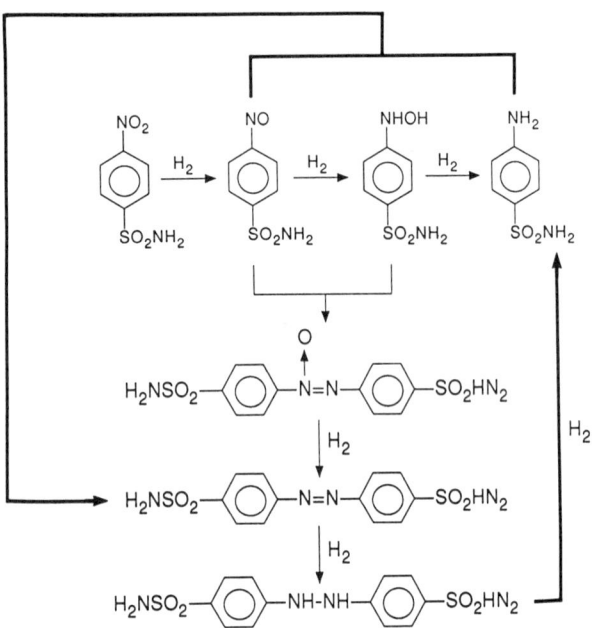

Figure 16.1 Reaction mechanism proposed by Haber [4] for the reduction of aromatic nitro compounds.

sulfonamide in Figure 16.1. There is experimental evidence to indicate that in many cases the reduction proceeds as a series reaction (see Fig. 16.2) and the parallel pathways are not important [5-7]. It is assumed here that the reduction can be represented by the following equations:

$$A + B \rightarrow E \tag{16.1}$$

$$A + E \rightarrow F \tag{16.2}$$

$$A + F \rightarrow G \tag{16.3}$$

where A is hydrogen, B is the nitro compound, E is the nitroso intermediate, F is the hydroxylamine intermediate, and G is the amine. Assuming that the power law model holds and neglecting the intraparticle diffusional resistance, the rate of change of concentrations of various species is given by:

$$-\frac{dB_1}{dt} = wk_1 A_s^{\beta_1} B_s^{\alpha_1} \tag{16.4}$$

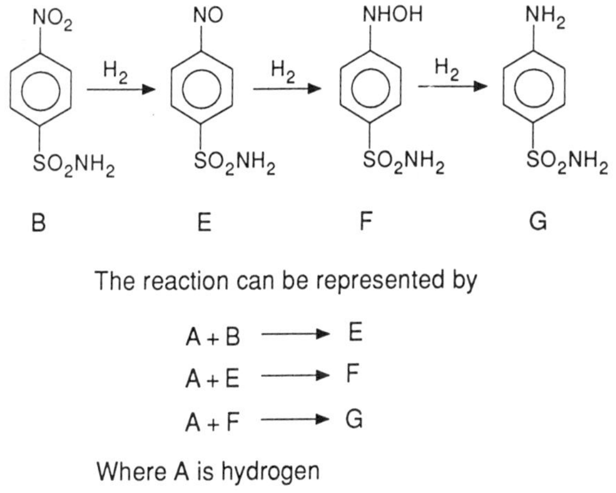

The reaction can be represented by

$$A + B \longrightarrow E$$
$$A + E \longrightarrow F$$
$$A + F \longrightarrow G$$

Where A is hydrogen

Figure 16.2   Simplified representation of aromatic nitro reduction.

$$-\frac{dE_1}{dt} = -wk_1 A_s^{\beta 1} B_s^{\alpha 1} + wk_2 A_s^{\beta 2} E_s^{\alpha 2} \tag{16.5}$$

$$-\frac{dF_1}{dt} = -wk_2 A_s^{\beta 2} E_s^{\alpha 2} + wk_3 A_s^{\beta 3} F_s^{\alpha 3} \tag{16.6}$$

The rate of change of product concentration G_1 can be obtained by the mass balance. The values of surface concentrations can be obtained by assuming pseudo-steady state and equating the rates of mass transfer and reaction:

$$(k_s a_p)_B (B_1 - B_s) = wk_1 A_s^{\beta 1} B_s^{\alpha 1} \tag{16.7}$$

$$(k_s a_p)_E (E_1 - E_s) = wk_2 A_s^{\beta 2} E_s^{\alpha 2} - wk_1 A_s^{\beta 1} B_s^{\alpha 1} \tag{16.8}$$

$$(k_s a_p)_F (F_1 - F_s) = wk_3 A_s^{\beta 3} F_s^{\alpha 3} - wk_2 A_s^{\beta 2} E_s^{\alpha 2} \tag{16.9}$$

$$M_A(A^* - A_s) = wk_1 A_s^{\beta 1} B_s^{\alpha 1} + wk_2 A_s^{\beta 2} E_s^{\alpha 2} + wk_3 A_s^{\beta 3} F_s^{\alpha 3} \tag{16.10}$$

where M_A is the overall mass transfer coefficient for hydrogen, which combines the resistances for gas-liquid and liquid-solid mass transfer; M_A is given by

$$M_A = \left[\frac{1}{K_L a_b} + \frac{1}{k_s a_p}\right]_A^{-1} \tag{16.11}$$

The system is defined completely by equations (16.4)-(16.10). The actual values of the rate constants and the mass transfer coefficients are usually not known a priori and are difficult to estimate. Thus it is advantageous to reformulate the governing equations in dimensionless form as follows:

$$- \frac{db_1}{d\tau} = a_s^{\beta_1} b_s^{\alpha_1} \tag{16.12}$$

$$- \frac{de_1}{d\tau} = -a_s^{\beta_1} b_s^{\alpha_1} + \psi_2 a_s^{\beta_2} e_s^{\alpha_2} \tag{16.13}$$

$$- \frac{df_1}{d\tau} = -\psi_2 a_s^{\beta_2} e_s^{\alpha_2} + \psi_3 a_s^{\beta_3} f_s^{\alpha_3} \tag{16.14}$$

$$\rho_B(b_1 - b_s) = a_s^{\beta_1} b_s^{\alpha_1} \tag{16.15}$$

$$\rho_E(e_1 - e_s) = \psi_2 a_s^{\beta_2} e_s^{\alpha_2} - a_s^{\beta_1} b_s^{\alpha_1} \tag{16.16}$$

$$\rho_F(f_1 - f_s) = \psi_3 a_s^{\beta_3} f_s^{\alpha_3} - \psi_2 a_s^{\beta_2} e_s^{\alpha_2} \tag{16.17}$$

$$\rho_A(1 - a_s) = a_s^{\beta_1} b_s^{\alpha_1} + \psi_2 a_s^{\beta_2} e_s^{\alpha_2} + \psi_3 a_s^{\beta_3} f_s^{\alpha_3} \tag{16.18}$$

The dimensionless concentrations are defined as follows:

$$a_s = \frac{A_s}{A^*}, \quad b_s = \frac{B_s}{B_{li}}, \quad e_s = \frac{E_s}{B_{li}}, \quad f_s = \frac{F_s}{B_{li}},$$

$$b_1 = \frac{B1}{B_{li}}, \quad e_1 = \frac{E1}{B_{li}}, \quad f_1 = \frac{F1}{B_{li}} \tag{16.19}$$

The dimensionless time is defined by

$$\tau = twk_1(A^*)^{\beta_1} B_{li}^{\alpha_1 - 1} \tag{16.20}$$

The dimensionless rate constants are given by:

$$\psi_i = \frac{k_i}{k_1} (A^*)^{\beta_i - \beta_1} B_{li}^{\alpha_i - \alpha_1} \qquad i = 2,3 \tag{16.21}$$

and the following equations give the definitions of the dimensionless mass transfer coefficients:

$$\rho_j = \frac{(k_s a_p)_j}{wk_1 A^{*\beta_1} B_{li}^{\alpha_1-1}} \qquad j = B, E, F \qquad\qquad (16.22)$$

$$\rho_A = \frac{M_A}{wk_1 (A^*)^{\beta_1-1} B_{li}^{\alpha_1}} \qquad\qquad (16.23)$$

By solving equations (16.12)-(16.18) simultaneously, one can predict the values of dimensionless concentrations in the reactor at any given time. Usual initial conditions are as follows:

$$b_1(0) = 1 \quad \text{and} \quad e_1(0) = f_1(0) = 0 \qquad\qquad (16.24)$$

Analytical solution to equations (16.12)-(16.18) is possible only in a few special cases. In general the equations have to be solved using numerical methods. The nonlinear algebraic equations (16.15-16.18) were solved using the Powell hybrid method [8], and the differential equations (16.12-16.14) were solved using the implicit Adams method [9]. Special safeguards had to be built into the algorithm for preventing concentrations from becoming negative when α_i was 1 or less. The numerical solution scheme worked very well when α_i was 1 or more and β_i was 1 or more. Numerical difficulties were encountered for α_i less than 1. As α_i became close to zero, it was progressively more difficult to solve the equations. The numerical solution scheme did not work when α_i was zero or less. When α_i were zero, the system of differential equations became stiff and, for α_i less than zero the right-hand sides of equations (16.13) and (16.14) were undefined at the initial conditions.

III. RESULTS AND DISCUSSION

The variation of concentrations of various organic species in the reaction mixture versus the dimensionless time for α_i, β_i, ρ_i, and ψ_i all equal to one is shown in Figure 16.3. Often in the reduction of nitro aromatic compounds the rate of hydrogen transport to the catalyst surface is the limiting factor [2]. Furthermore, the transport of hydrogen across the gas-liquid interface is frequently the slowest step [10]. The effect of changes in the value of the mass transfer coefficient of hydrogen on the concentrations of various organic species during the reaction is shown in Figures 16.4-16.7. For the curves in all the figures, α_i, β_i, and ψ_i are equal to 1.0 and ρ_B, ρ_E, and ρ_F are equal to 50.0. It can be seen from these curves that the overall reaction slows down as the mass transfer of hydrogen decreases and the intermediates, nitroso and hydroxylamine,

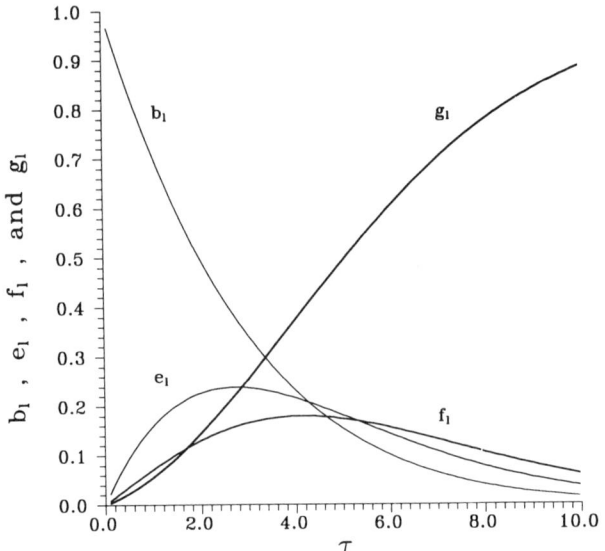

Figure 16.3   Simulated concentration profiles.

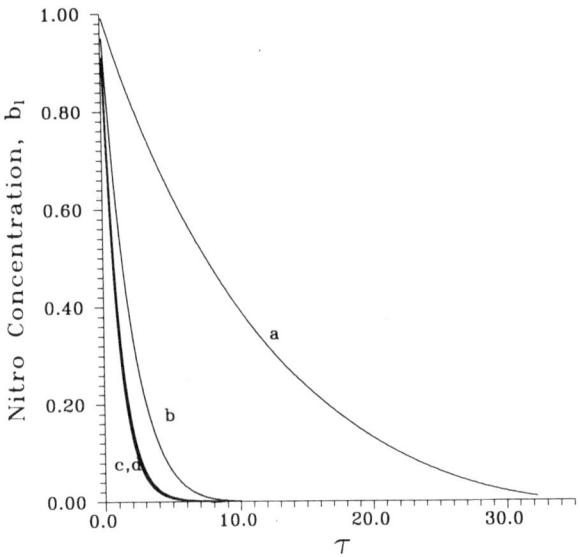

Figure 16.4   The effect of variation in the mass transfer of hydrogen on the disappearance of nitro compound. The respective P_A for each curve are (a) 0.1, (b) 1.0, (c) 10.0, and (d) 50.0.

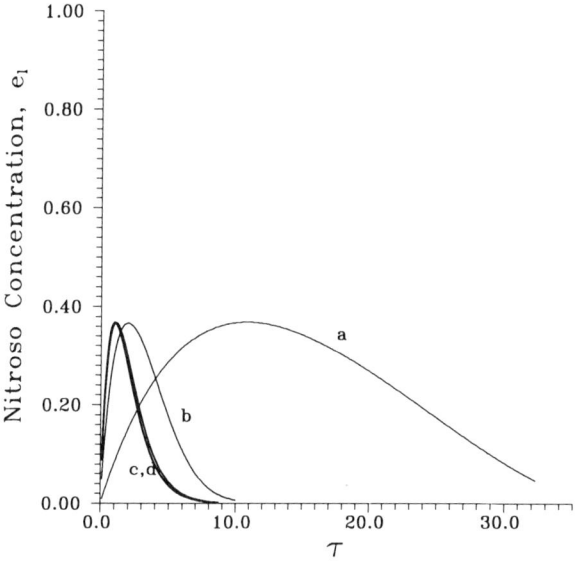

Figure 16.5 The effect of hydrogen availability on the concentration of nitroso intermediate in the reaction mixture. The respective P_A for each curve are (a) 0.1, (b) 1.0, (c) 10.0, and (d) 50.0.

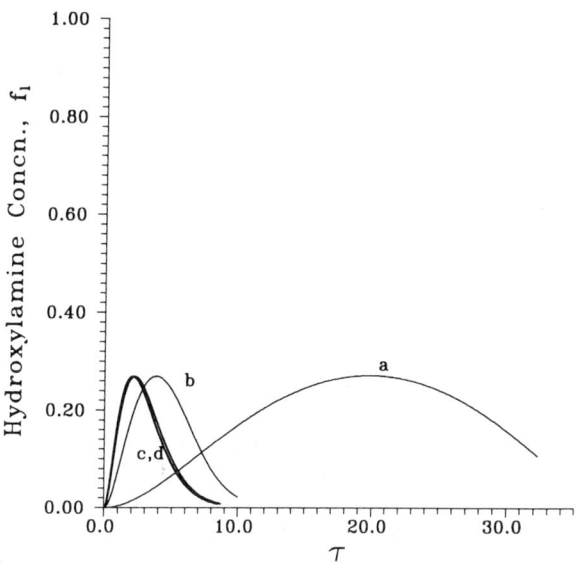

Figure 16.6 The effect of hydrogen availability on the concentration of hydroxylamine intermediate in the reaction mixture. The respective P_A for each curve are (a) 0.1, (b) 1.0, (c) 10.0, and (d) 50.0.

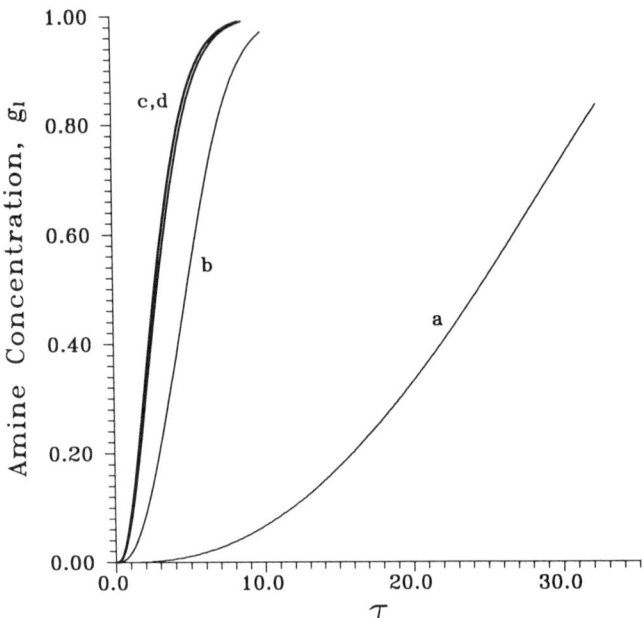

Figure 16.7 The effect of variation in the mass transfer of hydrogen on the formation of amine. The respective P_A for each curve are (a) 0.1, (b) 1.0, (c) 10.0, and (d) 50.0.

accumulate in the reaction mixture. From Figure 16.1 it is clear that the accumulation of these intermediates can lead to the formation of azo and azoxy compounds, which are often undesired impurities in the final product. Hence the "hydrogen-poor" conditions during the reduction can be detrimental to product quality. The effects of other parameters in the model on the reactor performance can be seen similarly.

IV.   CONCLUSIONS

A comprehensive model of the reduction of aromatic nitro compounds in a mechanically agitated semibatch reactor is presented. The model carefully accounts for various mass transfer resistances and their interaction with the kinetics of the reactions. The model clearly shows that the mass transfer of hydrogen can have a significant effect on the performance of the reactor. The simulation of the reactor behavior based on the model can be very useful in scale-up, troubleshooting, and optimization of reactor performance.

NOMENCLATURE

a_b	gas-liquid interfacial area per unit volume of slurry, cm^2/cm^3
a_p	external area of the catalyst particles per unit volume of slurry, cm^2/cm^3
a_s	dimensionless concentration of A as defined by eq. (16.19)
A^*	concentration of A in liquid in equilibrium with gas, mol/cm^3
A_s	concentration of A at the catalyst surface, mol/cm^3
b_1	dimensionless concentration of B in liquid as defined by eq. (16.19)
b_s	dimensionless concentration of B at the catalyst surface as defined by eq. (16.19)
B_{li}	initial concentration of B in liquid, mol/cm^3
B_1	concentration of B in liquid, mol/cm^3
B_s	concentration of B at the catalyst surface, mol/cm^3
e_1	dimensionless concentration of E in liquid as defined by eq. (16.19)
e_s	dimensionless concentration of E at the catalyst surface as defined by eq. (16.19)
E_1	concentration of E in liquid, mol/cm^3
E_s	concentration of E at the catalyst surface, mol/cm^3
f_1	dimensionless concentration of F in liquid
f_s	dimensionless concentration of F at the catalyst surface
F_1	concentration of F in liquid, mol/cm^3
F_s	concentration of F at the catalyst surface, mol/cm^3
k_1	rate constant for the reaction shown by eq. (16.1), $(cm^3/g)\,(cm^3/mol)^{\alpha_1 + \beta_1 - 1}\,s^{-1}$
k_2	rate constant for the second reaction, eq. (16.2), $(cm^3/g)\,(cm^3/mol)^{\alpha_2 + \beta_2 - 1}\,s^{-1}$
k_3	rate constant for the third second reaction, eq. (16.3), $(cm^3/g)(cm^3/mol)^{\alpha_3 + \beta_3 - 1}\,s^{-1}$
k_s	liquid-solid mass transfer coefficient, cm/s
K_L	gas-liquid film mass transfer coefficient, cm/s
M_A	overall gas-solid mass transfer coefficient for species A as defined by eq. (16.11), s^{-1}
w	catalyst mass per unit volume of slurry, g/cm^3
α_1	order of first reaction, eq. (16.1), with respect to compound B
α_2	order of the second reaction, eq. (16.2), with respect to compound E
α_3	order of the third reaction, eq. (16.3), with respect to compound F
β_1	order of the first reaction with respect to compound A

β_2 order of the second reaction with respect to compound A
β_3 order of the third reaction with respect to compound A
ρ_A dimensionless overall mass transfer of compound A as defined by eq. (16.23)
ρ_B dimensionless liquid-solid mass transfer coefficient of compound B as defined by eq. (16.22)
ρ_E dimensionless liquid-solid mass transfer coefficient of compound E as defined by eq. (16.22)
ρ_F dimensionless liquid-solid mass transfer coefficient of compound F as defined by eq. (16.22)
ψ_2 dimensionless rate constant of reaction 2 as defined by eq. (16.21)
ψ_3 dimensionless rate constant of reaction 3 as defined by eq. (16.21)
τ dimensionless time as defined by eq. (16.20)

REFERENCES

1. Kusunoki, K., Gondo, S., Nakano, K., Kawakami, K., and Arimitsu, Y., Selectivities for consecutive-parallel catalytic reactions in gas-liquid-solid semi-batch reactors and a gas-solid fixed bed reactor, *Kagaku Kogaku Ronbunshu, 1*, 286 (1975).
2. Roberts, G. W., The influence of mass and heat transfer on the performance of heterogeneous catalysts in gas/liquid/solid systems, in *Catalysis in Organic Synthesis 1976* (P. N. Rylander and H. Greenfield, eds.), Academic Press, New York, 1976, pp. 1-48.
3. Ramachandran, P. A., and Chaudhari, R. V., *Three Phase Catalytic Reactors*, Gordon & Breach, New York, 1983.
4. Haber, F., *Z. Elektrochem., 22*, 506 (1898).
5. Burge, H. D., Collins, D. J., and Davis, B. H., Intermediates in the Raney nickel catalyzed hydrogenation of nitrobenzene to aniline, *Ind. Eng. Chem. Prod. Res. Dev., 19*, 389 (1980).
6. Collins, D. J., Smith, A. D., and Davis, B. H., Hydrogenation of nitrobenzene over a nickel boride catalyst, *Ind. Eng. Chem. Prod. Res. Dev., 21*, 279 (1982).
7. Joshi, M. G., Pascoe, W. E., and Butterfield, D. E., Kinetics of hydrogenation of p-nitrobenzenesulfonamide in a three-phase slurry reactor, in *Catalysis of Organic Reactions* (D. W. Blackburn, ed.), Dekker, New York, 1990, pp. 71-45.
8. Powell, M. J. D., A hybrid method for nonlinear equations, in *Numerical Methods for Nonlinear Algebraic Equations* (R. Rabinowitz, ed.), Gordon & Breach, New York, 1970.

9. Hindmarsh, A. C., ODEPACK: A systematized collection of ODE solvers, in *Scientific Computing* (R. S. Stepleman, et al., eds.), North-Holland, Amsterdam, 1983, pp. 55-64.
10. Albal, R. S., Shah, Y. T., and Schumpe, A., Mass transfer in multiphase agitated contactors, *Chem. Eng. J.*, *26*, 61 (1983).

17

Solid Acid Catalysis of the Fries Rearrangement: Thermodynamic Limitations Based on Solvent Polarity

Kevin R. Lassila and Michael E. Ford

Air Products and Chemicals, Inc., Allentown, Pennsylvania

I. INTRODUCTION

Solid acid catalysis of classical organic transformations continues to attract considerable attention in both industrial and academic laboratories because of the potential of such technology to simplify processing and improve the economics of acid-catalyzed reactions [1]. In contrast to the mineral or Lewis acids utilized in conventional procedures, solid acids are readily removed at the completion of a reaction by a simple physical procedure such as filtration. Since the catalyst is not destroyed in this operation, it can be recycled and can, in some cases, remain in service for several years, reducing both catalyst cost and waste stream size. Furthermore, solid acids are noncorrosive, nontoxic, nonvolatile, and easy to handle, and their use makes continuous and batch processes virtually interchangeable.

From a practical standpoint, these properties introduce the potential for considerable economic advantage relative to conventional technologies. At least as important, however, are the dramatic selectivity enhancements and changes in reactivity that can occur under heterogeneous conditions. Often these changes arise because the chemistry occurs at a solid-liquid or solid-gas interface, introducing concentration or steric effects not present in homogeneous media. The biphasic nature of the chemistry can destroy simple extrapolations of results obtained under classical conditions, providing a fertile hunting ground for fundamental understanding and processes of potential commercial value.

In connection with a program dealing with the production of monomers for the advanced polymers market, we became interested in exploring the feasibility of utilizing a double Fries rearrangement

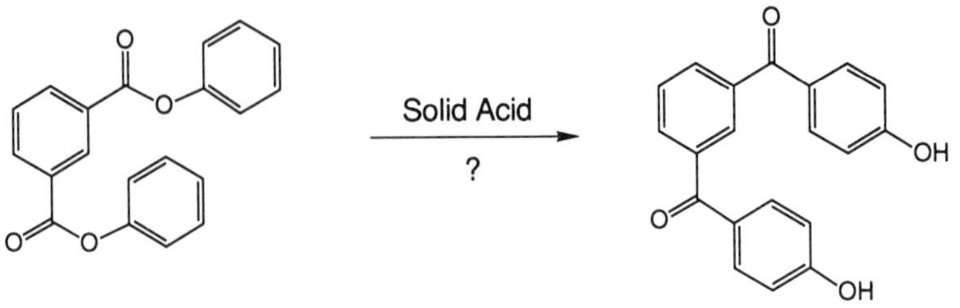

Figure 17.1 Double Fries rearrangement as a route to diphenol monomers.

as a means of producing diol monomers from diphenyl esters of aromatic dicarboxylic acids (Fig. 17.1). This chapter details our findings on the limitations of such chemistry.

II. RESULTS AND DISCUSSION

A. Literature Background

The rearrangement of phenyl esters to acyl-substituted hydroxy-arenes has been known since the late nineteenth century [2], but Fries was the first to recognize and exploit the generality of the transformation [3]. Traditionally, the reaction has been carried out in mineral acids [4] or in organic solvents employing stoichio-metric quantities of Lewis acids [5], even though the reaction is catalytic in acid. More recently, reports of solid acid catalyzed Fries rearrangements have appeared sporadically in the patent [6] and journal literature [7], but they describe only reactions of mono-functional systems, resulting in at best moderate yields of rearrange-ment products. High conversions and selectivities to para product were critical in our application, since we hoped to produce diphenols for high molecular weight polymers. Unreacted functionality would terminate polymer chains and o-hydroxy substituents would have deleterious effects on polymer properties.

B. Initial Experiments

Screening experiments to determine the relative activities and selec-tivities of representative solid acid catalysts for the Fries rearrange-ment were carried out by heating phenyl benzoate to reflux in the presence of 10 wt % catalyst for 17 hours. Phenyl benzoate rather than diphenyl isophthalate was used in these experiments to simplify analysis by decreasing the number of products formed and by in-creasing their volatility. The results summarized in Figure 17.2

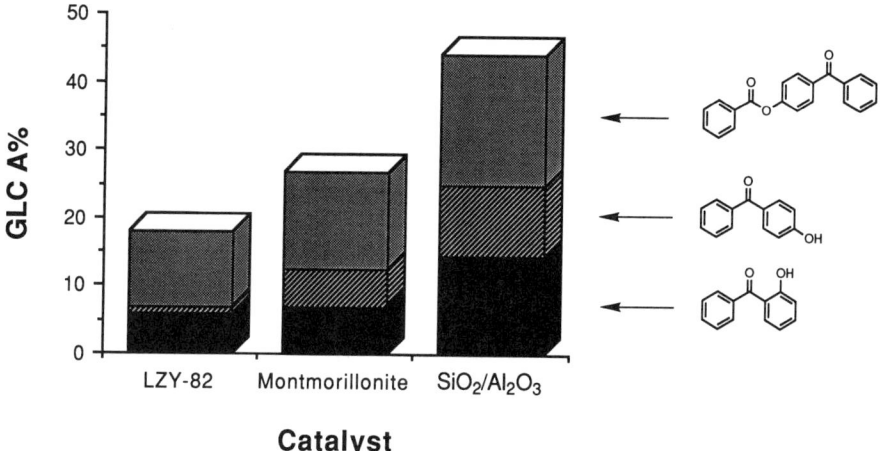

Figure 17.2 Comparison of catalyst activity for Fries rearrangement of phenyl benzoate (10 wt % catalyst in phenyl benzoate at 235°C/17 h).

show that a number of solid acid catalysts induce rearrangement and that the activity of the catalysts appears to be related to the accessibility of the acidic sites on the catalyst. SiO_2/Al_2O_3, an amorphous catalyst with virtually no restriction of access to the acidic sites, exhibits the highest activity; montmorillonite, which has a sheet like structure, exhibits the next highest activity, and LZY-82, a Y zeolite with a 7.4 Å pore opening, exhibits the lowest activity.*

The mere fact that any rearrangement at all was observed was encouraging, but we were concerned with the low selectivity to the desired para product. The low selectivity was a result of the formation of ortho product and 4-benzoylphenyl benzoate. Examination of the mechanisms of formation of these two by-products provided some insight into how the selectivity problem could be solved.

Formation of both the desired rearrangement product and the two side products is initiated by protonation of the starting material followed by heterolysis of the carbon-oxygen bond (Fig. 17.3).

---

*The SiO_2/Al_2O_3 was Davison SMR 7-6244 and contained 13% Al_2O_3 by weight. The montmorillonite was montmorillonite K-306 obtained from United Catalyst Company, Louisville, KY. LZY-82 is a Y zeolite manufactured by Union Carbide Corporation. All catalysts were finely ground with a mortar and pestle prior to use.

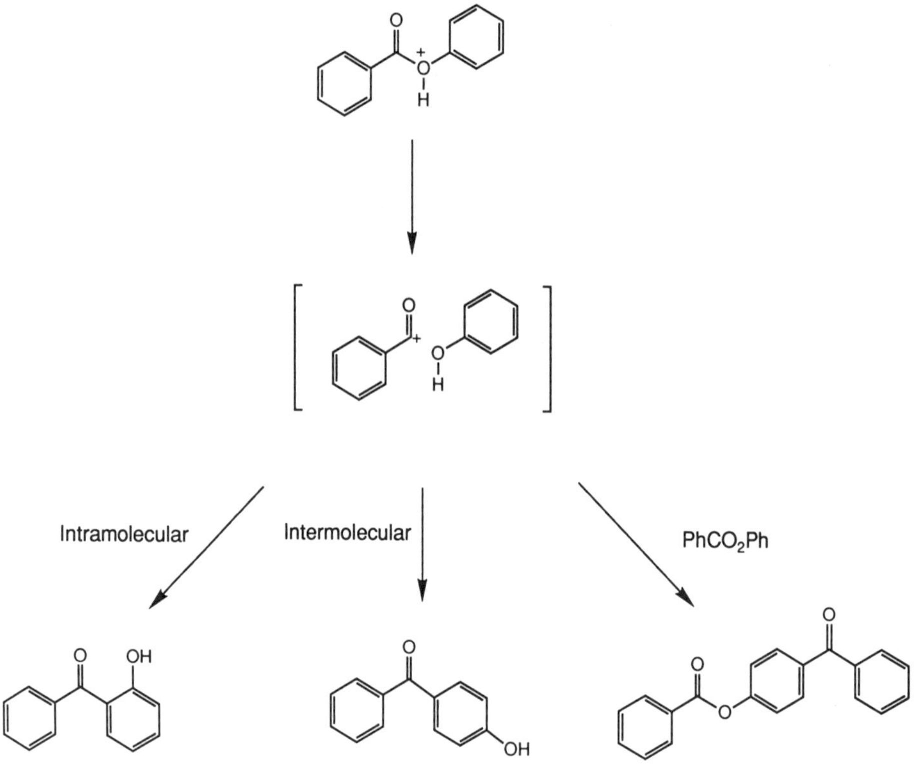

Figure 17.3   Mechanisms of by-product formation.

The partitioning of the resulting phenyl acylium ion determines
the product distribution. The ratio of *para-* to *ortho-*hydroxybenzo-
phenone is determined by the positional selectivity of the condensation with phenol. The experiments of Baltzly and coworkers [8]
have shown that *o*-hydroxybenzophenone is derived primarily from
recondensation of the acylium ion with the phenol originally attached
to the carbonyl carbon, whereas *p*-hydroxybenzophenone is derived
predominantly from condensation with another molecule of phenol.
Since the rate of the latter reaction is proportional to the concentration of phenol, the proportion of para product formed in the reaction
should increase if the reaction is performed in the presence of
phenol. The other by-product, 4-benzoylphenyl benzoate, is derived
from condensation of the acylium ion with starting material. The
rate of this reaction is proportional to the concentration of phenyl
benzoate, and again, addition of phenol should be beneficial because
it decreases the concentration of phenyl benzoate.

The catalyst also appears to affect the selectivity of the reaction. Particularly striking is the high selectivity to *o*-hydroxybenzophenone in reactions performed using LZY-82. Again, the results may be interpreted in terms of the molecularity of the rearrangement. Since there is only enough volume inside the catalyst pore of LZY-82 to accommodate one molecule of phenyl benzoate, the rate of intermolecular condensation is diminished and high selectivity to ortho product is observed.

This analysis provided a protocol for increasing the selectivity to para product. Addition of phenol should enhance selectivity to *p*-hydroxybenzophenone by increasing the relative proportion of intermolecular condensation and by decreasing the amount of condensation of the intermediate acylium ion with starting material. Use of an amorphous catalyst should increase the rate of intermolecular condensation and result in higher yields of para product. Experiments to test this hypothesis are summarized in the next section.

C. Effect of Reaction Stoichiometry

Several SiO_2/Al_2O_3-catalyzed rearrangements were performed in the presence of 3 equivalents of phenol to examine the effect of reaction stoichiometry on the selectivity of the reaction. The results in Figure 17.4 show that addition of phenol results in a substantial increase in the selectivity to *p*-hydroxybenzophenone. Most of this selectivity enhancement can be attributed to a decrease in the amount of self-condensation product formed. However, a substantial increase in the para/ortho ratio is also observed. These results are consistent with the analysis above and indicate that there is a significant degree of control of the product distribution by variation of the reaction stoichiometry.

Unfortunately, the increase in selectivity is not without cost. As seen in Figure 17.5, the conversion after 17 hours in the presence of phenol drops from about 45% to 10% with SiO_2/Al_2O_3 and from almost 30% to less than 10% with montmorillonite. The decrease in conversion may have several origins, but an obvious possibility is the decrease in reaction temperature encountered upon changing the reaction medium from phenyl benzoate (bp 235°C) to phenol (bp 180°C). Alternatively, the low conversions may be a result of catalyst deactivation. In either event, the low conversions could, in principle, be overcome by performing the reaction with another catalyst.

D. Catalysis by Nafion

Olah and coworkers have reported the use of Nafion as a catalyst for the Fries rearrangement [7b], obtaining a 73% yield (*p/o* = 2/1)

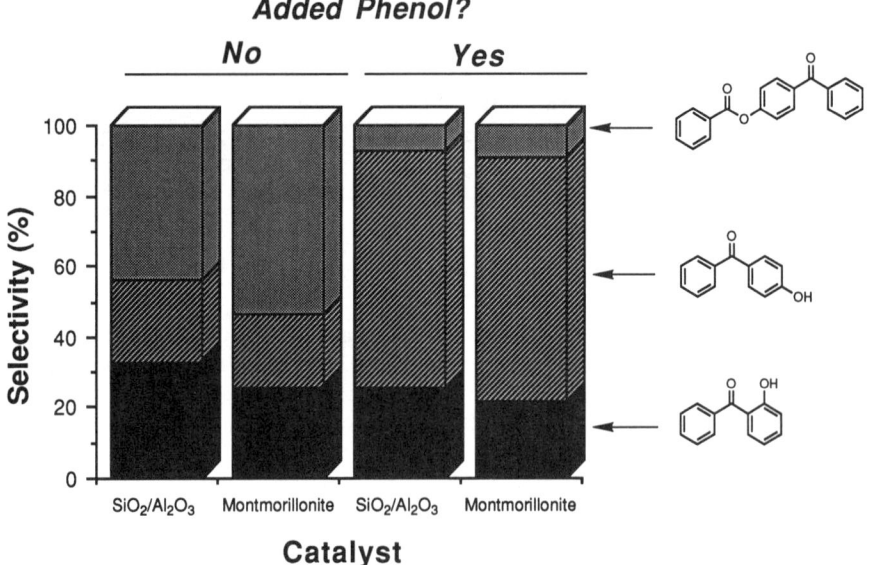

Figure 17.4   Effect of added phenol on selectivity (phenyl benzoate/ phenol, 1:3, with 10 wt % catalyst at 180°C/17 h).

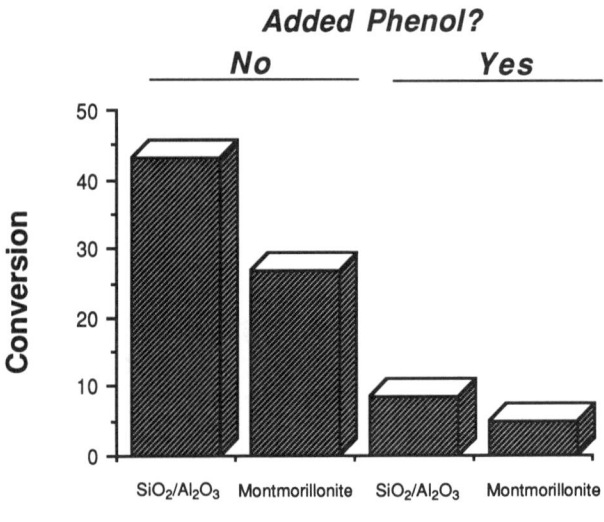

Figure 17.5   Effect of added phenol on conversion (phenyl benzoate/ phenol, 1:3, with 10 wt % catalyst at 180°C/17 h).

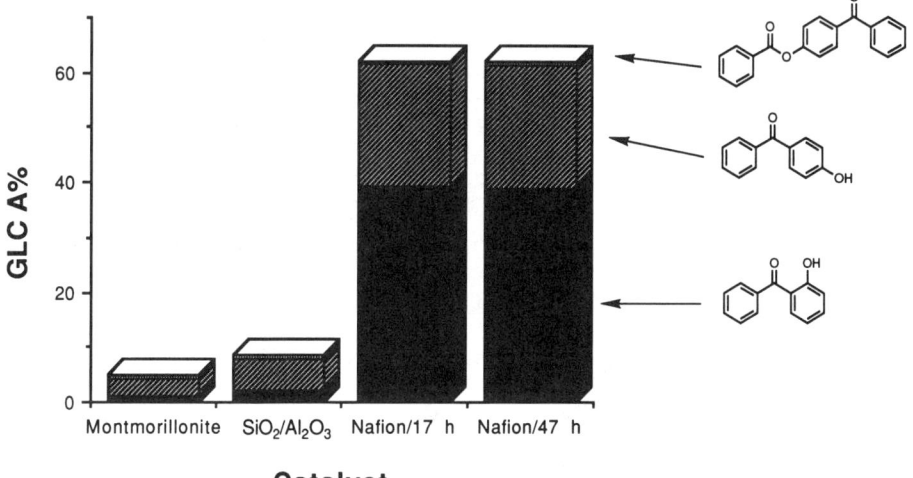

Catalyst

Figure 17.6 Nafion as a catalyst for phenyl benzoate rearrangement (phenyl benzoate/phenol, 1:3, with 10 wt % catalyst at 180°C/17 h).

in rearrangements of phenyl benzoate after only 12 hours. Although these workers emphasized that nitrobenzene was the only solvent suitable for the reaction, we examined catalysis by Nafion in phenol to determine whether their high rates would persist under conditions in which we had observed high selectivities to para product. The results in Figure 17.6 indicate that performing the reaction using Nafion rather than SiO_2/Al_2O_3 or montmorillonite increases conversions from about 10% to more than 60% after 17 hours. Although this result was encouraging despite the drop in para selectivity, a 70% conversion was insufficient for a double Fries rearrangement and we therefore continued heating the reaction mixture to increase the conversion. Surprisingly, after an additional 30 hours, the composition of the reaction mixture was virtually identical to that obtained after 17 hours. Although this result does not rigorously exclude the possibility of catalyst deactivation, it led us to at least consider the possibility that we were faced with a thermodynamic, rather than a kinetic, limitation.

E. Equilibrium Limitations

Thermodynamic calculations on the starting material and product in the Fries rearrangement were performed using the group additivity method of Benson [9]. The calculated enthalpies of formation for phenyl benzoate (-37.5 kcal/mol) and p-hydroxybenzophenone (-27.5 kcal/mol) are consistent with that obtained experimentally for phenyl benzoate (-36.5 kcal/mol) [10] and indicate that the reaction is 10 kcal/mol endothermic. The substantial endothermicity, which was compelling despite the crudity of the calculations, provided further incentive for examining the thermodynamic limitations of the reaction.

To determine whether the reaction was limited thermodynamically under the reaction conditions we had employed, a phenol solution of p-hydroxybenzophenone was heated at reflux in the presence of Nafion for 24 hours. At the end of this period, the reaction mixture contained, in addition to p-hydroxybenzophenone (27.7 GC FID area %), phenyl benzoate (43.0%) and o-hydroxybenzophenone (30.3%), demonstrating conclusively that the low conversions were a result of thermodynamic, and not kinetic, limitations. These results created an apparent contradiction, which required that the status of the Fries rearrangement as a classical organic transformation be reconciled with the reaction endothermicity and reversibility.

F.   Effect of Solvation

Since the calculations refer to the reaction in the ideal gas state, since our demonstration of reaction reversibility was done in phenol, and since Fries rearrangements are classically performed in homogenous acid media or in organic solvents with stoichiometric quantities of Lewis acids, it seemed that the contradiction could be resolved by considering the effect of solvation on the reaction. Consideration of the thermodynamic square in Figure 17.7 indicates that the free

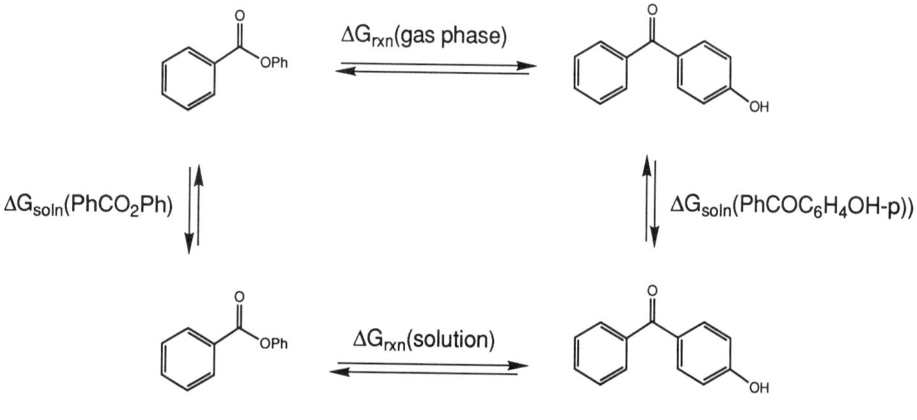

Figure 17.7   Effect of solvation on Fries rearrangement.

energy change for the reaction that occurs in solution differs from
that calculated in the ideal gas state by the difference in the free
energies of solvation of the starting material and product. Since
p-hydroxybenzophenone has both a larger dipole moment and more
solvatable functionality than phenyl benzoate, increased solvent
"polarity" should favor product formation. [An analogous argument
applies to reactions "catalyzed" by Lewis rather than Brønsted
acids if complexation with the Lewis acid is considered to be an
extreme form of solvation.]

A critical examination of the literature provides support for
this analysis. Excellent results have been obtained in homogeneous
acid media [4] such as HF, trifluoromethanesulfonic acid, phosphoric
acid, and methanesulfonic acid, or in organic media in the presence
of stoichiometric quantities of Lewis acids [5] such as $AlCl_3$, $HgCl_2$,
$SnCl_2$, and BF_3. Moderate yields have been obtained with Nafion
catalysis in nitrobenzene [7b]. Reactions performed in 1,2-
dichloroethane using catalytic quantities of trifluoromethanesulfonic
acid [11] produced an equilibrium mixture, and a gas phase reaction
of phenyl acetate in the presence of a zeolite catalyst [6a] was reported to provide traces (<5% total) of o- and p-hydroxyacetophenone.

This analysis indicates that solvent plays a central role in
driving the equilibrium and brings up the amusing speculation that
a solid acid catalyzed Fries rearrangement could be driven to completion if it were performed in mineral acid media. The question
of just how far the equilibrium could be driven in organic media
remained.

G. Determination of Equilibrium Compositions in Organic Media

Equilibrium compositions in a variety of solvents were determined
by performing reactions using, in separate experiments, either
phenyl benzoate or p-hydroxybenzophenone as starting materials,
and running the reactions until the compositions of the mixtures
were the same. The results in Figure 17.8 show that as the polarity
of the solvent increases, both the equilibrium conversion and the
selectivity to the para isomer increase. The least polar solvent
1,2-dichlorethane (ε = 10.4) provides an equilibrium conversion
of about 50% and a product distribution heavily favoring the ortho
isomer. Upon changing the solvent to phenol (ε = 9.8), which is
more polarizable and has the capability of hydrogen bonding, the
equilibrium conversion increases to 70% and the ratio of ortho to
para product decreases from about 7 to about 2. When the solvent
is changed to nitrobenzene, which has a high dielectric constant
(ε = 35), is highly polarizable, and is a hydrogen bond acceptor,

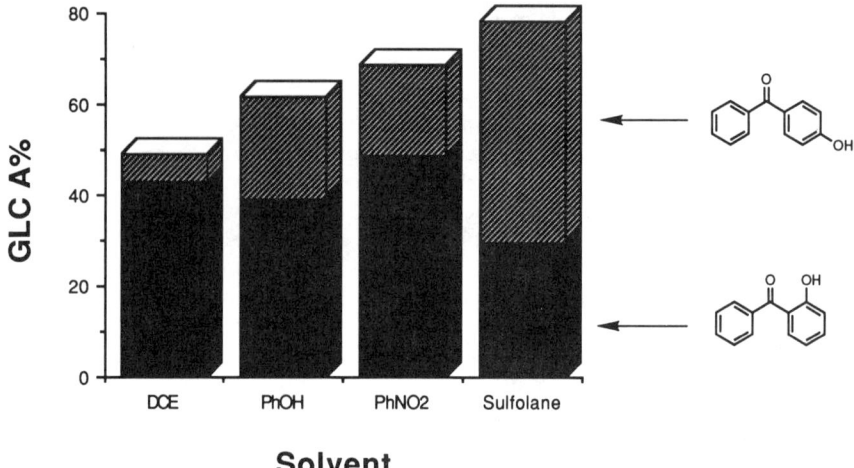

Figure 17.8   Effect of solvent on the equilibrium conversion in
Fries rearrangements of phenyl benzoate.

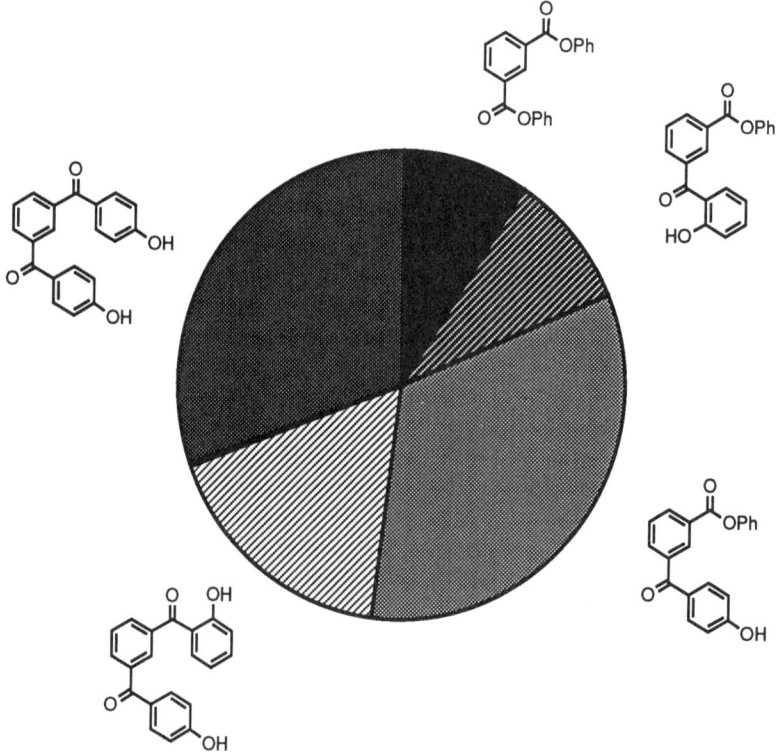

FIGURE 17.9   Anticipated product distribution from double Fries
rearrangement.

the conversion increases further, although the selectivity does
not differ significantly from that obtained in phenol. Sulfolane,
which has the highest dielectric constant ($\varepsilon = 44$) of any of the
solvents tested, provides the highest equilibrium conversion and
good selectivity to the para isomer. These results show rather nicely
the effect of solvent polarity on the outcome of the reaction.

Sulfolane is the most polar commonly available nonnucleophilic
solvent compatible with the cationic intermediate of the Fries rearrangement. To examine the feasibility of a double Fries rearrangement, we calculated the product distribution, assuming the ester
groups would react independently, and that we could, in a single
reaction, obtain the best observed equilibrium conversion and selectivity to para product, and completely eliminate side reactions.
The results presented in Figure 17.9 indicate that even under this
set of best conditions, the selectivity of a double Fries rearrangement to p,p' product is only about 30%. The complexity of the
reaction mixture means that substantial processing would be required
to isolate this material.

III. CONCLUSIONS

Although solid acids catalyze the rearrangement of phenyl benzoate
to hydroxybenzophenones, thermodynamic limitations arising from
the diminished polarity of the medium employed in the catalyzed
reaction preclude use of a solid acid catalyzed double Fries rearrangement in a commercial process for the production of diphenol monomers.
The results illustrate that the outcome of a solid acid catalyzed
reaction cannot be predicted by a simple extrapolation of results
obtained under classical conditions. Medium effects, which can be
at least as important as catalyst acidity and shape selectivity in
determining the course of an acid-catalyzed reaction, must be considered as well.

REFERENCES

1. Friedel-Crafts alkylations: (a) Laszlo, P., and Mathy, A.,
   Helv. Chim. Acta, 70, 577-586 (1987). (b) Clark, J. H., Kybett,
   A. P., Macquarrie, D. J., Barlow, S. J., and Landon, P.,
   J. Chem. Soc., Chem. Commun., 1353-1354 (1981). Friedel-
   Crafts acylation: (c) Gauthier, C., Chiche, B., Finiels, A.,
   and Geneste, P., *J. Mol. Catal., 50,* 219-229 (1989). (d) Corma,
   A., Climent, M. J., Hermengildo, G., and Primo, J., *Appl.
   Catal., 49,* 109-123 (1989). Arylamine alkylation: (e) Burgoyne,
   W. F., Dixon, D. D., and Casey, J. P. *Chemtech, 19,* 690-697
   (1989).

2. For reviews see: (a) Gerecs, A., in *Friedel-Crafts and Related Reactions*, Vol. III (G. A. Olah, ed.), Wiley-Interscience, New York, 1964, pp. 499-533. (b) Blatt, A. H., *Organic Reactions*, Vol. 1, Wiley, New York, 1942, pp. 342-369. (c) Dewar, M. J. S., and Hart, L. S., *Tetrahedron*, 973-1000, 1970. (d) Shine, H. J., *Aromatic Rearrangements*, Elsevier, New York, 1967, p. 72.

3. (a) Fries, K., and Finck, G., *Chem. Ber.*, *41*, 4271 (1908). (b) Fries, K., and Pfaffendorf, W., *Chem. Ber.*, *43*, 212 (1910).

4. (a) Snyder, H. R., and Elston, C. T., *J. Am. Chem. Soc.*, *77*, 364 (1955). (b) Darm, O., and Wylius, A., *Justus Liebigs Ann. Chem.*, *587*, 1 (1954). (c) Hocking, M. B., *J. Chem. Technol. Biotechnol.*, *30*, 626-641 (1980). (d) Norell, J. R., *J. Org. Chem.*, *38*, 1924-1928 (1973).

5. (a) Martin, R., Gros, N., Boehmer, V., and Kaemmerer, H., *Monatsh. Chem.*, *11*, 81-92 (1980). (b) Chakravarti, D., Chakravarti, A., and Sarker, U. *J. Indian Chem. Soc.*, *48*, 1017-1019 (1971). (c) Martin, R., Gros, N., Boehmer, V., and Kaemmerer, H. *Monatsh. Chem.*, *110*, 1057-1066 (1979), and references therein.

6. (a) Landis, P. S., and Venuto, P. B., U.S. Patent 3,354,221. (b) Fujita, T., Ishiguro, M., Takahata, K., and Saeki, K. Jpn. Kokai Tokkyo Koho JP 60/152444. (c) Onoda, T., and Wada, K., Jpn. Kokai JP 51/8231.

7. (a) Sekiguchi, M., and Tanaka, M., *Nippon Kagaku Kaishi*, 742-746 (1985). (b) Olah, G. A., Arvanaghi, M., and Krishnamurthy, V. V., *J. Org. Chem.*, *48*, 3359-3360 (1983). (c) Cundy, C. S., Higgins, R., Kibby, S. A. M., Lowe, B. M., and Paton, M. R., *Tetrahedron Lett.*, *30*, 2281-2284 (1989).

8. Baltzly, R., Ide, W. S., and Phillips, A. P., *J. Am. Chem. Soc.*, *77*, 2522 (1955).

9. Benson, S. W., *Thermochemical Kinetics*, 2nd ed., Wiley, New York, 1976.

10. Adams, G. P., Fine, D. H., Gray, P., and Laye, P. G., *J. Chem. Soc. B*, 720-722 (1967).

11. Effenberger, F., and Gutmann, R., *Chem. Ber.*, *115*, 1089-1102 (1982).

18

Superacid-Catalyzed One-Pot Synthesis of Aromatic Sulfoxides from Arenes: Mechanistic Aspects and Synthetic Utility

K. Laali and D. S. Nagvekar

Department of Chemistry, Kent State University, Kent, Ohio

I. INTRODUCTION

Pioneering work by Olah et al. led to the generation of and detailed studies on a great variety of stable arenium ions in superacid media [1,2]. Whereas low temperature protonation of benzene, fluorobenzene, and fluorotoluenes with $HF \cdot SbF_5$ or $FSO_3H \cdot SbF_5$ "magic acid" in SO_2ClF cleanly furnishes the arenium ions, addition of SO_2 as cosolvent leads to rapid formation of protonated sulfinic acid [3-5]. The formation of this acid was explained either by nucleophilic attack of SO_2 on the arenium ion or by electrophilic attack of SO_2-SbF_5 complex (a potent sulfinating agent) on the arene itself, with SO_2 promoting arenium ion deprotonation [5]. Ring sulfination was also observed by Winstein et al [6] when studying O-protonated aromatic alcohols in magic acid/SO_2. Synthetic utility of O-protonated sulfinic acids (and the derived $ArSO^+$) in electrophilic chemistry, however, remains little explored.

In our earlier work on protosolvated onium ions Me_3S^+, Me_3Se^+, and Me_3Te^+ in superacid media and their alkylation ability toward aromatics [7], an "unexpected" product, p-tolyl sulfoxide, was formed in the reaction of toluene with $FSO_3H \cdot SbF_5/SO_2$. We suggested that it might be formed by a sequence of fluorosulfonation, ionization, and condensation of toluene itself. Such a process should initially give a sulfone, which is presumably reduced in situ to sulfoxide! In view of the importance of sulfoxides as organic synthons (chirons) in synthesis [8,9], a high yield, one-pot approach from simple arenes is highly desirable. Syntheses of diaryl sulfoxides from arenes/$SOCl_2$/$AlCl_3$ [8] and from arene/arenesulfinyl chloride/$AlCl_3$ [10] are usually limited to activating substituents and often give low yields [8]. The $ArMgX/SOCl_2$ system is more promising but is limited to symmetrical sulfoxides [8].

II. EXPERIMENTAL

When a cold homogeneous solution of $FSO_3H \cdot SbF_5$ (1:1) diluted
in SO_2 was added slowly to excess toluene (8-fold) in Freon/SO_2
under nitrogen at dry-ice acetone temperature, an orange-green
solution was formed which, on warming to $-30°C$, became dark blue.
The temperature was slowly raised to $0°C$ (purple solution). Quench-
ing and workup furnished p-tolyl sulfoxide in 75% isolated yield.
In addition, a minor product (6%), identified as the sulfide, was
confirmed by independent reduction of the sulfoxide ($PhTMS/I_2$)
[11] and gas chromatographic coinjection with the reaction mixture
(Fig. 18.1).
   Whereas a wide variety of methods are available for sulfoxide →
sulfide conversion [9,12] reduction in a "highly oxidizing" superacid
medium appears to be unconventional. To establish the potential
role of protonated sulfoxides in the reduction step, we studied
the low temperature reaction of several functionalized sulfoxides
with magic acid/SO_2 under stable ion conditions.
   Aliphatic sulfoxides are protonated on sulfur in superacid media,
and the S—H signal is observed at 5-7 ppm [13] with parent diphenyl
sulfoxide, protonation is accompanied by ring sulfonation. Ring
sulfonation can be avoided by protonation in $HF \cdot SbF_5$, where
a sulfur-protonated onium ion (SH^+ at 5.03 ppm) is observed [13].

III. RESULTS

We found that unlike the parent diphenyl sulfoxide, substituted
diaryl sulfoxides are O-protonated in magic acid/SO_2. Derived stable
ions were probed by multinuclear (^1H, ^{13}C, ^{19}F) NMR spectroscopy.
   Based on a number of control experiments, it was established
that sulfoxonium ions were the key intermediate en route to reduction. Our suggested mechanism for the reduction is shown in
Figure 18.2.
   The proposed mechanism for the arene/superacid/SO_2 system
involves sulfination of the arenium ions, O-protonation of the resulting sulfinic acid, dehydration of the oxonium ion "$ArSO^+$," and
arylation (Fig. 18.3). In the absence of SO_2, the fluorosulfonation-
ionization-arylation path becomes dominant.

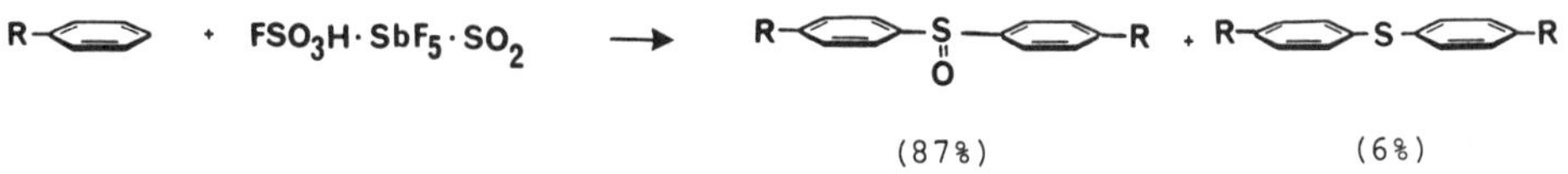

Figure 18.1

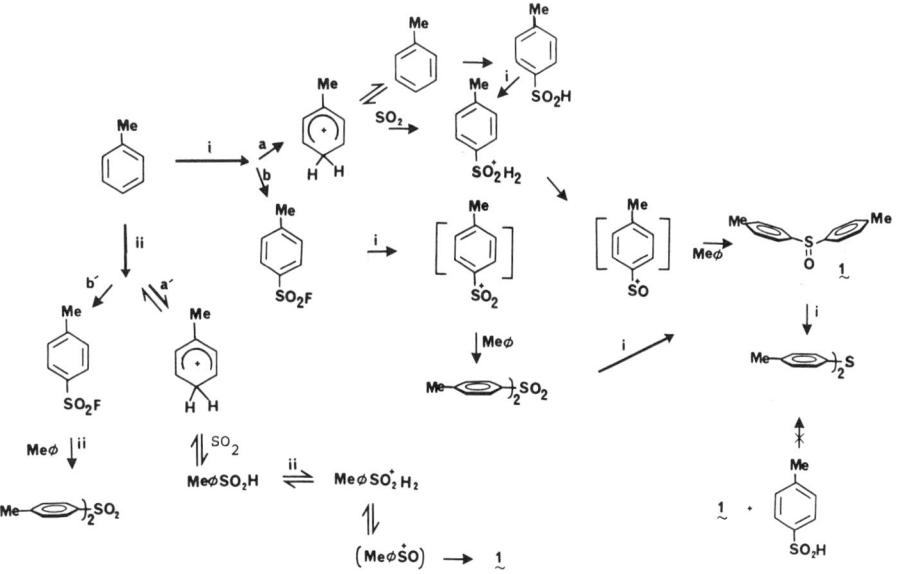

Figure 18.2

Figure 18.3

i: $FSO_3H \cdot SbF_5$ (1:1)
ii: FSO_3H
a,a': SO_2 solvent
b,b': Freon solvent

REFERENCES

1. For comprehensive reviews on monoarenium ions, see: Olah,
   G. A., Surya Prakash, G. K., and Sommer, J., in *Superacids,*
   Wiley, New York, 1985, pp. 99–101; Brouwer, D. M., Mackor,
   E. L., and MacLean, C., in *Carbonium Ions*, Vol. 2 (G. A.
   Olah and P. v. R. Schleyer, ed.), Wiley, New York, Chap. 20.
2. For representative examples of diarenium ions, see: Pagni,
   R. M., *Tetrahedron*, 40, 4161 (1984).
3. Olah, G. A., and Kiovsky, T. E., *J. Org. Chem.*, 89, 5692
   (1967).
4. Olah, G. A., and Kiovsky, T. E., *J. Org. Chem.*, 90, 2583
   (1968).
5. Olah, G. A., Schlosberg, R. H., Kelly, D. P., and Mateescu,
   G. D., *J. Am. Chem. Soc.*, 92, 2546 (1970).
6. Brookhart, M., Anet, F. A. L., and Winstein, S., *J. Am.
   Chem. Soc.*, 88, 5657 (1966).
7. Laali, K., Chen, H. Y., and Gerzina, R. J., *J. Organomet.
   Chem.*, 348, 199 (1970).
8. Drabowicz, J., Kielbasinski, P., and Mikolajczyk, M., in *The
   Chemistry of Sulfones and Sulfoxides* (S. Patai, Z. Rappoport,
   and C. J. M. Stirling, ed.), Wiley, New York, 1988, Chap. 8.
9. Madesclaire, M., *Tetrahedron*, 44, 6537 (1988).
10. Olah, G. A., and Nishimura, J., *J. Org. Chem.*, 39, 1203
    (1974).
11. Olah, G. A., Balaram Gupta, B. G., and Narang, S. C.,
    Synthesis, 533 (1977).
12. Groosert, J. S., in *The Chemistry of Sulfones and Sulfoxides*
    (S. Patai, Z. Rappoport, and C. J. M. Stirling, eds.), Wiley,
    New York, 1988, Chap. 20.
13. Olah, G. A., Ku, A. T., and Olah, J. A., *J. Org. Chem.*,
    35, 3904 (1970).

19

Dehydrogenation of Cyclic Alcohols in the Presence of ZSM-5 Zeolites

Cv. Bezouhanova and Yu. Kalvachev
Department of Chemistry, University of Sofia, Sofia, Bulgaria

H. Lechert
University of Hamburg, Institute of Physical Chemistry, Hamburg, Germany

I. INTRODUCTION

At the present time several different types of synthetic zeolites are known, and many of them find application in organic syntheses [1,2]. The catalytic activity of the zeolites is usually related to their acid-active sites (Brønsted or Lewis) or to the presence of specific metal cations. Their selectivity is determined mainly by the crystal structure, and the channel shapes and dimensions.

The concept of the relation between the conjugated acids and bases applied to zeolites reveals several peculiarities [3]. The basic properties of the zeolites are due to the presence of oxygen atoms species O^{2-}, AlO_4^-, or $-OH$ and are also influnced by the countercation.

The zeolites are able to activate dehydrogenation reactions in the absence of the usually necessary transition metal, as well. Especially interesting from this point of view are the so-called pentasils or H-ZSM-5 zeolites, which activate the conversion of methanol into aromatics or olefins [4]. In some cases the simultaneous participation of acidic and basic zeolite lattice sites has been postulated to initiate the methanol reactivity [5].

The ring or chain alkylation of toluene is influenced by the acidity and electronegativity of the zeolites. It has been found that methanol alkylates toluene to styrene, and ethylbenzene, through formaldehyde on highly basic alkaline metal-modified exchanged X and Y zeolites [6]. The dehydration of 2-propanol on alkali metal cation-modified exchanged X and Y zeolites [7], accompanied the dehydrogenation at high temperature (700 K). The maximal yield of acetone was found to be 9.3 mol % on CsY.

Recently we have observed [8.9] that zeolites of type ZSM-5 in acidic or sodium form were able to dehydrogenate unsaturated alcohols or cyclohexanol in very mild conditions—in an IR cell as catalytic reactor at room temperature, under reduced pressure. Our investigations were extended to other alcohols (1- and 2-butanol, 3-methylcyclohexanol, cyclopentanol, benzylalcohol, etc.). The results showed that the yield of the corresponding carbonyl compound depended not only on the composition of the zeolite catalyst, but also on the structure and geometry of the starting alcohol.

II. RESULTS AND DISCUSSION

A. Carbonyl Compounds from Cyclic Alcohols

In the IR spectra of cyclohexanol, 3-methylcyclohexanol, cyclopentanol, and benzylalcohol, adsorbed on H-ZSM-5 or Na-ZSM-5 zeolites heated in vacuo (1.3×10^{-2} Pa), a new band with a position between 1724 and 1765 cm^{-1} for the different alcohols, characteristic for carbonyl group, appeared after several minutes of contact at room temperature. The dependence of the carbonyl band intensity on the composition of the H-ZSM-5 zeolites used for cyclohexanol (Figs. 19.1 and 19.2) and 3-methylcyclohexanol (Fig. 19.3) showed that the degree of dehydrogenation increased with higher aluminum content. Heating at higher temperatures diminished the quantity of dehydrogenation product. In the interaction of the foregoing zeolites with cyclopentanol (Fig. 19.4) and benzylalcohol (Fig. 19.5), the dependence of the dehydrogenation from the aluminum content

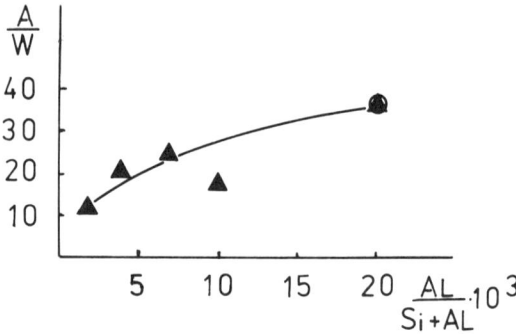

Figure 19.1   Dependence of the absorbance A at 1735 cm^{-1} per gram of zeolite W from the aluminum (Al) molar fraction in H-ZSM-5 zeolites by cyclohexanol adsorption at room temperature.

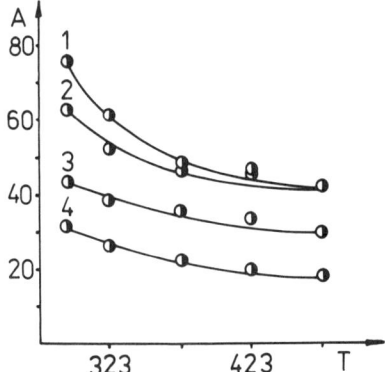

Figure 19.2 Changes of the intensity of the band at 1735 cm^{-1} with temperature by the contact of cyclohexanol with H–ZSM-5 zeolites with different Si/Al ratios: 1, 50; 2, 100; 3, 250; and 4, 500.

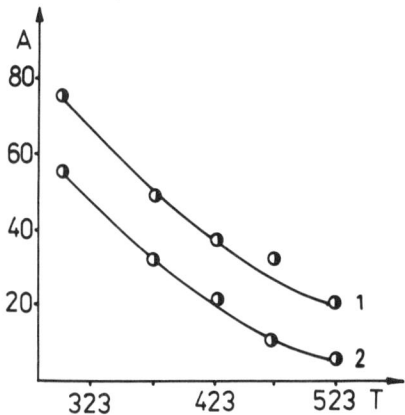

Figure 19.3 Changes of the carbonyl band (1736 cm^{-1}) intensity with temperature by the contact of 3-methylcyclohexanol with H–ZSM-5 zeolites: 1, Si/Al = 50; and 2, Si/Al = 140.

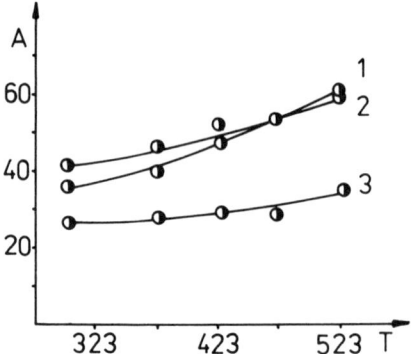

Figure 19.4   Absorbance at 1765 cm^{-1} by interaction of cyclopentanol with H–ZSM-5 zeolites at different temperatures and different Si/Al ratios: 1, 50; 2, 100; and 3, 140.

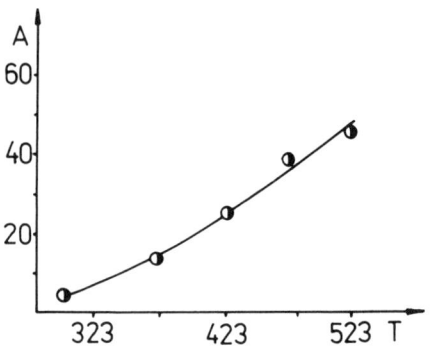

Figure 19.5   Changes of the carbonyl band (1724 cm^{-1}) intensity with temperature by the contact of benzylalcohol with H–ZSM-5 (Si/Al = 50).

was the same, but in both cases the quantity of the carbonyl compound product was increased by heating up to 523 K. It was difficult to understand why the influence of the temperature had an opposite effect on the dehydrogenation activity of alcohols with six- and five-membered rings.

The catalytic experiments with cyclohexanol in a flow reactor (Table 19.1) showed that, on the acidic H–ZSM-5 zeolites at higher temperatures (according to the IR spectra, > 373 K), the main

Table 19.1 Conversion of Cyclohexanol[a]

Catalyst[b]	Temperature (K)	Unreacted alcohol (wt %)	Ketone (wt %)	Cyclohexene	Cyclohexane	C-6 Aliphatic compound
H-ZSM-5(50)	523	Traces	4.9	76.2	18.9	
H-ZSM-5(140)	523	16.0	6.1	48.7	29.1	
H-ZSM-5(250)	523	21.6	5.7	53.6	15.6	3.5
H-ZSM-5(500)	523	35.8	1.5	50.2	9.1	3.3
H-ZSM-5(1000)	523	62.7	2.8	24.8	4.2	5.5
Na-ZSM-5(50)	473	91.8	8.2			
Na-ZSM-5(100)	473	91.9	8.1			
Na-ZSM-5(140)	473	92.5	7.5			
Na-ZSM-5(250)	473	96.0	4.0			
Na-ZSM-5(500)	473	98.2	1.8			
Na-ZSM-5(1000)	473	98.4	1.6			

[a]LHSV = 1 hour^{-1}, 1 hour on stream.
[b]Ratio of Si to aluminum in parentheses.

reaction was dehydration to cyclohexene in accordance with the
concentration of the Brønsted acid sites on the catalyst. Benzylalcohol is unable to dehydrate, and the higher temperature favors
the dehydrogenation. The difference in the reactivity of cyclohexanols
and cyclopentanol has probably thermodynamic reasons—the calculated
equilibrium constant for cyclohexanol dehydration is almost twice
as much as that for cyclopentanol (data for liquid state) [10].

As seen from Table 19.1, cyclohexane was also obtained. The
experiments were performed in the absence of added hydrogen.
Before the experiment, the catalyst was flushed with argon. Thus
cyclohexane could be formed at the expense of the hydrogen released
by dehydrogenation. The suggestion that the dehydration and dehydrogenation on H–ZSM-5 zeolites proceed as concurrent reactions
on active sites of two different types—acid and basic—was confirmed
by the experiments with Na–ZSM-5 zeolites (Table 19.1 and Fig. 19.6).

The maximal yield of carbonyl compound (Table 19.1) was 8%
at 473 K. Evidently the reduced pressure favors dehydrogenation.
It was interesting also to check the activity of NaY in cyclohexanol
dehydrogenation. The carbonyl band appeared at room temperature,
but its intensity diminished considerably at 323 K and disappeared
at 473 K; there was severe coking of the catalyst, and the catalyst
turned black.

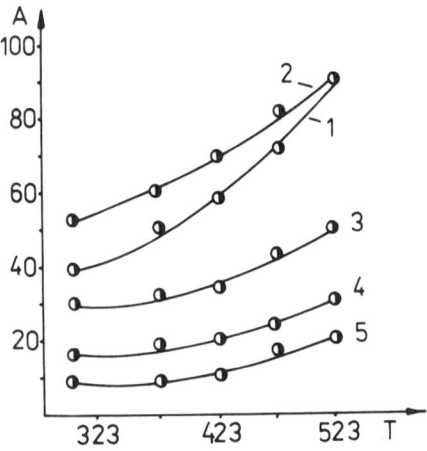

Figure 19.6   Dependence of the intensity of the band at 1765 cm^{-1}
on temperature by the interaction of cyclopentanol with Na–ZSM-5
zeolites with different Si/Al ratios: 1, 50; 2, 100; 3, 140; 4, 250;
and 5, 500.

B. Behavior of Noncyclic Alcohols

It is natural to expect similarities in the chemical properties of cyclic and noncyclic alcohols. According to some references [6,7,11], methanol and propanol also dehydrogenate on ZSM-5 zeolites to carbonyl compounds.

Our experiments with 1- and 2-propanol and 1- and 2-butanol showed that H-ZSM-5 zeolites activated only the dehydration to the corresponding olefin. In addition, isomerization of butenes occurred. A carbonyl band (1732 cm^{-1}) was observed for the adsorbed species after interaction of *n*-butanols with H-ZSM-5 (Si/Al = 50) at room temperature. The band disappeared by heating at 373 K while the olefin bands appeared. The situation was very similar to the phenomenon observed by Lercher et al. [11] who suggested adsorption of propane-2-ol on Brønsted acid sites and alkali cations.

In the interaction with Na-ZSM-5 above 400 K, a new band appeared at 1717 cm^{-1}, reached a maximum at 473 K, and disappeared upon further heating, giving rise to another band at 1680 cm^{-1}. Meanwhile the only reaction products detected were propene and water. The conclusion was that the dehydrogenated species (acetone) had decomposed to propene. If an enolization of the ketone is accepted, the formation of olefin is understandable:

$$CH_3-C-CH_3 \rightleftharpoons CH_3-C=CH_2 \longrightarrow CH_3-CH=CH_2 + H_2O$$
$$\underset{O}{\|} \qquad \underset{OH}{|} \quad +H_2$$

Probably this is the case with all the alcohols producing carbonyl compounds.

III. CONCLUSIONS

The existence on the zeolites of two types of active site determines their polyfunctionality as catalysts in alcohols conversion. The Brønsted acid sites activate dehydration to olefins and their isomerization, while the oxygen atoms from the zeolite lattice act as hydrogen acceptors. Both activities depend on the countercation. The zeolites containing alkali metals as countercations are selective for alcohol dehydrogenation.

Another factor influencing zeolite catalytic activity is geometrical conformity between the zeolite structure and the reacting molecules. The appropriate combination of all these requirements permits the selective dehydrogenation of some alcohols to carbonyl compounds under reduced pressure at room temperature.

REFERENCES

1. Venuto, P. B., *Catalysis in Organic Syntheses*, Academic Press, New York, 1977, p. 67.
2. Van Bekkum, H., and Kouwenhoven, H. W., in *Heterogeneous Catalysis and Fine Chemicals* (M. Guisnet et al., eds.), Elsevier, Amsterdam, 1988, p. 45, and Hoelderich, W. F., *ibid.*, p. 93.
3. Barthomeuf, D., Coudurier, G., and Vedrine, J. C., *Mater. Chem. Phys.*, *18*, 559-575 (1988).
4. Chang, C. D., *Catal. Rev. Sci. Eng.*, *25*, 1 (1983).
5. Lin, L., Tobias, R. G., McLaughlin, K., and Anthony, R. G., in *Gas and Alcohols to Chemicals* (R. G. Herman, ed.), Plenum Press, New York, 1984, p. 323, and references therein.
6. Giordano, N., Pino, L., Cavallaro, S., Vitarelli, P., and Rao, B. S., *Zeolites*, *7*, 131 (1987).
7. Yashima, T., Suzuki, H., and Hara, N., *J. Catal.*, *33*, 486 (1974).
8. Bezouhanova, Cv., Dimitrov, Chr., Lechert, H., Nenova, V., and Krusteva, M., in *Proceedings of the 6th International Symposium on Heterogeneous Catalysis*, Sofia, Bulgaria, July 13-18, 1987, Part 2 (D. Shopov et al., eds.), Publishing House of the Bulgarian Academy of Sciences, p. 216.
9. Bezouhanova, Cv., Dimitrov, Chr., Nenova, V., Kalvachev, Yu., and Lechert, H., in *Zeolites: Facts, Figures, Future* (P. A. Jacobs and R. A. Van Santen, eds.), Elsevier, Amsterdam, 1989, p. 1223.
10. Stull, D. R., Westrum, E. F., and Sinke, G. C., *The Chemical Thermodynamics of Organic Compounds*, Wiley, New York, 1969.
11. Lercher, J. A., Warecka, G., and Derewinski, M., *Proceedings of the 9th International Congress on Catalysis*, Calgary, Ontario, Canada, 1988, p. 364.

20

Selectivity Dependence on the Acidity and Metal Area of Copper-Alumina Catalysts in the Reaction of Cyclohexanol

P. Kanta Rao, C. Sivaraj, S. T. Srinivas, and V. Nageshwar Rao

Catalysis Section, Indian Institute of Chemical Technology, Hyderabad, India

I. INTRODUCTION

Catalytic dehydrogenation of cyclohexanol to cyclohexanone is of industrial importance in the manufacture of nylon. More than 90% of all the cyclohexanone output is used either to make caprolactum for nylon 6 or adipic acid for nylon 66. Cyclohexanone is also used as a solvent and as a building block in the synthesis of many organic compounds such as pharmaceuticals and insecticides. Industrially cyclohexanone is manufactured either by the dehydrogenation of cyclohexanol [1-4], which is produced by catalytic air oxidation of cyclohexane [5], or by catalytic hydrogenation of phenol [6]. The catalytic dehydrogenation of cyclohexanol has gained much importance in recent years [7-10]. For example, Sideltseva and Erofeev [7] have studied the dependence of the metal deposition procedure on the specific surface and catalytic properties of Cu/MgO catalysts. Petrova et al. [8] have examined the promotional role of cadmium in copper-alumina ($Cu-Al_2O_3$) alloy catalysts. They achieved about 87% cyclohexanone selectivity at a reaction temperature of 370°C. Pridman et al. [10] have postulated that the $CuCrO_4$ is a precursor of the most active catalytic site of copper-chromium catalysts for dehydrogenation of cyclohexanol. They have shown that the dehydration of cyclohexanol to cyclohexene occurs in the Cr_2O_3 formed on the surface of the reduced catalysts. More recently, copper-based catalysts have been studied in the dehydrogenation [11] and oxidative dehydrogenation [12] of cyclohexanol. Kanta Rao and coworkers [11] have shown that $Cu-ZnO-Al_2O_3$ catalyst prepared via a novel deposition-precipitation method is highly selective (> 99%) for the dehydrogenation of cyclohexanol to cyclohexanone at a reaction temperature of 250°C. Wang and coworkers [12] have

shown that the activity and stability of a commercial CuO–ZnO
catalyst in the oxidative dehydrogenation of cyclohexanol to cyclohexanone depends on reaction temperature and oxygen partial
pressure. Both the activity and stability of the catalyst could be
improved by modifying the catalyst with palladium oxide or heteropoly
acid. Cyclohexene is one of the by-products of the reaction [11,13].
However, the characteristics of the catalysts that govern the selectivity of the reaction of cyclohexanol to cyclohexanone and cyclohexene are not known. This chapter presents results indicating
the dependence of the selectivity to cyclohexanone or cyclohexene
in the reaction of cyclohexanol on the acidity and copper area of
highly active copper-alumina catalysts prepared by a novel coprecipitation method.

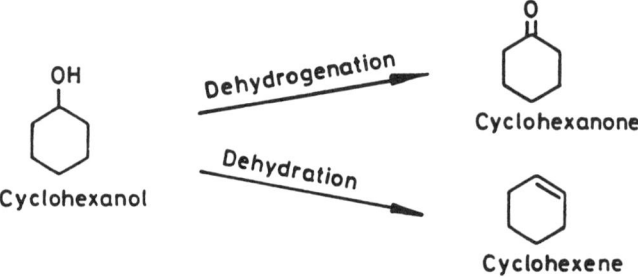

The acidity of the catalysts was measured by chemisorption of
ammonia at room temperature, and the copper area was determined
by adsorptive decomposition of nitrous oxide, also at room temperature. The copper-alumina catalysts were also characterized by their
surface area and X-ray diffraction. Copper-alumina catalysts of
varying composition were prepared using a urea hydrolysis procedure [14,15].

II. EXPERIMENTAL METHODS

A. Materials

Copper(II) nitrate (Fluka), aluminum(III) nitrate (Sarabhai), and
urea (Sarabhai) were of analytical reagent (AR) grade. Hydrogen,
nitrous oxide, and nitrogen gases were of IOL (Indian Oxygen
Limited) AR grades, and ammonia was from the U.S. vendor Matheson;
all materials were 99.9% pure.

B. Preparation

An aqueous solution (2 L) containing copper(II) nitrate and aluminum(III) nitrate of the required concentrations, as well as urea (1.6 M), was placed in a 3-liter round-bottomed glass reactor and heated to 90-95°C with constant stirring. The pH of the solution was measured at different intervals of time using a digital pH meter. When the precipitation was complete (pH 7-7.5), the solution was filtered in a Büchner funnel, and the precipitate washed with distilled water and dried at 110°C for 24 hours. The oxide catalysts were then obtained by calcining the catalyst precursors at 400°C for 24 hours. The copper content of the calcined copper-alumina catalysts varied between 4 and 34 wt%.

C. X-Ray Diffraction (XRD), Gas Chromatography (GC), and Atomic Absorption Spectrometry (AAS)

X-Ray diffractograms were recorded on a Philips model PW 1051 diffractometer using nickel-filtered Cu Kα radiation. The reaction products—cyclohexanone, cyclohexene, and unreacted cyclohexanol— were analyzed by gas chromatography using a 2 m × 3 mm stainless steel column packed with 10% Carbowax 20 M maintained at 120°C. A flame ionization detector was used. The copper content of the catalysts was determined using a Perkin-Elmer model 5000 atomic absorption spectrometer, after the samples had been digested in a mixture of hydrochloric acid and nitric acid and diluted to obtain the required concentration range.

D. Surface Area, Ammonia Chemisorption, and Copper Metal Area

A conventional all-glass high vacuum system was used to measure the Brunauer-Emmett-Teller (BET) surface areas by nitrogen (0.162 nm^2) adsorption at -196°C. The same system, having a stationary background vacuum of 10^{-6} torr (1 torr = 133 Pa) was employed to carry out NH$_3$ chemisorption measurements. In a typical experiment, about 0.5 g of the catalyst sample was placed in a glass adsorption cell and evacuated at 110°C (10^{-6} torr) for 2 hours, and then cooled to room temperature (25°C) under vacuum prior to NH$_3$ adsorption measurements. The first adsorption isotherm, representing both reversible and irreversible ammonia adsorption, was generated, allowing 20 minutes equilibration time at each pressure. Then the catalyst was evacuated at 25°C for 1 hour to remove reversibly adsorbed ammonia. After this, a fresh second isotherm representing only the reversibly adsorbed ammonia was generated in an identical manner.

From the difference between the first and second adsorption isotherms, the irreversibly chemisorbed ammonia was calculated. Helium was used to calculate the dead volume. Specific copper area of the catalysts was determined by adsorptive decomposition of nitrous oxide on the copper surface at room temperature, using a chromatographic pulse technique according to the procedure described by Evans et al. [16] and modified by Denise et al. [17]. The standard procedure employed was the reduction of the catalyst sample (about 0.5 g) by flowing hydrogen (40 cm^3 min^{-1}) at 250°C for 5 hours, followed by cooling to 25°C in a helium flow for 1 hour before the nitrous oxide decomposition experiment. The nitrous oxide decomposes on the copper surface according to the following reaction:

$$N_2O(gas) + 2Cu_s \longrightarrow (Cu_s)_2O + N_2(gas)$$

The copper metal areas were determined using the following equation [18]:

$$S_H = n_m^s X_m n_s^{-1}$$

where S_H is the total metallic surface area, n_m^s is the total nitrous oxide molecules decomposed, X_m is chemisorption stoichiometry at monolayer coverage, and n_s^{-1} is the number of copper metal atoms per unit area of surface (1.47×10^{19} m^{-2}).

E. Activity Measurements

Activity measurements were made in a fixed bed flow microreactor operating under normal atmospheric pressure. For each run, about 0.5 g of the catalyst was loaded in a borosilicate glass reactor of 10 mm i.d. fitted with a thermowell and an electrically heated vertical tubular furnace. The average particle size (0.6 mm) was chosen to eliminate mass transfer effects. The catalyst was first reduced at 250°C for 5 hours in hydrogen flow and then the activities were measured at 250°C on the reduced catalyst maintaining a cyclohexanol flow rate of 0.25 mol h^{-1}. The reaction products analyzed were mainly cyclohexanone and cyclohexene.

III. RESULTS AND DISCUSSION

When the solution containing $Cu(II)(NO_3)_2$, $Al(III)(NO_3)_3$, and urea is heated to 90-95°C, ammonium hydroxide is generated uniformly throughout the solution by urea hydrolysis coprecipitating Cu^{2+} and $Al(OH)_3$:

$$CO(NH_2)_2 + 3H_2O \longrightarrow 2NH_4OH + CO_2$$

The pH-versus-time curve in a typical experiment is represented
in Figure 20.1. The pH of the solution rises slowly in the beginning,
from about 2.7 to 3.5, and then rises at a faster rate, indicating
the commencement of precipitation. The pH remains almost constant
after about 4.5 hours of heating time, attaining a value of about 7.
The precipitation can be considered to be complete at this stage.
Upon filtration, the filter cake and the solution appear blue, apparently because some soluble copper(II) tetramine complex has formed.
The coprecipitated catalysts were first dried at 110°C and then
calcined at 400°C. All the copper values reported in this study
are based on calcined samples.

A.   Characterization of Catalysts

1.   X-Ray Diffraction

The XRD data of calcined catalysts are given in Table 20.1. Catalysts
with copper content of less than 23.6 wt % did not show any crystalline

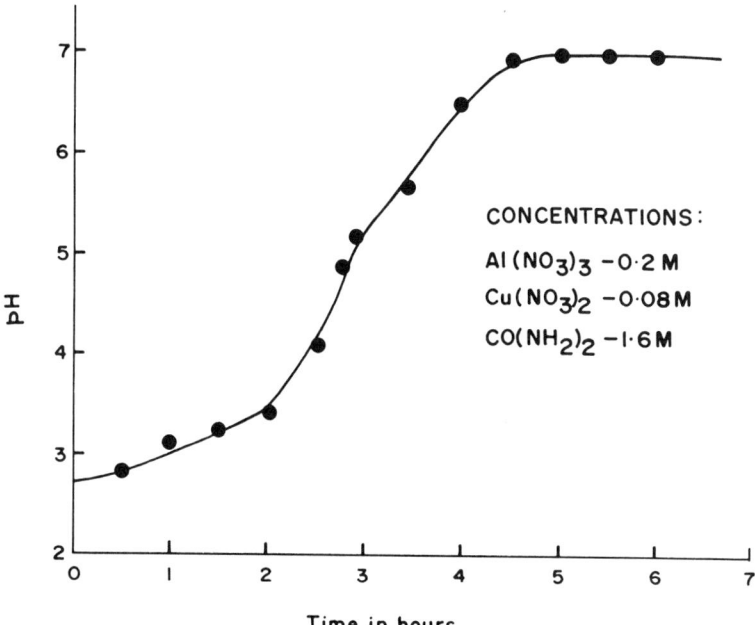

Figure 20.1   The pH as a function of time in the coprecipitation
of copper(II) and aluminum hydroxide from homogeneous solution.

Table 20.1 Catalyst Composition, BET Surface Area, XRD, Copper Area, Ammonia Chemisorption, Activity and Selectivities of Copper–Alumina Catalysts

S. No.	Copper content, wt.%	BET surface area m²/g. cat.	XRD	Copper area m²/g. cat.	NH_3 uptake, μ mol/m² . cat.	Reaction rate mol h⁻¹/ g . cat × 10	% Selectivity Cyclo-hexanone	% Selectivity Cyclo-hexene
1.	12.0	250	Amorphous	3.9	1.06	0.202	70.2	29.8
2.	17.0	220	–do–	7.9	0.79	0.231	75.4	24.6
3.	21.2	200	–do–	10.2	0.57	0.252	80.0	20.0
4.	23.6	150	CuO, $CuAl_2O_4$, $\gamma-Al_2O_3$	12.0	0.60	0.275	82.5	17.5
5.	25.4	128	–do–	9.2	0.62	0.260	83.8	16.2
6.	30.7	75	–do–	7.0	0.47	0.207	93.8	6.2

phase, indicating that copper is present in a highly dispersed or amorphous state in these catalysts. However, γ-Al$_2$O$_3$, CuO, and CuAl$_2$O$_4$ phases were present in all the catalysts containing 23.6-30.7 wt % copper. This suggests that copper oxide may be distorting or suppressing the formation of γ-Al$_2$O$_3$ phase in catalysts containing $\leq$ 21 wt % Cu calcined at 400°C. It may be noted that Kanta Rao and coworkers [14] have shown that γ-Al$_2$O$_3$ was well crystallized from bayerite even at a calcination temperature of 300°C. While the copper-alumina catalysts prepared by coprecipitation and calcined at 400°C have shown the formation of CuAl$_2$O$_4$ phase, the catalysts prepared by deposition-precipitation on suspended γ-Al$_2$O$_3$ calcined at the same temperature have not shown any copper-aluminate phase [15]. According to Wolberg and Roth [19], the direct formation of a CuAl$_2$O$_4$ surface phase can occur even at 300°C, when there is an appropriate combination of copper concentration and alumina surface area. It appears that copper content is one of the factors in determining the extent of metal oxide support interaction.

2. Surface Area, Ammonia Chemisorption, and Free Copper Area

Composition, BET surface area, ammonia uptake, and copper areas of calcined catalysts are given in Table 20.1. The surface areas of the catalysts decrease continuously with increase in copper content. This may be due to the gradual filling up of the pores of Al$_2$O$_3$ by copper species. The copper metal areas, expressed in meters per gram of square catalyst, increased with copper content up to 23.6 wt % and then decreased with further increase in copper content. The maximum value at 23.6 wt % copper indicates that at this loading, the maximum number of copper species is available on the catalyst surface. The decrease in copper metal area with further increase in copper content in the catalyst may be due to the buildup of large copper crystallites, as well as pore-mouth blocking.

Ammonia uptake as a function of copper loading for various catalysts is shown in Figure 20.2. The ammonia uptake decreases with increases in copper content in the catalysts. Recently Corma et al. [20] have demonstrated that the acidity of aluminas could be measured by ambient temperature NH chemisorption. Hirschler has reported that adding metal ion to silica-alumina replaced strong acid sites with weak acid sites [21]. From a Fourier transform-infrared spectroscopic study, Busca has demonstrated that the CuO surface displays a basic character [22], which was deduced from the relevant stability of the surface carbonate species either deriving from the relative adsorption of carbon oxides or being residual from preparation procedures. In the present study a large amount of ammonia chemisorption at the lowest Cu loading studied

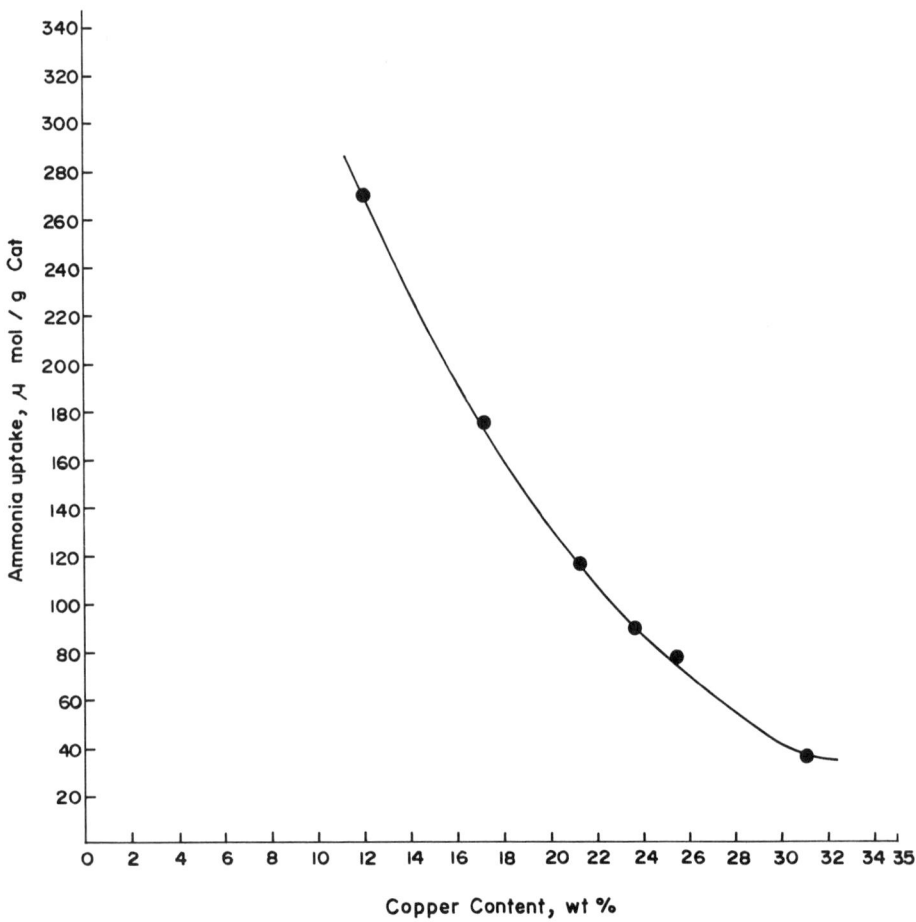

Figure 20.2   Ammonia uptake as a function of copper loading.

decreased with increase in copper content in the catalysts. The
results on ammonia uptake capacities of these catalysts, which offer
a measure of their acidity, are in agreement with the findings that
CuO surface displays a basic character. As the copper loading
in the catalyst increases, the acidic sites of alumina are neutralized
by the basicity of CuO, as evidenced by the decrease in ammonia
uptake.

B.   Catalytic Activity

Activity and selectivities of the catalysts are also given in Table 20.1.
Activity of the catalysts increased with increase in copper loading
up to 23.6 wt % and then declined with further copper loading.
The cyclohexanone selectivity increased with copper loading, while
the cyclohexene selectivity decreased. The cyclohexanone selectivity
varied between 70.2 to 93.8% depending on the copper content in
the catalyst and, correspondingly, the cyclohexene selectivity varied
between 29.8 and 6.2%. These coprecipitated $Cu-Al_2O_3$ catalysts
have shown higher cyclohexanol conversion than the $Cu-ZnO-Al_2O_3$
catalysts with maximum conversions of 55 and 42%, respectively,
under similar experimental conditions [11]. However, the latter
catalysts have shown higher selectivities for cyclohexanone. Ammonia
uptake and the selectivities of cyclohexanone and cyclohexene for
various catalysts as a function of copper loading are shown in
Figure 20.3, which indicates that the ammonia uptake decreases

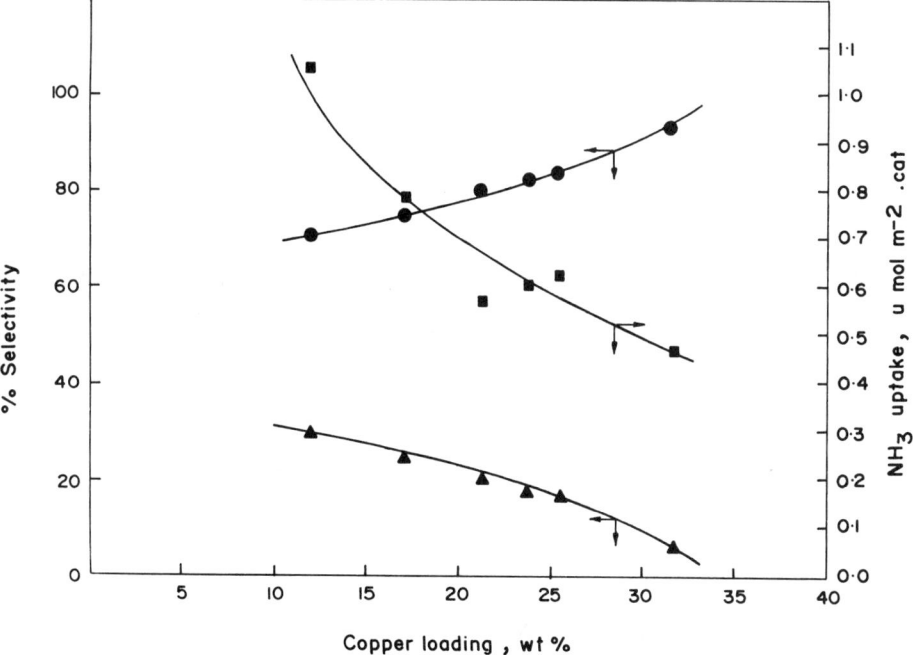

Figure 20.3   Effect of copper loading on ammonia uptake and selectivity to cyclohexanone and cyclohexene: (■) ammonia uptake, (●)
selectivity to cyclohexanone, and (▲) selectivity to cyclohexene.
Reaction temperature, 250°C; feed rate of cyclohexanol, 0.25 mol h^{-1}.

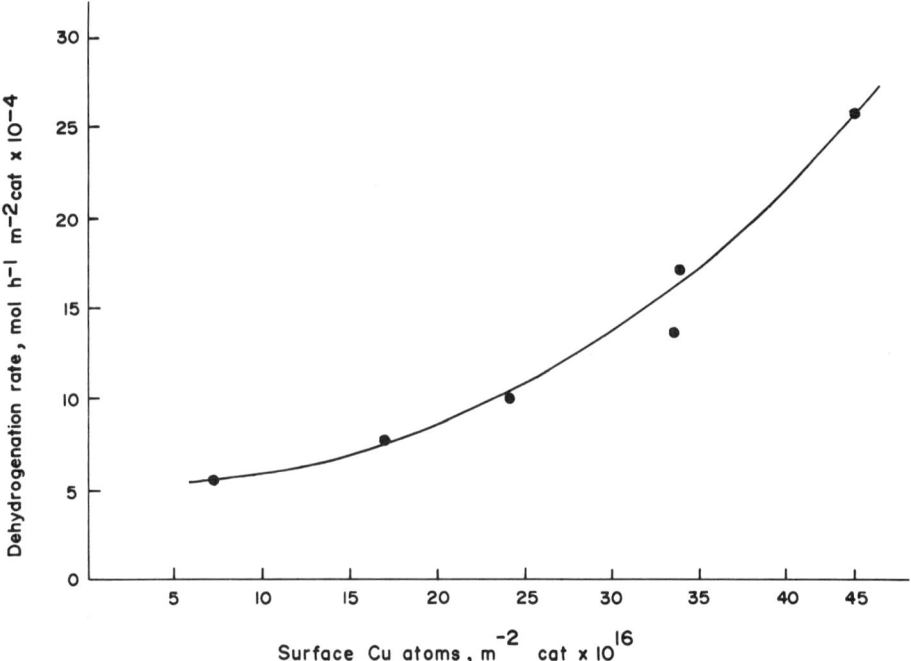

Figure 20.4 Dependence of the rate of dehydrogenation of cyclohexanol on the number of surface Cu atoms per unit area of the
copper-alumina catalysts.

with increase in copper loading. While the selectivity to cyclohexanone
increases with increase in the Cu content, the selectivity to cyclohexene decreases. The dependence of the dehydrogenation rate
of cyclohexanol on the number of surface Cu atoms per unit area
is depicted in Figure 20.4: clearly the dehydrogenation rate increases
with increase in the number of surface Cu atoms.

The dehydrating activity of γ-Al_2O_3 is well known [14,23].
When copper is added, the dehydrogenation activity arises, and
this activity has been shown to depend on copper content. Alumina
produces mainly cyclohexene by the dehydration of cyclohexanol
on the acidic sites of the catalyst [13]. The correlation observed
in Figure 20.3 clearly demonstrates that the selectivity to cyclohexene
depends on the acidity of the catalyst. Furthermore, the selectivity
to cyclohexanone increases with decrease in the acidity of the catalyst. However, the activity of the catalysts depends on the Cu
loading, showing a maximum at 23.6 wt % Cu. Interestingly, at
this Cu loading, CuO, $CuAl_2O_4$, and γ-Al_2O_3 phases have been

detected by XRD, while the catalysts with lower copper content do not show (by XRD) the presence of any crystalline phases. It was shown that the metallic copper species is the probable active site for the dehydrogenation activity of copper-based catalysts [11,15,24]. From Figure 20.4 it is clear that the activity of Cu-Al$_2$O$_3$ catalysts for the reaction of cyclohexanol to cyclohexanone is dependent on the number of surface Cu atoms per unit BET area.

IV. CONCLUSIONS

1. Cu-Al$_2$O$_3$ catalysts prepared by coprecipitation through urea hydrolysis are highly active for the reaction of cyclohexanol to cyclohexanone and cyclohexene at a reaction temperature of 250°C.

2. The selectivity to cyclohexene and cyclohexanone depends on the acidity of the catalysts determined by room temperature (25°C) NH$_3$ chemisorption.

3. The selectivity to cyclohexanone is dependent on the number of surface Cu atoms per unit BET area.

4. The catalyst with 23.6 wt % Cu showed the maximum activity for the reaction of cyclohexanol.

5. Room temperature ammonia chemisorption and/or adsorptive decomposition of N$_2$O appear to be promising techniques for assessing the Cu-Al$_2$O$_3$ catalysts for their relative selectivity to cyclohexanone and cyclohexene in the dehydrogenation of cyclohexanol.

REFERENCES

1. Emelyanov, N. P., Belskaya, R. I., and Semyachko, R. Ya., U.S. Patent 3,652,460 (1972).
2. Gibbson, C. A., U.S. Patent 3,998,884 (1976).
3. Jorda, L. G., Garcia Ochog, F., Aracil, J., and Munoz, V., *Afinidad*, *42*, 181 (1985).
4. Kozlov, N. S., Osonovik, E. S., and Yanchuk Vestsi, A. F., *Akad. Navuk BSSR, Ser. Khim. Nauk, 84* (1978); *Chem. Abstr.*, *90*, 5960f.
5. Doder, D. J., U.S. Patent 2,223,494 (1940).
6. Phelix, P., U.S. Patent 3,305,586 (1967).
7. Sideltseva, M. A., and Erofeev, B. V., *Bull. Acad. Sci. USSR, Div. Chem. Soc.*, *2*, 30 (1986).
8. Petrova, V., Orizavski, I., and Draganov, A., *Khim. Ind. (Sofia)*, *9*, 401 (1983).
9. Dobrovolszky, M., Tetenyi, P., and Paal, Z., *J. Catal.*, *74*, 31 (1982).

10. Pridman, V. Z., Bedina, L. N., and Petrov, I. Ya., *Kinet. Katal.*, *29*, 621 (1988) (Engl. Transl., 1988, 535).
11. Sivaraj, C., Mahipal Reddy, B., and Kanta Rao, P., *Appl. Catal.*, *45*, L11 (1988).
12. Lin, Yu Ming, Wang, Ikai, and Yeh, Chuim-Tih, *Appl. Catal.*, *41*, 53 (1988).
13. Uemichi, Y., Sakai, T., and Kanzuka, T., *Chem. Lett.*, 777 (1989).
14. Sivaraj, C., Prabhakara Reddy, B., Rama Rao, B., and Kanta Rao, P., *Appl. Catal.*, *24*, 25 (1986).
15. Sivaraj, C., and Kanta Rao, P., *Appl. Catal.*, *45*, 103 (1988).
16. Evans, J. W., Wainwright, M. S., Bridgewater, A. J., and Young, D. Y., *Appl. Catal.*, *7*, 75 (1983).
17. Denise, B., Sneeden, R. P. A., Beguin, B., and Cherifi, O., *Appl. Catal.*, *30*, 353 (1987).
18. Shelef, M., Otto, K., and Gandhi, H., *J. Catal.*, *12*, 361 (1968).
19. Wolberg, A., and Roth, J. F., *J. Catal.*, *15*, 250 (1969).
20. Corma, A., Fornes, U., and Ortega, E., *J. Catal.*, *92*, 284 (1985).
21. Hirschler, A. E., *J. Catal.*, *2*, 428 (1963).
22. Busca, G., *J. Mol. Catal.*, *43*, 225 (1987).
23. Knozinger, H., Buhl, H., and Kochloefl, K., *J. Catal.*, *24*, 57 (1972).
24. Pepe, F., Angeletti, C., De Rossi, S., and Lo Jacono, M., *J. Catal.*, *91*, 69 (1985).

21

Selective Dehydrogenations and Dehydrocyclizations over Novel Copper Aluminum Borate Catalysts

Patrick E. McMahon and Larry C. Satek

Amoco Research Center, Amoco Chemical Company, Naperville, Illinois

I. INTRODUCTION

Catalytic dehydrogenation is a core technology in the chemical industry. Hydrocarbon applications include re-forming, synthesis of aromatics from aliphatics and cycloaliphatics, synthesis of aliphatic olefins, production of styrene and substituted styrenes, and synthesis of isopropenyl aromatics (α-methylstyrenes) [1]. Most hydrocarbon dehydrogenations are nonoxidative, the coproduct formed being molecular hydrogen. Oxidative dehydrogenation, being carried out in the presence of a hydrogen acceptor, can provide a thermodynamic and, depending on the mechanism, a kinetic advantage. The term "oxidative dehydrogenation" generally does not distinguish among possible mechanisms.

Dehydrogenation of alcohols to aldehydes and ketones is an important reaction for intermediates synthesis in a number of industries. The majority of methods use stoichiometric oxidants; a smaller portion proceed through a catalytic cycle with a stoichiometric co-oxidant [2]. Nonoxidative heterogeneous catalytic dehydrogenations, which have large-scale advantages, are rare, and catalysts now used are restricted to oxides of copper, chromium, and zinc. Large-scale industrial catalytic examples (with or without a hydrogen acceptor) include methanol to formaldehyde, cyclohexanol to cyclohexanone, and 2-butanol to 2-butanone [1].

Dehydrocyclization can most simply be described as a ring formation reaction with accompanying dehydrogenation. Substrates such as aliphatics, olefins with varying degrees of unsaturation, and alkyl- or alkenyl-substituted ring systems are converted to corresponding carbocyclic and heterocyclic rings or fused ring systems. Dehydrocyclization, by this general description, can proceed by a variety of mechanisms.

Linear aliphatics are converted to aromatics by the standard re-forming catalysts: combinations of platinum, rhenium, tin, and iridium on an acidic support [3-5]. A model reaction such as the conversion of n-hexane to benzene has been extensively studied, for example, over platinum or other Group VIII metals supported on nonacidic zeolites [6] or tellurium-NaX zeolites [7]. Fused rings can be synthesized from alkyl and alkenyl aromatics through side-chain ring closure on the parent ring. Noncatalytic electrocyclic ring closure leads to a variety of fused ring systems such as 1,2-dihydronaphthalenes from phenyl-1,3-butadienes [8], 5,6- and 7,8-dihydroquinolines from pyridinyl-1,3-butadienes [9] and 2-acenaphthones and 2-naphthoate esters from various butadienyl aromatics [10]. The related catalytic dehydrocyclization occurs over platinum on alumina to give indan and indene (plus a variety of other products) from n-propylbenzene [11]. Formation of dimethyl-naphthalenes by dehydrocyclization of methyl-4-(p-tolyl)butane, -butene, and -butadiene occurs over supported chromium and rhenium oxides [12].

Dehydrocyclization can occur across neighboring side chains in o-disubstituted alkyl aromatics—for example, in the synthesis of indene and various other products from o-ethyltoluene over supported platinum catalysts [13]. Benzofurans from o-alkenyl-substituted phenols can be prepared in low yields over platinum on alumina neutralized by alkali metal salts [14]. An oxidative dehydrocyclization is used to form indoles from 2-allyl- or 2-ethenylanilines catalyzed by palladium chloride and benzoquinone [15]. Although other general methods for the synthesis of oxygen and nitrogen heterocycles do not proceed through dehydrocyclization, related reactions such as solution phase iron or palladium reductive cyclization of o-nitrophenyl enamines [16] and palladium cyclization of 2-alkynlanilines [17] can be used to form a variety of indole derivatives.

Copper aluminum borate and copper/cometal aluminum borates are novel compositions of matter, possessing a unique crystal structure [18]. These materials are shown to function as mild and active catalysts for high temperature nonoxidative dehydrogenations and dehydrocyclizations with little loss of reactant to degradative pathways [19]. They are particularly useful for reactions of substrates that show sensitivity to cracking and other by-product formation, easily form coke, or tend to poison noble metal catalysts. Substitution of lattice copper by metals such as palladium, platinum, and nickel, or substitution of aluminum by trivalent cations, gives this catalyst system a high degree of flexibility for reaction optimization. Synthesis of isopropenylbenzenes and substituted naphthalenes, formation of fused ring carbocyclics and heterocycles, and dehydrogenation of alcohols to aldehydes and ketones are applications described in this chapter. For many of these applications, the catalysts based on copper aluminum borate are superior to any other known catalyst.

II. EXPERIMENTAL

A. Catalysts

The synthesis of copper aluminum borate and copper/cometal aluminum borates can be accomplished by a variety of techniques [19]. The synthesis and structure of these materials is described in detail elsewhere in this book [18]. Catalysts that are listed as containing additional metals are, in general, given as atom percent replacement of copper by the specific metal. In some cases, noted in the tables, the catalysts were prepared by impregnation, in which case percentage metal is a weight percent. Potassium oxide modification was always through impregnation.

B. Reactions

All high temperature reactions were carried out in gas phase reactors. Screening units varied in size from laboratory reactors holding a 10-30 g catalyst charge to small (200-1000 g) pilot plants. Products were analyzed by standard gas chromatographic and gas chromatographic/mass spectrometric techniques. When on-line gas chromatographs were not available, only condensable reactor products were analyzed. Liquid phase alcohol dehydrogenations were performed in standard glassware setups.

III. RESULTS AND DISCUSSION

A. Dehydrogenation of Alkyl Aromatics

1. Ethyl Group Dehydrogenation: Synthesis of *p*-Methylstyrene

Synthesis of styrene from ethylbenzene is a standard dehydrogenation reaction in the chemical industry. Consequently, the dehydrogenation catalysts based on iron oxide have been thoroughly optimized for this and similar ethyl aromatic feedstocks. A specific application for a new catalyst system was discovered in connection with the synthesis of *p*-methylstyrene by dehydrogenation of *p*-ethyltoluene, in the presence of the ethylene-toluene alkylate stream.

The presumed industrial synthesis of *p*-methylstyrene proceeds through a highly selective, molecular sieve catalyzed ethylation of toluene to form the desired para isomer. Following a separation section, the pure ethyltoluene feed is dehydrogenated in the presence

Table 21.1 Dehydrogenation of p-Ethyltoluene over Copper
Aluminum Borate

Time on stream (h)	Temperature (°C)	WHSV	Molar diluent[a] ratio	Conversion (%)	Selectivity (%)
50	625	0.2	12:1	63	90
150	625	0.2	12:1	55	90
200	625	0.2	12:1	48	90
50	625	0.5	12:1	50	90
50	625	1.0	12:1	40	90
50	625	2.0	12:1	36	90
50	625	0.2	10:1	50 ± 3	90
50	625	0.2	5:1	50 ± 3	88
50	625	0.2	1:1	50 ± 3	85

[a]Diluents: nitrogen + solvent.

of steam diluent to p-methylstyrene. A process advantage would
accrue if the alkylate stream could be sent directly to the dehydrogenation reactor without an intermediate separation, and with the
excess toluene (plus ethylene) as the diluent.

   Some of the earliest work with the copper aluminum borate catalyst system examined the dehydrogenation of p-ethyltoluene in the
absence of steam, and with toluene and toluene/nitrogen as diluents.
Table 21.1 summarizes some of the findings. In general, the approximately 90% dehydrogenation selectivity with hydrocarbon or nitrogen
diluents is similar to the steam-based iron oxide catalysts. Initial
conversions (50 h) at various space velocities, at least for our
earliest catalyst versions, are below what can be accomplished with
the iron oxides. Although the catalyst was shown to function without
steam, slow deactivation was seen over 1-2 weeks. Selectivity for
this reaction is tolerant of very low molar diluent ratios. Further
work with an optimized copper aluminum borate based catalyst led
to a pilot plant study.

2.   Isopropyl Dehydrogenation: Synthesis of Substituted
     α-Methylstyrenes

α-Methylstyrene is a commercially available product, used as a monomer for homo- and copolymer applications in acrylonitrile-butadiene-
styrene (ABS) resins, polyester resins, and coatings. Ring-methyl
substituted α-methylstyrenes can be used as comonomers in styrene

resins and in adhesive formulations. These compounds are often synthesized by dehydrogenation of the corresponding isopropyl-benzene derivative (eq. 21.1).

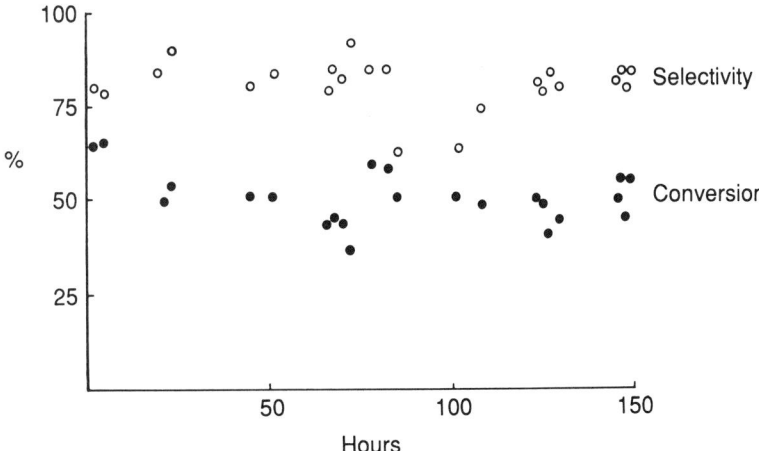

$$(21.1)$$

Early catalyst versions were examined for the dehydrogenation of cumene (Fig. 21.1). Over most potassium-modified copper aluminum borates, conversion can be maintained at approximately 50% at weight hourly space velocities (WHSVs) of 0.6-0.7, with selectivities in the 90-95% range. (In this chapter, "weight hourly space velocity" is defined as the total weight of reactant hydrocarbon per weight catalyst per hour.) As for the p-methylstyrene case, the catalyst is tolerant of low diluent ratios. No deactivation over 15 days was observed with a molar diluent ratio of 9:1 nitrogen and only 4:1 steam.

Dehydrogenation over copper aluminum borates contrasts with the reaction over iron oxide based catalysts. These commercial catalysts deactivate rapidly (1-2 days), generally require a high molar ratio of steam to hydrocarbon ($\geq 20:1$), and operate at modest

Figure 21.1 Selectivity and conversion versus time for copper aluminum borate: WHSV = 0.65, 9 moles of N_2 per mole of cumene, 4 moles of steam per mole of cumene at 600°C.

housely space velocities (0.3-0.8). Under similar conditions, copper
aluminum borates achieve an approximately equivalent selectivity
and similar conversion, but with longer times between regeneration
and at a lower diluent level than is required for the iron oxide
based catalysts.

Dehydrogenation of p-cymene over copper aluminum borates
prepared by different techniques has recently been examined in
detail. Preparation methods can have a remarkable effect on the
activity of the catalyst. Further work is described in Chapter 22.
Conversions of near 50% and selectivities in excess of 98% are
achievable at WHSVs of 0.5. The catalyst has been run for up to
2 weeks with essentially no deactivation. This contrasts with an
iron oxide based catalyst; under the same conditions, conversions
eventually rose to a comparable value after 5 days, but selectivities
were in the mid-90% range.

A limited amount of work has been done with 3,4-dimethylcumene.
A potassium-modified copper aluminum borate, prepared similarly
to the p-cymene catalyst, gave conversions in the 40% range at
a WHSV of 0.5. Selectivity to the substituted styrene was greater
than 99% at a relatively high steam-to-hydrocarbon molar diluent
ratio of 50:1. Figure 21.2 shows results of a long-term study. Slow
deactivation is observed, which must be corrected by regeneration of
the catalyst by steaming. Another technique was to increase the level
of potassium doping by adding potassium carbonate to the catalyst bed.
Figure 21.2, arrow #2 attempts to increase conversion by increasing

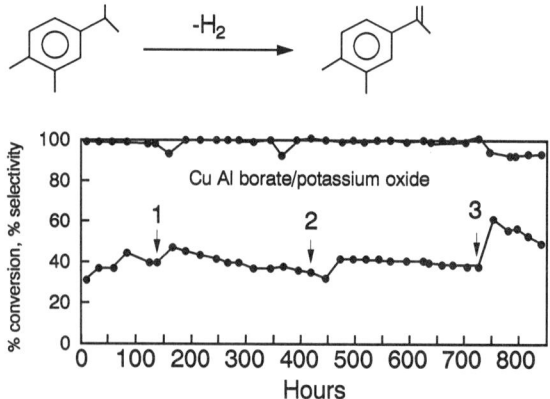

Figure 21.2 Dehydrogenation of 3,4-dimethylcumene.

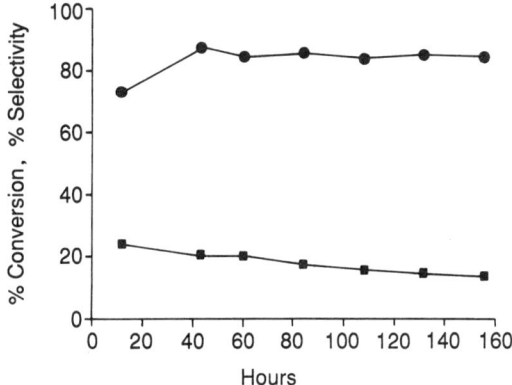

Figure 21.3   Conversion (■) and selectivity (●) versus time for iron oxide based catalyst.

temperatures resulted in irreversible damage to the catalyst as shown by the lower selectivity to the styrene product. In Figure 21.2, arrow #3, these results indicate that it may be necessary to optimize the copper aluminum borate catalysts for each specific dehydrogenation. Under the same steam-to-hydrocarbon ratio, temperature and flow conditions, and catalyst particle size, iron oxide based catalysts gave lower conversions (20% dropping to 15% after about 4 days), as well as lower selectivities of about 85% (Fig. 21.3).

3.   Dehydrogenation of Mono- and Diisopropylnaphthalene

Isopropenylnaphthalenes can be used as a comonomer in a variety of olefinic polymers (similar to the corresponding α-methylstyrenes), and as a precursor for isopropylidene- or indan-based monomers for condensation polymers. The diisopropenylnaphthalenes (e.g., 2,6-diisopropenylnaphthalene) can be used as crosslinking agents in olefin-based polymers, as precursors for carbon fibers, or for preparing bisisopropylidene monomers.

The dehydrogenation of isopropylnaphthalenes presents difficulties in addition to the ease of cracking of the isopropyl group. The naphthalene ring is particularly susceptible to coke formation at the required high temperatures, leading to rapid catalyst deactivation. The copper aluminum borate systems provide a mild reaction surface, which allows dehydrogenation to occur with minimum activation of cracking and coking pathways.

The gas phase dehydrogenation of 2-isopropylnaphthalene over our standard catalyst proceeds smoothly at 600°C (WHSV of 0.2; 50:1 diluent ratio with toluene, nitrogen, and/or steam) to give conversions of about 90% at a selectivity of 92%. Lower molecular weight alkyl-naphthalenes from cracking reactions make up the balance. A slow deactivation with time on stream was noted.

The conversion of 2,6-diisopropylnaphthalene was studied in more detail. We examined a number of catalyst preparations at different conditions, usually with a nitrogen/hydrogen diluent. In general, good conversion of the starting material was achieved, with small losses to cracking products. Total conversion of both isopropyl groups to form the diolefin was difficult; usually an approximately equal split between the monoolefin (2-isopropenyl-6-isopropylnaph-thalene) and diolefin (2,6-diisopropenylnaphthalene) was formed. Starting material losses occurred through the typical cracking at the isopropyl side chains.

Table 21.2 shows the results for the gas phase dehydrogenation of 2,6-diisopropylnaphthalene with a 20% potassium modified copper aluminum borate catalyst. The activity of this catalyst is not quite as high as the unmodified catalyst; however, the added potassium lowered the cracking activity. (Selectivity increased by approximately 10%.) At the lowest space velocity (0.05), continuous runs proceeded for 200-400 hours before conversion dropped below 90%, at which time net yield of olefin products was about 80%. A 6-month run was completed with catalyst regeneration at the 80% yield point.

Platinum-on-alumina catalysts will perform this dehydrogenation. Under similar conditions, initial selectivity is to the monoolefin plus some diolefin at 85-90% selectivity at 530°C. However, all these catalysts rapidly deactivate, sometimes with a concomitant loss in

Table 21.2 Dehydrogenation of 2,6-Diisopropylnaphthalene over Potassium-Modified Copper Aluminum Borate[a]

WHSV	Tempera-ture (°C)	Conver-sion (%)	Selectivity (%)		
			Total olefin	(monoolefin)	(diolefin)
0.05	575	94	91	48	43
0.05	595	98	85	37	48
0.1	575	81	91	60	31
0.1	595	97	80	35	45
0.2	575	62	86	67	19
0.2	595	85	84	52	32

[a]Molar diluent ratio: 20-40:1 solvent + nitrogen + hydrogen.

selectivity. The best preparation was a tin-passivated platinum on a lithium ion modified alumina [20]. At 570°C, for example, initial conversion of 90%, with 86% selectivity to olefins, was found. Under fixed conditions (WHSV 0.1, 20:1 molar diluent ratio of hydrogen plus nitrogen to hydrocarbon), this catalyst deactivated at a rate of 0.5-1.0% conversion loss for each hour on stream.

4. Dehydrogenation of Dixylylethane

Dixylylethane, 1,1-bis(3,4-dimethylphenyl)ethane, is a conveniently prepared feedstock for the nitric acid oxidation to the monomer benzophenone tetracarboxylic acid. As part of a study for homogeneous Co/Mn/Br oxidations, we attempted the dehydrogenation of this material to the corresponding olefin, dixylylethylene (eq. 21.2).

$$(21.2)$$

This molecule represents a difficult problem for high temperature dehydrogenation in that hydrocracking at the benzylic position is a very favorable competing pathway. Cleavage of the terminal methyl group from the 1,1-disubstituted ethane, leaving the dixylyl-methyl fragment, and parallel or subsequent cleavage of the phenyl bridging bonds, producing xylyl and ethylbenzene fragments, would be favored radical (or acidic) reactions.

Both the olefin and dixylylmethane were considered as possible oxidation feeds. Thus there was some incentive to confine any cracking reactions to formation of dixylylmethane, eliminating irreversible loss of material due to deeper fragmentation. Work by our colleagues showed that noble metal catalysts, even those with a passivated support, gave severe cracking, with the formation primarily of xylyl and substituted ethylbenzene fragments.

We examined this reaction with some of the early copper aluminum borate catalysts; both gel and solid state preparations performed about equally. The data in Table 21.3 show that, contrary to the results from supported noble metals, the synthesis of dixylylethylene proceeds with a selectivity of approximately 80% at moderate conversions (55-60%) at 590°C. The catalyst most studied for this reaction was a 1% Mo/0.2% Pd modified copper aluminum borate catalyst, originally optimized for another reaction. Under equivalent conditions, the nonmodified catalyst performed similarly. Higher temperatures gave correspondingly lower selectivity; however, the major cracking product was dixylylmethane. This reaction demonstrates the extremely weak cracking tendency of these materials.

Table 21.3 Dehydrogenation of Dixylylethane over Copper Aluminum Borate-Based Catalysts[a]

Catalyst description	Temperature (°C)	Conversion (%)	Selectivity (%)		Benzyl cleavage products[b]	Others[a]
Unmodified	640	85	60	20	9	11
0.2% Pd	590	76	63	19	7	11
1% Mo; 0.2% Pd	590	57	80	12	4	4
1% Mo; 0.2% Pd	640	73	60	22	9	9
1% Mo; 0.2% Pd	670	87	40	33	11	16

[a]WHSV, 0.2; nitrogen diluent at 28:1 molar ratio.
[b]Xylenes, dimethylethylbenzenes.
[c]Arrylmethyl cracking products; isomerizations; heavies.

B. Dehydrocyclization of Alkylaromatics: Side-Chain Closure to Ring

Dehydrocyclization of alkyl aromatics provides a useful method for constructing multiple-ring systems. Although often catalytically demanding, building up of the ring system, when possible, has several advantages. The desired ring structure may not be obtainable in available feeds; or isolation and purification (from coal tar extract or appropriate hydrocarbon streams) may be difficult and costly. Introduction of stereoselectivity in substituted ring systems may be more easily accomplished by synthesizing an appropriate dehydrocyclization precursor—for example, for preparing multisubstituted naphthalenes from corresponding substituted benzenes.

Dehydrocyclizations can proceed though a number of mechanisms, mainly distinguished by the timing of the dehydrogenation steps versus the ring closure step, and the mechanism of ring closure. Concerted electrocyclic or electrophilic aromatic substitution, or other nucleophilic, electrophilic, or radical reactions, can all be involved in ring closure. Aliphatic to aromatic conversion over reforming catalysts is considered to go through a bifunctional process whereby the metal(s) dehydrogenate the paraffin and ring closure proceeds by a cationic olefin substitution [4]. Monofunctional catalysts, such as platinum or other Group VIII metals supported on nonacidic (alkali metal exchanged) zeolites are proposed to dehydrocyclize by using the zeolitic cage structure to provide appropriate orientation of the precursor [6]. Noncatalytic electrocyclic ring closure of the linear hexatriene formed by successive dehydrogenation of hexane is implicated for tellurium-NaX zeolites [7]. Noncatalytic electrocyclic ring closure, involving a diene side chain and a parent aromatic ring, has been demonstrated in the syntheses of a variety of fused ring systems [8-10].

For catalytic dehydrocyclization of substituted aromatics beginning with a saturated alkyl side chain, the catalyst must activate the appropriate carbon through dehydrogenation. If ring closure proceeds through an electrocyclic mechanism, then catalyst function is restricted to dehydrogenation. Formation of the carbon-carbon bond may also be catalytic if, for example, ring closure is perceived as going through a forming intermediate olefin (bound or free) followed by cationic or radical aromatic substitution. Catalyst requirements must then include activation of a pathway for carbon-carbon bond formation (e.g., by acid catalysis), without also activating similar, but undesirable, pathways such as hydrocracking, isomerization, and coke formation. Potential dehydrocyclization precursors and suitable catalyst materials will be dependent on equilibrium and kinetic constraints.

1. Dehydrocyclization of Alkylbenzenes

The dehydrocyclization of a series of *n*-alkyl-substituted benzenes
was studied using the copper aluminum borate based catalysts.
One general observation is that product distribution in these reactions is extremely sensitive to catalyst modification and preparation
techniques. For example, the results for the reaction of *n*-
propylbenzene (eq. 21.3) over two unmodified catalysts are shown
in Table 21.4. The formation of the five-membered indene fused
ring system is not particularly favored under these conditions,
and the primary products arise from simple dehydrogenation. Significant isomerization to the stable isopropenyl olefin is also seen.

$$\qquad\qquad \xrightarrow{-_nH_2} \qquad + \qquad + \qquad \qquad (21.3)$$

Product distribution changes as a function of time on stream, with
straight dehydrogenation increasing at the expense of fused ring
formation. Hydrocracking of the feed remains approximately constant
and is dependent on small preparation differences. Since an electrocyclic mechanism is not possible in this case, the catalyst has a
role in ring closure. Product changes with time can be explained
if coking is selectively deactivating sites for the ring closure reaction. However, since activation must involve prior or concerted
dehydrogenation, the nature of the catalytic site or sites is not
clear.
    A different case is presented by the dehydrocyclization of amyl-
benzene, (eq. 21.4). Formation of the fused six-membered ring
is more favored under reaction conditions, and thus the primary
products are methylnaphthalenes and methyltetralins in 85-90% selectivity. The lengthening of the alkyl side chain dramatically increases
the susceptibility of the substrate to cracking reactions at the
temperatures (550-600°C) employed for propylbenzene. Because
the base case copper aluminum borate is not sufficiently active
for the reaction at lower temperatures, a 1% palladium modified
catalyst was used. The data are shown in Table 21.5.

$$\qquad\qquad \xrightarrow{-_nH_2} \qquad + \qquad \qquad (21.4)$$

Conversion falls slowly over the course of the 100-hour run, but
the product distribution changes little. Hydrocracking accounts for

Table 21.4 Dehydrocyclization of n-Propylbenzene over Copper Aluminum Borate Catalysts[a]

Catalyst	Time on stream (h)	Conversion (%)	Selectivity (%)				
			Cracking products[b]	(isopropylbenzene + α-methylstyrene)	(allylbenzene)	(propenylbenzene)	(indan)[c]
Cu Al borate, pH 5.7	7	86	25	18	1	5	51
	41	91	15	36	4	21	24
	9	87	12	34	4	18	32
Cu Al borate, pH 5.9	47	91	13	36	4	19	28
	150	85	12	35	5	27	21
	205	84	9	35	6	30	20
	280	65	8	30	8	41	11

[a]WHSV, 0.02; molar diluent ratio, 40:1; temperature, 550–600°C.
[b]Toluene, ethylbenzene, styrene.
[c]Sometimes contained small percentage of indan.

Table 21.5 Dehydrocyclization of Amylbenzene over 1% Pd/Copper Aluminum Borate[a]

Time on stream (h)	Temperature (°C)	Conversion (%)	To dehydrogenated products	Selectivity (%)			
					[b]		Others[c]
6	525	71	81	1	10	70	19
12	550	84	74	0	10	64	26
8	500	59	90	1	5	84	10
20	500	43	93	3	7	83	7
56	500	35	92	4	10	78	8
100	500	30	89	5	12	72	11

[a]WHSV, 0.05; molar diluent rations, hydrogen 6:1, nitrogen + toluene, 12:1.
[b]Contains some methyldihydronaphthalene.
[c]Cracking products; other ring systems.

only about 10% of the product and remains essentially constant. A small shift toward incomplete aromatization (methyltetralins vs. methylnaphthalenes) accompanies the deactivation. The only olefin found, in small amounts, is the ring-conjugated product. Because ring closure proceeds through carbon 4 of the pentyl chain, the remaining thermodynamic and kinetic favored olefins (or forming olefins) directly activate the required side-chain position for ring attack. The significant methyltetralin formation, combined with the lack of olefin intermediates, and the loss of aromatization activity with deactivation suggests initial rate-limiting activation (dehydrogenation) of the side chain followed by rapid carbon-carbon bond formation to form a methyltetralin intermediate. Subsequent dehydrogenation yields methylnaphthalene.

2. Dehydrocyclization of 1-Ethylnaphthalene

Acenaphthylene can be used as a pure material precursor to form carbon fibers. It and the reduced acenaphthene are also intermediates in the synthesis of 1,8-naphthalene dicarboxylic acid. We examined the dehydrocyclization of 1-ethylnaphthalene (eq. 21.5) over a number of catalysts, shown in Table 21.6.

$$- n H_2 \qquad\qquad\qquad\qquad (21.5)$$

The dehydrogenation-based products of the reaction are 1-vinylnaphthalene, acenaphthene, and acenaphthylene. Side products include methylnaphthalene, naphthalene, and other substituted naphthalenes derived from cracking, isomerization, and transalkylation. Copper aluminum borates provide good conversions with very little in the way of destructive side reactions (selectivities to the dehydrogenated products, 90-96%). Iron oxide based styrene catalysts and straight platinum on alumina re-forming catalysts are not suitable for this reaction, showing poor selectivity. The addition of rhenium to the platinum re-forming catalyst improves the dehydrogenation selectivity but does not catalyze ring formation.

The major product in the reaction over unmodified copper aluminum borate is 1-vinylnaphthalene; only small amounts of the bridged five-membered, fused ring products are found. Contrary to the case for amylbenzene, the activation of the appropriate side-chain carbon during or after olefin formation is not sufficient to drive the ring-closure reaction.

Table 21.6 Dehydrocyclization of 1-Ethylnaphthalene over Selected Catalysts[a]

Catalyst[b]	Temperature (°C)	Conversion (%)	To dehydrogenated products	Selectivity (%)			
				(vinylnaphthalene)	(acenaphthene)	(acenaphthylene)	Other
CuAB; gel prep	640	94	90	85	2	5	10
CuAB; solid prep	640	91	92	84	2	6	8
CuAB; 0.1% Pd	630	75	96	68	18	10	4
CuAB; 0.1% Pd	670	89	92	61	13	18	8
CuAB; 0.2% Pd[c]	630	87	91	76		15	9
CuAB; 2% Pd[c]	610	92	90	68	14	8	10
CuAB; 10% K$_2$O[c]	630	80	83	59	10	14	17
CuAB; 30% K$_2$O[c]	660	35	84	39	20	25	16
CuAB; 10% Ni:							
21 reactor hours	620	79	89	62	9	18	11
71 reactor hours	620	39	80	20	26	34	20
27 reactor hours	660	84	73	50	1	22	27
77 reactor hours	660	78	80	18	36	26	20
Iron oxide	630	93	40	34		6	60
Pt/alumina	580	55	40	33		7	60
Pt-Re/alumina	580	54	80	76		4	20

[a] WHSV, 0.1; 80:1 molar diluent ratio, solvent + nitrogen.
[b] CuAB: copper aluminum borate; iron oxide: commercial catalyst.
[c] Prepared by impregnation; metals are not lattice substituted.

Replacement of 0.1% lattice copper by palladium increases the yield of ring product to approximately 30% while maintaining very high selectivities to the dehydrogenation-based products. Palladium was also added to the catalyst by impregnation, at a 0.2 and a 2 wt % loading. Yield of cyclized product also increased while maintaining overall selectivity, although the lattice-substituted catalyst was slightly better. Impregnation by potassium oxide, used at times for steam reactions, increased the yield of ring products but at the expense of lower activity and selectivity to dehydrogenation-derived products.

Replacement of 10% lattice copper by nickel also increases the ring product. These catalysts showed an unusual product distribution dependence with time on stream. As the catalyst aged in the reactor, formation of acenaphthene plus acenaphthylene continued to increase (for the length of the run), eventually reaching approximately 60% of the total product. This may be due to a change in catalyst structure (e.g., metal atom migration), or a role for carbon or coke in the catalytic ring closure, independent of dehydrogenation. Barring an unusual coke-metal interaction, the latter explanation implies a free olefin intermediate.

C. Dehydrocyclization of 1,2-Disubstituted Aromatics: Closure Across Side Chains

Catalytic dehydrocyclization of ortho-substituted aromatics represents another strategy for building up fused ring systems. Both five- and six-membered rings can be formed over copper aluminum borate based catalysts. Equation (21.6) gives a general description of the five-membered, fused ring synthesis.

$$X = C, N, O$$

Benzofuran synthesis from o-alkenyl-substituted phenols over platinum on passivated alumina [14] and oxidative dehydrocyclization to form indoles from 2-allyl- or 2-ethenylanilines with palladium chloride/benzoquinone [15] require some unsaturation in a side chain. This is also true for the related iron or palladium reductive cyclization

of *o*-nitrophenyl enamines [16] and palladium cyclization of 2-alkynyl-
anilines [17]. There are few examples of reactions that begin with
saturated alkyl side chains, especially under nonoxidative conditions
[13]. This is due to stringent catalyst requirements, which must
effect activation of at least one of the appropriate side chains by
dehydrogenation, at sufficiently high temperatures to provide favorable kinetics, but with little degradation of the starting molecule.
Ring closure must also be catalyzed without activating (to any great
degree) related cracking and coking pathways.

1.   Dehydrocyclization of *o*-Ethyltoluene

The formation of indene from *o*-ethyltoluene exemplifies synthesis
of a carbocyclic fused ring. Indene and substituted indenes can
be used as comonomers in various olefinic polymers; the homopolymer
can be used as a precursor for carbon fibers. Results for the synthesis of this molecule by a one-step dehydrocyclization of *o*-
ethyltoluene over two copper aluminum borate preparations are
shown in Table 21.7. The major products of the reaction are *o*-
vinyltoluene and indene, with a few percent indan. The by-products
xylene, ethylbenzene, and styrene are formed through cracking.
Selectivities to the dehydrogenation-based products over these
catalysts are excellent, in the 90-94% range.
    Conversions at a variety of space velocities at 600°C are shown
in Table 21.7. The highest conversion achieved at the lowest space
velocity on a fresh catalyst was 77%. Rough calculations suggest
that this may be approaching thermodynamic limits. Data for this
reaction point out the subtle catalyst preparation differences that
can affect results. One of the catalysts shows deactivation, a conversion loss of about one-third, over 200 hours on stream. The
other does not show deactivation during an approximately equivalent
time under the same conditions.
    With decreasing conversion, either through deactivation or be-
cause of increases in space velocity, the product distribution changes
to favor simple dehydrogenation instead of dehydrocyclization. This
suggests that the olefin is an intermediate in the cyclization, although
more than one pathway at potentially more than one catalytic site
is a possibility. Recycle experiments with the crude reactor effluent
(no separations) demonstrate that vinyltoluene can be converted
to indene under reaction conditions. A mixture containing a 16:84
indene-to-vinyltoluene product split (not counting the other products) was converted to a mixture with approximately a 50:50 ratio
after one recycle. With continued recycle, a maximum product split
of approximately 66:33 indene (plus some indan) to vinyltoluene
was reached. This may represent the thermodynamic limit under
reaction conditions.

Table 21.7 Dehydrocyclization of o-Ethyltoluene over Copper Aluminum Borate[a]

Catalyst	Time on stream (h)	WHSV	Conversion (%)	Selectivity (%)			
				To dehydrogenated products	Cracking products[b]	(propenyl-benzene structure)	(indane structure)
Cu Al borate, pH 5.7	30	0.05	77	88	12	27	60
	189	0.05	51	93	7	67	26
	58	0.1	66	93	7	41	52
	38	0.2	37	89	11	53	36
	60	0.3	21	91	9	74	17
	82	0.4	16	91	9	80	11
Cu Al borate, pH 5.9	23	0.05	45	93	7	57	36
	200	0.05	51	93	7	53	40
	58	0.01	62	92	8	45	47
	112	0.05	51	93	7	51	42
	38	0.1	36	90	10	58	32
	35	0.2	28	90	10	62	26
	62	0.4	22	89	11	70	19

[a]Temperature, 600°C; molar diluent ratios (nitrogen plus solvent), 4-40:1.
[b]Xylene, toluene, ethylbenzene.

2. Synthesis of Fused-Ring Heterocycles

More favorable dehydrocyclization is found for the synthesis of the
corresponding fused five-membered heterocycles. For example, the
free energy for the dehydrocyclization of o-ethylaniline to indole
is negative at 630°C. Indole and substituted derivatives are used
in fragrances, and as an intermediate in pharmaceutical and agricultural products. Indole is a precursor in a biosynthetic process
to form the amino acid tryptophan. Supplies of indole currently
derive from coal tar extraction. A synthetic process, dehydrocyliza-
tion of o-ethylaniline using the copper aluminum borate catalyst
system, has the potential to produce indole for a cost far below
the current extraction-based selling price.

The data for this reaction are summed up in Table 21.8. The
major product of the reaction is indole, in total yields of 85-90%
under most conditions for our best catalyst. Loss of starting material
to by-products such as aniline or toluidine can be kept below 5-8%.
Over the most active catalysts, conversion to the final ring product
is nearly complete, with only traces of o-vinylaniline or indolene
formed, except at the highest space velocities. The favorable free
energy allows the reaction to be pushed to 100% conversion, given
sufficient rates. At the lower space velocities for our base case
catalyst, conversion remains at 100% for up to 1 week on stream.
Conversion can be kept above 90% for 2-3 weeks on stream at WHSVs
of 0.05-0.2 with continued high selectivity. Regeneration of the
catalyst completely restores high indole yields with no detected
limit on the number of possible regenerations.

The course of this study exemplifies the progress of continued
catalyst development. The first catalysts were prepared by a gel
technique. This method was supplanted by solid state synthetic
techniques, which were operationally simpler. However, further
improvements in the gel preparation have produced a catalyst that
is both more active and more selective than the previous best solid
state preparation.

Based on the work so far, copper aluminum borate is uniquely
suited for this dehydrocyclization. Iron oxide based catalysts give
poor conversion and selectivity, and they deactivate rapidly. Of
the supported noble metal catalysts, a platinum-alumina re-forming
catalyst did give good initial selectivity to indole. However, deactivation was very rapid, and conversion dropped by 50% in less
than 2 days, with a severe loss in selectivity. Moreover, the catalyst was not regenerable using standard techniques; activity could
not be restored by attempted coke removal by heating the catalyst
in air at various temperatures and oxygen concentrations. Consistent
with this observation, modification of the copper aluminum borates
with platinum or palladium lead to poorer results compared to

Table 21.8 Dehydrocyclization of *o*-Ethylaniline over Selected Copper Aluminum Borate Based Catalysts[a]

Catalyst[b]	WHSV	Conversion (%)	Selectivity (%)				
			To dehydrogenated products	Cracking products[c]	(vinyl-NH_2)	(indoline)	(indole)
CuAB gel prep[d]	0.05	100	94	6			93
	0.1	98	93	7	1	2	90
	0.2	95	92	8		3	89
	0.3	91	93	7		6	87
	0.7	78	91	9		18	73
CuAB, solid prep	0.1	98	89	11	1	1	87
	0.2	79	85	15	2	7	76
	0.3	59	86	14	2	16	68
	0.5	54	85	15	2	20	63
CuAB; 0.1% Pd	0.1	67	81	19	2	15	64
CuAB; 10% Ni	0.1	74	84	16	2	24	58

[a]Data taken for samples at 30–40 hours on stream; molar diluent ratio, 10–20:1 nitrogen + solvent.
[b]CuAB: copper aluminum borate.
[c]Aniline, toluidine.
[d]Temperature, 600°C; other catalysts at 630°C.

Table 21.9   Dehydrocyclization of o-Ethylphenol over Copper Aluminum Borate Catalyst[a]

Temperature (°C)	Conversion (%)	Selectivity (%)				
		To dehydrogenated products	Cracking products[b]	[o-ethylphenol structure]	[benzofuran structure]	[benzofuran structure]
600	31	86	14	13	2	71
630	57	86	14	7	1	78
650	82	83	17	2	1	80

[a]Data taken for samples at 50-60 hours on stream; WHSV, 0.1; molar diluent ratio, 20:1 nitrogen + solvent; CuAB catalyst, solid prep.
[b]Phenol, cresol.

unmodified catalysts. It is not known whether this is due specifically to poisoning by the amine reactant or to a more general mechanism for deactivation.

The feasibility of extending the dehydrocyclization reaction to the oxygen-based analogue is shown in Table 21.9. Benzofuran can be synthesized from o-ethylphenol over copper aluminum borate. Yields are not as high compared to the indole system for equivalent conditions, although only one catalyst was examined for the reaction.

D.   Dehydrogenation of Alcohols

Aldehydes and ketones are important intermediates in pharmaceuticals, flavors, fragrances, and dyes. Where appropriate, dehydrogenation of a corresponding alcohol represents a useful synthetic route. Nonoxidative heterogeneous catalysts have advantages, especially for industrial processes, since there are no problems due to heavy metal contamination, use of co-oxidants, recovery of homogeneous catalysts, or overoxidation. Catalysts for this reaction type are generally confined to oxides of copper, chromium, and zinc; industrial products include cyclohexanone, 2-butanone, and formaldehyde (oxidative) [1].

Laboratory examples using heterogeneous nonoxidative dehydrogenation of alcohols are scarce. This may be due, in part, to the frequent requirement for a gas phase reactor system; however, liquid phase reactions are often possible. Supported copper-chromium oxide converts various aliphatic alcohols to the respective aldehyde or ketone at temperatures of 300-350°C with poor to moderate yields (30-60%) [21]. Liquid phase reactions by copper chromite have also been reported for higher boiling monoalcohols (250°C) and glycols [22]. Raney nickel can be used for dehydrogenations at lower temperatures with a hydrogen acceptor, although high concentrations of catalyst are required to give 30-35% yields (24 h) of phenyl-substituted ketones [23]. Based on experiments to date, copper aluminum borate catalysts provide higher activity at lower temperatures, and usually with higher selectivity, than other known heterogeneous nonoxidative catalysts for liquid and gas phase alcohol reactions.

The data in Tables 21.10 and 21.11 summarize the results for a common test reaction, the gas phase conversion of 2-propanol to acetone. Comparison runs were performed for a number of copper aluminum borates and other catalysts. The basis of the experiment was to fix the liquid hourly space velocity (LHSV) at 6 and the conversion at approximately 50%. The temperature was then adjusted for each catalyst to achieve this fixed activity. No attempt was made to find optimum conditions for any catalyst, nor are the runs corrected for catalyst density and surface area. However, the comparisons can be used to indicate the general effectiveness of catalysts for secondary alcohol dehydrogenation.

The copper aluminum borates are extremely active for 2-propanol dehydrogenation, achieving the target conversion at temperatures lower than any other tested material. Selectivity to dehydrogenation is essentially 100%; condensation products consume 1-2% of the acetone formed. These results are in good agreement with the observed mild dehydrogenation activity for hydrocarbons. Copper chromite is also an excellent catalyst, as expected from its use industrially for alcohol dehydrogenations.

The relative tendency to catalyze dehydrogenation versus dehydration is often used to gain information about catalyst surface characteristics [24]. Dehydration to propylene is equated with surface acidity, although base catalysis is also a possibility. Our data confirm at least a qualitative agreement between propylene formation and the observed tendency of a catalyst to catalyze other nominally acidic-type reactions in hydrocarbons such as cracking and isomerization. The assignment of dehydrogenation to basic sites is less certain. Given the variety of pathways available for dehydrogenation, it is probably prudent to restrict this interpretation to catalysts for which reaction mechanisms are well understood [25].

Dehydrogenation of primary alcohols to aldehydes is less favorable, thus reaction of 1-propanol is a similar, but more discriminating test system (Table 21.12). Temperatures approximately 100°C higher are required to achieve initial 50% conversion. The catalysts tested

Table 21.10   Dehydrogenation of 2-Propanol over Selected Catalysts
Normalized to LHSV 6 at Approximately 50% Conversion

Catalyst	Time on stream (h)	Tempera- ture (°C)	Conver- sion (%)	Selectivity (%) ⌒	○ ⌄	Condensa- tion products
MnO$_2$/alumina	27	350	50.4	78.7	18.1	3.2
98% MgO	60	450	49.9	18.1	73.6	8.3
Platinum/alumina	142	225	50.8	88.2	0.1	11.7
Pt-Re/alumina	150	235	48.5	88.0	1.0	11.0
1.2% Pt/carbon	40	230	51.4	5.2	92.0	2.8
0.5% Pt/alumina	57	255	50.3	87.0	4.8	8.2
0.5% Pd/alumina	21	225	55.3	6.9	90.2	2.9
	104	220	49.3	18.1	80.0	1.9
4.6% Pd/carbon	30	395	48.3	45.6	54.1	0.3
	50	395	19.7	89.3	10.3	0.4
1.1% Ru/carbon	23	225	48.0	28.8	69.6	1.6
1.2% Ir/carbon	45	220	42.1	10.4	86.7	2.9
20% Iron(III) oxide/alumina	32	325	49.5	86.9	8.3	4.8
CuCO$_3$/silica- (6% Cu)	45	275	50.0	0.8	99.2	0
	73	275	35.0	0.9	99.1	0
Copper chromite	71	180	51.1	0	99.6	0.4
	150	180	47.7	0	99.6	0.4

show almost no propylene formation. The copper aluminum borates
are much more active for formation of condensation products and
thus give lower yields of propionaldehyde. Modified materials have
an advantage, however, as demonstrated for both propanol conversions. A specially modified copper aluminum borate shows essentially
no deactivation over the course of several hundred hours on stream.
Other catalysts deactivate rapidly. Copper chromite deactivates
rapidly for 1-propanol conversion.

Reactions of higher boiling alcohols can be performed in the
liquid phase at or below reflux. For example, at 150°C, 3% catalyst
loading, sec-phenethyl alcohol (neat) is converted to acetophenone
in 80-85% yield with a standard copper aluminum borate. Conversion
is quantitative after approximately 18 hours, driven to completion
by hydrogen loss from the system. Yield is lower than 100% as
a result of the formation of the dehydration compound styrene,

Table 21.11 Dehydrogenation of 2-Propanol over Copper Aluminum Borate (CuAB) Based Catalysts, Normalized to LHSV 6 at Approximately 50% Conversion

Catalyst	Time on stream (h)	Temperature (°C)	Conversion (%)	Selectivity (%)		
				⌒	⌾	Condensation products
CuAB, pH 7.5	42	150	48.2	1.3	97.2	1.5
	118		47.2	1.9	95.4	2.7
CuAB, pH 9.5	17	145	49.1	0	95.9	4.1
	100		46.7	0.2	97.3	2.5
	200		37.5	0.2	98.8	1.0
	300		33.1	0	99.3	0.7
CuAB, modified	30	160	50.1	0	98.8	1.2
	100		47.7	0	98.4	1.6
	200		50.4	0	98.7	1.3
	280		49.1	0	98.8	1.2
CuAB/1% Ni	31	160	59.3	0.2	94.8	5.0
	59	150	52.9	0.4	95.7	3.9
CuAB/5% Ni	37	160	50.6	0.5	97.5	2.0
CuAB/10% Ni	30	155	52.5	0	97.1	2.9
CuAB/15% Ni	47	180	49.1	1.0	97.8	1.2
CuAB/1% Pt	64	155	50.9	1.9	94.2	3.9
CuAB/1% Pd	34	160	50.3	1.5	93.8	4.7

which is immediately hydrogenated in the solution to ethylbenzene. Studies of observed completion time versus catalyst load have shown that in this case equilibrium is rapidly established and the rate of the reaction is controlled by loss of hydrogen from solution. Copper chromite is not as effective for this reaction; yields greater than 50% could not be achieved during a 48-hour run.

The gas phase synthesis of cyclohexanone from cyclohexanol was examined over a base case copper aluminum borate (Table 21.13). Below 200°C, the reaction is 96-99% selective for cyclohexanone, with condensation products making up the balance. Above 200°C, small amounts of cyclohexene and phenol are also formed. The constant conversion at 165°C for a wide range in WHSVs indicates that equilibrium conversion is maintained on this catalyst at least up to a WHSV of 6.6. Although conversion measurements were less precise at very high flow rates because of temperature control problems, equilibrium conversion at space velocities above 10 may be possible.

Table 21.12   Dehydrogenation of 1-Propanol over Selected Catalysts
Normalized to LHSV 6 at Approximately 50% Conversion

Catalyst	Time on stream (h)	Temperature (°C)	Conversion (%)	Selectivity (%)		
				⟋⟍	⟋⟍ CHO	Condensation products
CuAB, pH 7.5	10	255	47.5	0.5	83.5	16.0
	80		20.9	2.0	71.5	26.5
CuAB, pH 9.5	20	260	49.4	0.5	84.6	14.9
	70		39.2	1.8	91.3	6.9
CuAB, modified	30	245	50.4	3.1	58.2	38.7
	100		48.3	0.5	61.1	38.4
	150		45.8	0.4	66.3	33.3
	200		46.3	0.3	73.7	26.0
Copper/chromite	50	265	49.4	0	97.7	2.3
	60		40.8	0	97.4	2.6
	70		38.3	0	98.3	1.7

Table 21.13   Dehydrogenation of Cyclohexanol over Copper
Aluminum Borate

Temperature (°C)	WHSV	Conversion (%)	Selectivity (%)			
			⬡=O	⬡	⬡-OH	Condensation products
165	0.6	25	96			4
165	0.9	24	96			4
165	1.2	27	99			1
165	2.4	28	96			4
165[a]	6.6	26	93	2	2	3
185	1.2	27	97	1		2
205	1.2	36	92	1	< 1	6
225	1.2	50	88	3	< 1	8
220[a]	13.0	42	90	4	< 1	5

[a]Approximate temperature; control was difficult at these WHSVs.

IV. CONCLUSION

Copper aluminum borates are a new family of catalytic materials, possessing a novel composition of matter with a unique crystal structure. They provide a mild active surface for selective nonoxidative hydrogenations and dehydrocyclizations. The catalysts are particularly suited for substrates that are susceptible to degradation and coking. Replacement of lattice copper by metals such as palladium, platinum, or nickel and lattice aluminum by trivalent cations permits fine-tuning of activity and selectivity for a variety of applications. Dehydrogenation of isopropylbenzenes and substituted naphthalenes, formation of certain fused ring carbocyclics and heterocycles, and synthesis of aldehydes and ketones from precursor alcohols have been demonstrated.

ACKNOWLEDGMENTS

We gratefully acknowledge the work of the following technical experts on this project: Bill Dangles, John Featherstone, Chris Lipa, Pat Mayo, Pat Shope, and Jim Smith. Among our many colleagues who contributed to this program are Rich DeSimone, Jim Kaduk, Marc Kullberg, John Melquist, Eric Moore, and Alex Zletz. This chapter is dedicated to the memory of Chris Lipa.

REFERENCES

1. Information on industrial processes can be found in: (a) Lowenheim, F. A., and Moran, M. K., *Industrial Chemicals*, Wiley, New York, 1975. (b) Wittcoff, H. A., and Reuben, B. G., *Industrial Organic Chemicals in Perspective*, Wiley, New York, 1980. (c) Grayson, M., ed., *Kirk-Othmer Encyclopedia of Chemical Technology*, Wiley, New York, 1978.
2. A good review of alcohol oxidations can be found in Haines, A. H., *Methods for the Oxidation of Organic Compounds*, Academic Press, New York, 1988.
3. Ciapetta, F. G., and Wallace, D. N., *Catal. Rev.*, 5, 67 (1972).
4. Pines, H., and Nogueira, L., *J. Catal.*, 70, 391 (1981).
5. Nogueira, L., and Pines, H., *J. Catal.*, 70, 404 (1981).
6. (a) Rees, L. V. C., ed., *Proceedings of the Fifth International Conference on Zeolites*, Heyden and Sons, London, 1980. (b) Lambert, S. L., et al., U.S. Patent 4,614,834 (1986). (c) Derouane, E. G., and Vanderveken, D. J., *Appl. Catal.*, 45, L15 (1988).

7. (a) Price, G. L., Ismagilov, Z. R., and Hightower, J. W., *J. Catal.*, *73*, 361 (1982). (b) Price, G. L., Ismagilov, Z. R., and Hightower, J. W., *J. Catal.*, *81*, 369 (1983).

8. (a) Valkovich, P. B., Conger, J. L., Castiello, F. A., Brodie, T. D., and Weber, W. P., *J. Am. Chem. Soc.*, *97*, 901 (1975). (b) Radcliffe, M. M., and Weber, W. P., *J. Org. Chem.*, *42*, 297 (1977).

9. Rosen, B. I., and Weber, W. P., *J. Org. Chem.*, *42*, 47 (1977).

10. (a) Zoeller, J. R., U.S. Patent 4,783,548 (1988). (b) Zoeller, J. R., *Tetrahedron Lett.*, *30*, 1457 (1989). (c) Zoeller, J. R., and Summers, C. E., Jr., *J. Org. Chem.*, *55*, 319 (1990).

11. Shephard, F. E., and Rooney, J. J., *J. Catal.*, *3*, 129 (1964).

12. Taniguchi, K., et al., U.S. Patent 3,931,348 (1976).

13. Csicery, S. M., *J. Catal.*, *110*, 348 (1988).

14. Illingworth, G. E., and Loucvar, J. J., U.S. Patent 3,285,932 (1966).

15. (a) Harrington, P. J., Hegedus, L. S., and McDaniel, K. F., *J. Am. Chem. Soc.*, *109*, 4335 (1987). (b) Hegedus, L. S., Allen, G. F., Bozell, J. J., and Waterman, E. L., *J. Am. Chem. Soc.*, *100*, 580 (1978). (c) Harrington, P. J., and Hegedus, L. S., *J. Org. Chem.*, *49*, 2657 (1984).

16. (a) Ponticello, G. S., and Baldwin, J. J., *J. Org. Chem.*, *44*, 4003 (1979). (b) Kozikowski, A. P., Ishida, H., and Chen, Y., *J. Org. Chem.*, *45*, 3350 (1980).

17. Rudisill, D. E., and Stille, J. K., *J. Org. Chem.*, *54*, 5856 (1989).

18. For a complete description of these materials, see Satek, L. C., Kaduk, J. A., and McMahon, P. E., Characterization, structure, and active site hypothesis for novel copper aluminum borate dehydrogenation and dehydrocyclization catalysts, Chap. 22 of this book.

19. (a) Satek, L. C., U.S. Patent 4,590,324 (1986). (b) Zletz, A., Satek, L. C., and Miller, J. T., U.S. Patent 4,645,753 (1987). (c) Zletz, A., U.S. Patent 4,729,979 (1988). (d) Hussman, G. P., and McMahon, P. E., U.S. Patent 4,740,647 (1988). (e) DeSimone, R. E., Moore, E. J., and Rosen, B. I., U.S. Patent 4,755,497 (1988). (f) McMahon, P. E., U.S. Patent 4,891,462 (1990). (g) Satek, L. C., U.S. Patent 4,913,886 (1990).

20. (a) Dautzenberg, F. M., Helle, J. N., Biloen, P., and Sachtler, W. H., *J. Catal.*, *63*, 119 (1980). (b) Burch, R., and Garla, L. C., *J. Catal.*, *71*, 360 (1981). (c) Beltramini, J., and Trimm, D. L., *Appl. Catal.*, *31*, 113 (1987).

21. (a) Dunbar, R. E., *J. Org. Chem.*, *3*, 242 (1938). (b) Dunbar, R. E., and Arnold, M. R., *J. Org. Chem.*, *10*, 501 (1945).

22. (a) Kyrides, L. P., and Zienty, F. B., *J. Am. Chem. Soc.*, *68*, 1385 (1946). (b) Schniepp, L. E., and Geller, H. H., *J. Am. Chem. Soc.*, *69*, 1545 (1947). (c) Halasz, A., *J. Chem. Educ.*, *33*, 624 (1956).
23. (a) Kleiderer, E. C., and Kornfeld, E. C., *J. Org. Chem.*, *13*, 455 (1948). (b) Mitra, M. N., and Elliott, W. H., *J. Org. Chem.*, *33*, 175 (1968). (c) Mitra, M. N., and Elliott, W. H., *J. Org. Chem.*, *33*, 2814 (1968).
24. Hathaway, P. E., and Davis, M. E., *J. Catal.*, *119*, 497 (1989).
25. Bowker, M., Petts, R. W., and Waugh, K. C., *J. Catal.*, *99*, 53 (1986).

22

Characterization, Structure, and Active Site Hypothesis for Novel Copper Aluminum Borate Dehydrogenation and Dehydrocyclization Catalysts

Larry C. Satek, J. A. Kaduk, and Patrick E. McMahon

Amoco Research Center, Amoco Chemical Company, Naperville, Illinois

I. SYNTHESIS

Copper aluminum borate, $Cu_2Al_6B_4O_{17}$, can be prepared by a process of the sol/gel type, as shown in equation (22.1)

$$2Cu(NO_3)_2 \cdot 6H_2O + 3Al_2O_3(sol) + 4H_3BO_4 \xrightarrow{\text{pH 4-10}} \xrightarrow{\text{air-dry}}$$

$$\xrightarrow{\text{vacuum dry}} \xrightarrow{\text{calcine}} Cu_2Al_6B_4O_{17} \qquad (22.1)$$

The calcination temperature is determined by the minimum temperature required for compound synthesis and a desire for significant surface area. The differential thermal analysis scan, shown in Figure 22.1, identifies the peak synthesis temperature at about 710°C, with the synthesis onset occurring at about 680°C. Properly prepared and dried gels will allow synthesis at these temperatures of pure copper aluminum borate samples with surface areas near 200 m^2/g.

The preparation of a good gel—that is, one that does not have significant separation of components through the precalcination stage—is a nontrivial feat. Use of organic cosolvents and organic bases can dramatically affect the copper coordination sphere in the gel, as can be observed by the color of the gel. Furthermore, the appropriate preparation can affect the catalytic activity. Figures 22.2 and 22.3 show conversions and selectivities of two catalyst runs that were duplicated several months later; the catalysts differ only in the various catalyst preparation parameters. The reproducibility of the catalyst runs and the differences observed in catalyst performance show that catalyst preparation is extremely important.

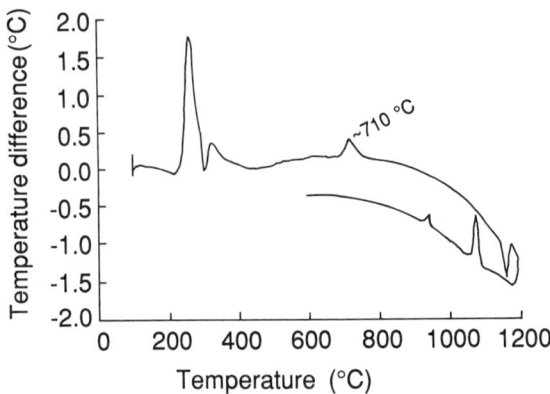

Figure 22.1   DTA scan of copper aluminum borate vacuum-dried precursor.

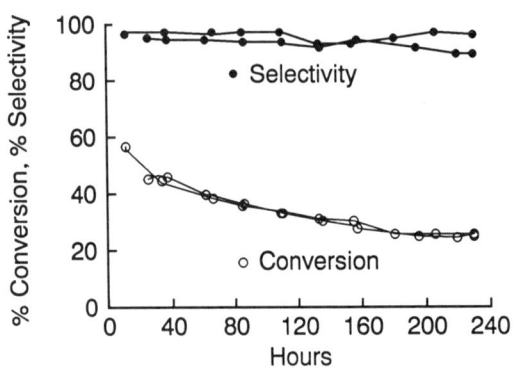

Figure 22.2   Conversion and selectivity data for a copper aluminum borate gel preparation with NH_4OH and 780°C.

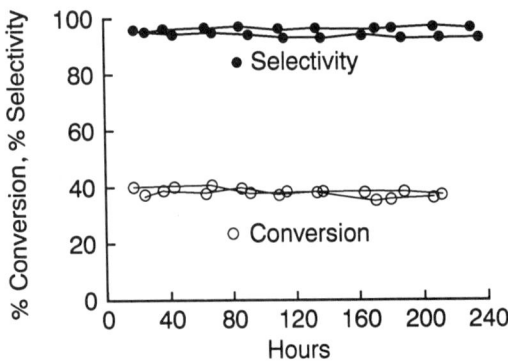

Figure 22.3   Conversion and selectivity data for a copper aluminum borate gel preparation with MeOH, tetramethylammonium hydroxide, 830°C, and K_2CO_3.

II. STRUCTURE

The structure was redetermined by neutron and X-ray powder diffraction (XRD) methods using Rietveld refinement techniques to
obtain a higher precision description of the structure [1,2]. The
compound crystallizes in the tetragonal space group I_4/m with a =
10.5736 (7) and c = 5.6703 (6). The structure is made up of edge-
sharing chains of octahedral Al atoms parallel to the c-axis. These
chains are joined in the a- and b-directions by trigonal planar
BO_3 groups. There is a trigonal bipyramidal site, occupied by both
Cu and Al, which shares a face with the Al octahedron. These
trigonal bipyramidal sites share equatorial corners at a square planar
oxygen atom. The structure contains eight ring channels approximately 4 Å in diameter parallel to the c-axis.

Figure 22.4 contains an overview of the structure, depicting
the square planar oxygen atom and the trigonal bipyramidal disordered Cu/Al site. Important bond lengths and angles are shown

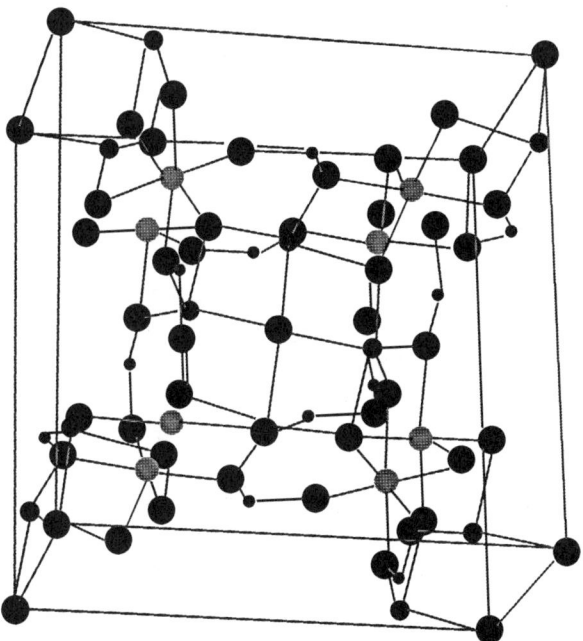

Figure 22.4 Overview of copper aluminum borate.

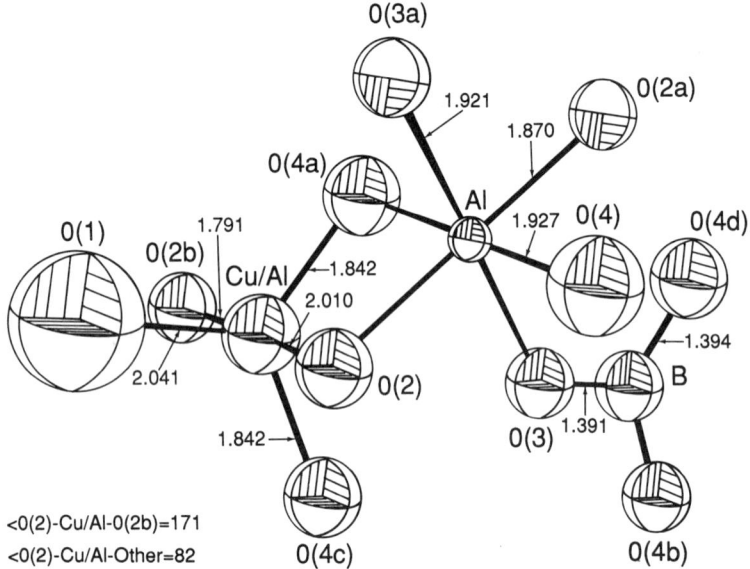

Figure 22.5 One half formula unit of copper aluminum borate showing significant bond lengths (Å) and angles.

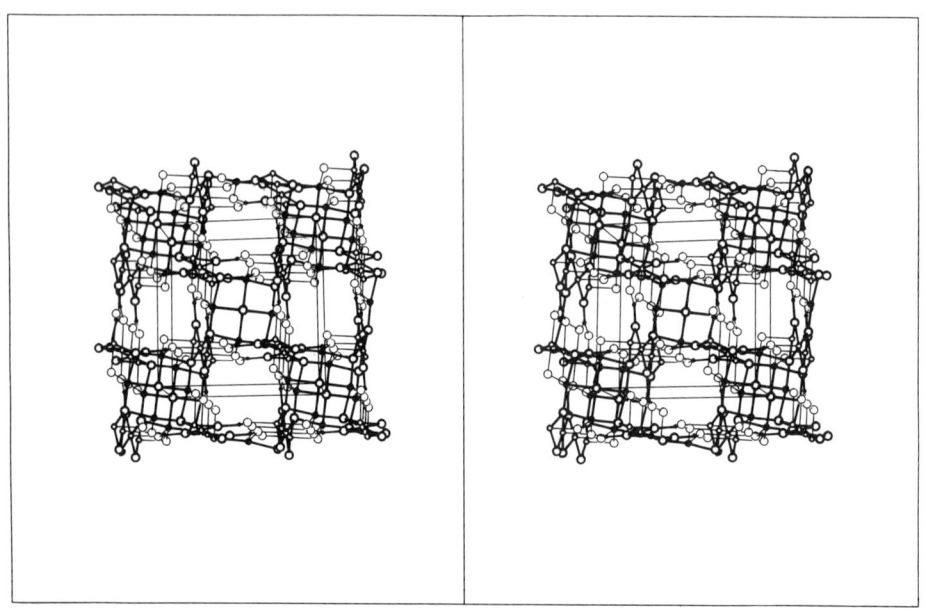

Figure 22.6 Stereo pair of three unit cells of copper aluminum borate.

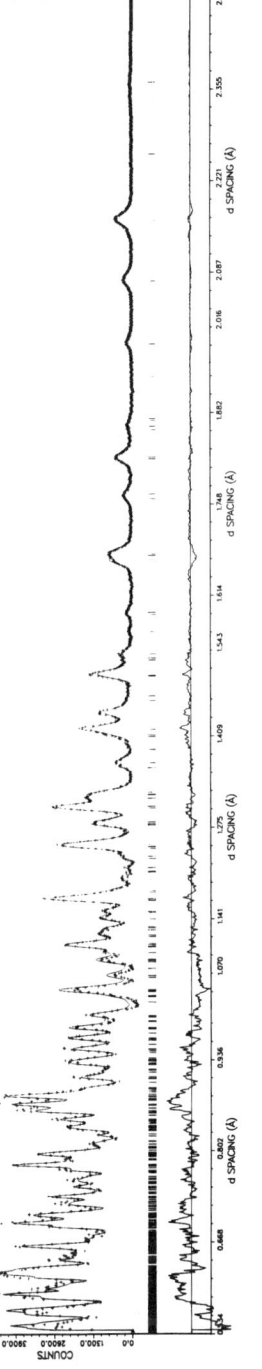

Figure 22.7 Neutron powder diffraction difference map: calculated and observed values.

in Figure 22.5. The bond lengths to the square planar oxygen are essentially normal (2.04 Å).

Figure 22.6 is a stereo pair showing approximately three unit cells, so that the channels can be observed. Figure 22.7 is the neutron difference map, showing the agreement of the fit.

III. ACTIVE CATALYST

Used catalysts in many of the catalytic applications described show that significant chemical changes have occurred in copper aluminum borate. The XRD powder patterns for a new and used catalyst are shown in Figure 22.8. The used powder pattern more closely resembles that of $Al_4B_2O_9$ with copper metal, and one might expect that equation (22.2), showing the reaction of copper aluminum borate with a reducing agent, would describe the active catalytic species.

$$2Cu_2Al_6B_4O_{17} + 4[RED] \longrightarrow 3Al_4B_2O_9 + 4Cu + B_2O_3 + 4[RED-O]$$

$$(22.2)$$

where RED is a reducing agent.

There are a few differences, however, notably lines at 29, 32, 45, 50, and 55° 2θ. Indeed, even the XRD powder patterns of $Cu_2Al_6B_4O_{17}$ and $Al_4B_2O_9$ are quite similar.

It is remarkable that the $Al_4B_2O_9$-like phase can be synthesized under these conditions, since the synthesis temperature of the pure

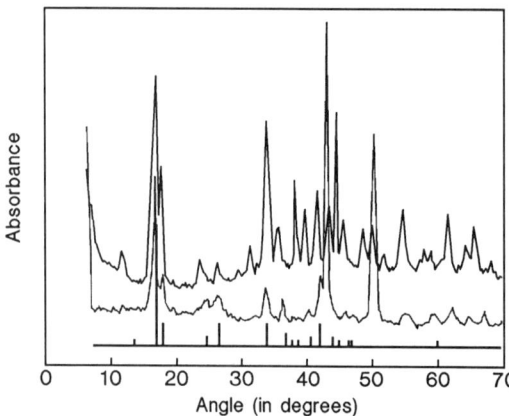

Figure 22.8 X-Ray diffraction powder pattern of new and used catalyst.

Table 22.1   Analysis of a Used Catalyst Extracted Twice with Dilute Aqua Regia at 70°C, and Calculated Metal Oxide and Material Balance Values[a]

Solids		Filtrate	
From ICP[b]	Calculated	First extraction	second extraction
4.1% Cu $\longrightarrow$	5.1% CuO	1.52% Cu	0.103% Cu
36.1% Al $\longrightarrow$	68.2% Al_2O_3	0.084% Al	0.141% Al
8.6% B $\longrightarrow$	27.7% B_2O_3	0.073% B	0.039% B
	100.9%		

[a]Overall material balance: 100 × (wt solids + calculated wt of extracts)/(initial wt of solids) = 101.6%.
[b]Inductively coupled plasma spectroscopy.

phase is approximately 900°C. Clearly the active catalyst never sees temperatures higher than about 600°C, and it is observed in reactions as low as 500°C; additionally, since the copper aluminum borate is generally prepared at temperatures 100-200°C below the aluminoborate's synthesis temperature, at no time during the active catalyst's preparation does the temperature approach the synthesis temperature of the pure phase.

Further insight into the active catalyst, for *p*-ethyltoluene dehydrogenation, can be obtained by leaching the used catalyst with warm dilute aqua regia to remove the elemental copper.

Table 22.1 describes the results of a material balance on multiple leachings of a used catalyst. Approximately 20% of the total copper is not removed, hence is not reduced out of the catalyst as copper metal or a copper oxide. Note that essentially 100% of the material is accounted for by the experiment. Successive leaching experiments result only in the dissolution of small quantities of the remaining solids, not selective removal of copper.

Figure 22.9 plots conversion and selectivity of the catalyst, before and after leaching, for *p*-ethyltoluene dehydrogenation run under identical conditions [3]. The reproducibility is quite good, demonstrating that the elemental copper is not involved in the catalysis, but that the residual copper ions in the aluminoborate solid are probably responsible for the catalytic activity. That no more of the copper is reduced can be seen in Figure 22.10, where no difference is observed in the leached catalyst before and after it is placed in the reactor.

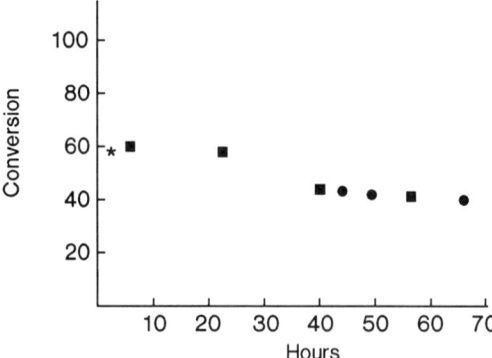

Figure 22.9 Selectivity and conversion of the dehydrogenation of p-ethyltoluene with a fresh copper aluminum borate catalyst versus a leached catalyst; (■) leached and (●) fresh.

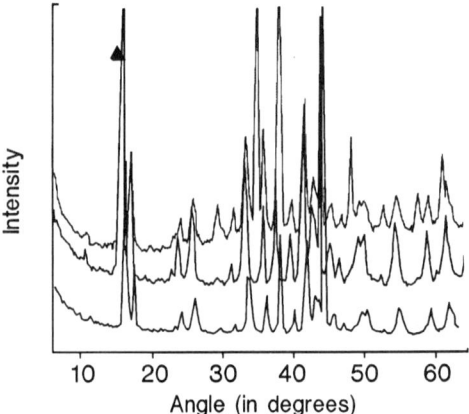

Figure 22.10 X-Ray diffraction powder pattern of a used unleached catalyst, a used leached catalyst, and a catalyst that had been used before and after leaching.

More information can be gained when one examines the used catalyst from low temperature dehydrogenations of alcohols. The XRD pattern is unchanged after up to 10 days operation. Thus the active site of the catalyst is not dependent on the reduction event alone. Furthermore, there is no activation time observed, although poorer catalysts will deactivate with some feeds.

Given the similarity of the XRD powder patterns of copper aluminum borate, aluminum borate, and the active catalyst, we believe that the active site must look something like the copper/aluminum sites in the starting copper aluminum borate—that is, the square planar oxygen, which is bonded to the trigonal bipyramidal disordered copper/aluminum sites.

REFERENCES

1. Richter, L., Synthese und Strukturuntersuchungen von Eisen-und Kupfer-Aluminium-Boraten, Technische Hochschule, Aachen, Feb. 25, 1977. Details of the structure reported here are essentially correct, but with only 137 data on a single crystal, and a final R = 9.7%. This served as a model in the Rietveld refinement. A preliminary report of the neutron diffraction data was presented at the 194th meeting of the American Chemical Society in New Orleans, Sept. 1, 1987.

2. Subsequent work has been done on a single crystal (R = 4%). While the structure is in good agreement with this powder diffraction study, the data are more precise. Satek, L. C., Hriljac, J. A., Brown, R. D., Kaduk, J. A., and Cheetham, A. K., Synthesis, structure, and catalytic activity of the novel aluminoborate $Cu_2Al_6B_4O_{17}$, manuscript in preparation.

3. Moore, E. J., unpublished results.

23

Homogenous Catalytic Oxidations Using Metalloporphyrin Complexes with Emphasis on Ruthenium Systems

Brian R. James

Department of Chemistry, University of British Columbia, Vancouver, British Columbia, Canada

I. INTRODUCTION

Partly because of environmental concerns with the use of classical reagents such as permanganate and dichromate for the oxidation of organics, there has been increasing interest in the use of O_2 (or air) as oxidant, which is cheap as well as environmentally harmless [1-3]. There are, of course, several commercial hydrocarbon O_2 oxidation processes that currently use transition metal salts as catalysts; indeed, this class of reactions represents the largest scale application of homogeneous catalysis [4]. Examples include processes for production of terephthalic acid (a polyester precursor) from p-xylene using Co(II)/Mn(II) acetates/bromides catalysts, and for cyclohexanone/cyclohexanol (nylon precursors) from cyclohexane using Co(II) salts [4]. However, these O_2 oxidations suffer from low selectivity because they operate via radical chain processes involving an alkyl hydroperoxide, ROOH; the major function of the metal complex is catalytic decomposition of hydroperoxides, and any interaction between the metal complex and O_2 is largely incidental [1,4].

Selective O_2 oxidations are accomplished by certain enzyme systems in which direct interaction of the metal center and O_2 is a critical step in the overall process, and in recent years there have been notable advances in selective O_2 oxidations emanating from attempts to mimic the increasingly well-understood enzyme chemistry using model, protein-free metal complexes [3,5-9]. Some enzymes of the mono- and dioxygenase types, where one or both oxygen atoms of O_2, respectively, are incorporated into a substrate, function via an iron-porphyrin center.

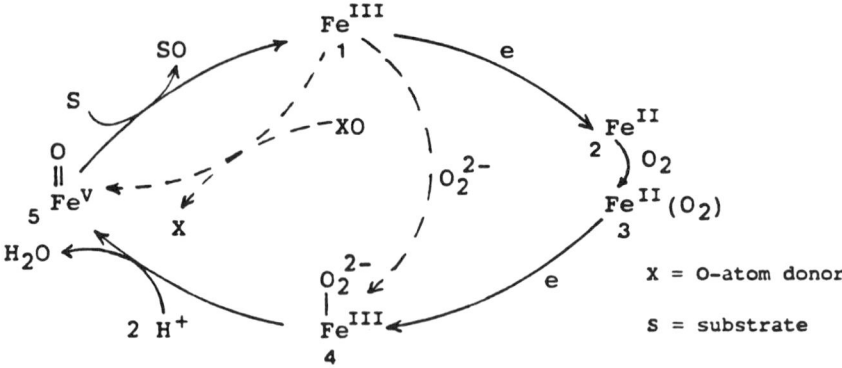

Figure 23.1   Basic features of the cytochrome P-450 mechanism, including shunt pathways (see text).

It is instructive to consider in more detail one monooxygenase, cytochrome P450, because many recent major advances in oxidation chemistry have stemmed from increased understanding of this enzyme and related model (biomimetic studies) [3,5-9]. Figure 23.1 illustrates the key steps in the catalytic cycle that operates by reductive activation of O_2, whereby one O atom is reduced to H_2O and the second O atom becomes available, within a high-valent metal-oxo species for the oxidation process (eq. 23.1).

$$\text{substrate(S)} + O_2 + 2H^+ + 2e \longrightarrow S(O) + H_2O \qquad (23.1)$$

There are five steps in the overall mechanism.

1.   Addition of substrate S to the resting state of the enzyme [low spin Fe(III)] to give a high spin enzyme-substrate complex, 1; crystallographic data for a P450 enzyme that hydroxylates camphor have revealed that the camphor is bound by the protein in close proximity to the Fe(III) center [10].

2.   Addition of one electron to give a high spin Fe(II) species, 2, that subsequently binds O_2 to give the low spin oxygenated species 3.

3.   Addition of a second electron to give species 4, which, from data on model studies [6,11] is written as an Fe(III)-peroxide [although this is formally the same electronically as Fe(I)(O_2), Fe(II)(O_2^-), or Fe(V,($O^{2-})_2$)].

4. Species 4 undergoes electrophilic attack by proton(s) to generate, with loss of H_2O, the putative, high valent oxo species 5, which is written as $Fe^V=O$, although model studies favor a $O=Fe^{IV}$ (porp$^+$), porphyrin cation-radical species [5-7].

5. Species 5 "attacks" substrate S with a resulting net O atom transfer to give S(O) product with regeneration of the starting Fe(III) form of the enzyme. The porphyrin of the P450 enzyme is PpIX as in myoglobin, but, in contrast to this protein that "simply" binds O_2 reversibly (cf. 2 $\rightleftharpoons$ 3), the ligand trans to the bonded O_2 is thiolate of a cysteine amino acid residue (vs. the histidine imidazole in myoglobin).

The P450 enzymes can effect transformations, for example, of saturated hydrocarbons to hydroxylated products, olefins to epoxides, and thioethers to sulfoxides; the mechanistic details of the net O-atom transfer processes from metal-oxo complexes, more generally, are currently subject to intense scrutiny [3,9,12-15].

The dashed lines in Figure 23.1 represent pathways that, in principle, "shunt through" the two one-electron steps via the use of oxygen at the peroxide level (1 → 4), and indeed the enzyme can utilize peroxide [5], and oxygen atom donors (1 → 5). There is a vast oxidation literature on model systems considered to operate by a 1 → 5, metal-oxo type of pathway, using O atom donors (per-acids, alkyl hydroperoxides, amine oxides, PhIO, NaOCl, IO_4^-, etc.), with a diverse range of metal complexes. Obvious models for the P450 system are Fe(porp)X complexes (where porp = a porphyrin dianion and X = an anionic ligand), and these and many other metalloporphyrin complexes have been used as precursors [6-9,15,16] including photochemically activated systems [17], as well as other macrocyclic ligand complexes, polyoxometallate species, phosphine-containing complexes, and other coordination compounds [18-23].

In the P450 systems, the reducing equivalents (eq. 23.1, Fig. 23.1) are provided by NADPH via various coupled reductase systems [8,24]. Model studies using metal complexes, with O_2 as oxidant, require the addition of two electrons (cf. 1 → 2 → 3 → 4) or some other "sacrificial" coreductant. Again, both metalloporphyrin complexes [7,8] and nonporphyrin complexes [25,26] have been used, with the necessary reducing equivalents being provided by, for example, BH_4^-, ascorbic acid, H_2, or a Pt/H_2 electrode [7,8, 25,26]. Such model systems have been studied less intensively than the O atom transfer systems and are complicated by competing redox reactions between O_2 and the added coreductant.

With the aim of mimicking Fe-porphyrin oxygenase systems, studies were initiated some 15 years ago in this department on the second-row Ru-porphyrin analogues. This work included both protein and nonprotein systems, and some catalytic oxidation chemistry has been developed [6,14,27-32]. This chapter updates our studies on O atom transfer from iodosylbenzene (PhIO) using Ru-porphyrins [6], and studies on O_2 oxidations catalyzed by *trans*-Ru(porp)(O)$_2$ complexes, which serendipitously turned out to exhibit *dioxygenase* activity [14,32-34], although they were originally synthesized via a "monooxygenase-type" recipe using an O atom donor (cf 1 → 5, Fig. 23.1). Catalysis via the dioxo species using O_2 is summarized first because of the possible relevance of suggested intermediates to the oxidations using PhIO.

II. O_2-OXIDATIONS CATALYZED BY
   trans-Ru(porp)(O)$_2$ COMPLEXES

A. The Dioxo Species

Incorporation of Ru into a free base porphyrin H_2(porp) is achieved via reaction of the base with $Ru_3(CO)_{12}$, which yields Ru(porp)(CO) complexes, 7 [35]. Reaction of these with O atom donors (cf. 1 → 5, Fig. 23.1) yields μ-oxo species of the type Ru—O—Ru [6,36] unless the porphyrin bases are designed to prevent sterically the formation of such dinuclear species, which are inactive, thermodynamic sinks in Ru-porphyrin-O_2 chemistry. Use of 5,10,15,20-tetramesitylporphyrin (H_2TMP) or 5,10,15,20-tetra(2,6-dichlorophenyl)porphyrin (H_2OCP), however, prevents μ-oxo formation, and treatment of these Ru(porp) (CO) species with *m*-chloroperbenzoic acid gives the *trans*-Ru(porp) (O)$_2$ complexes 6 [14,37,38]; the OCP derivative has been characterized structurally [39]. The CO of the precursor Ru(porp)(CO) perhaps reacts with the peracid and is "burned off" as CO_2, leaving the four-coordinate Ru(porp) to subsequently add two further O atoms; an alternative mechanism could involve initial formation of O=Ru(porp)(CO), labilization of CO, and subsequent addition of a further O atom.

   The surprising formation of the dioxo species proved to be highly significant, because *both* O atoms can be transferred to substrates (see below) and, furthermore, can be replenished using

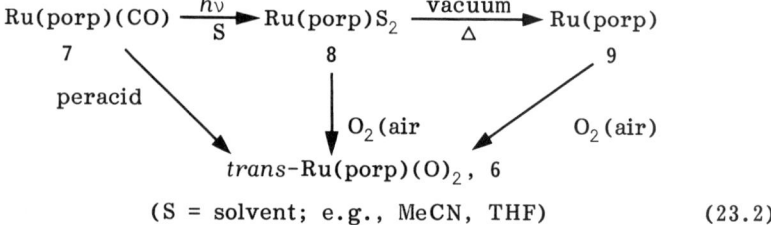

$$(S = solvent;\ e.g.,\ MeCN,\ THF) \qquad (23.2)$$

O_2 or air. Indeed, it soon became evident that 6 is readily formed by all the procedures outlined in equation (23.2) [33,34,40].
It should be noted that the 14-electron, "bare" Ru(TMP), 9, has been isolated and, with N_2, readily forms Ru(TMP)(N_2)$_2$ [40,41]. Reaction of 9 with air, however, preferentially forms 6. For use in practical O_2 oxidations, 6 may be generated from 7 using a stoichiometric amount of the peracid, or from 8 using O_2 (or air) [32,33, 37,40]. The precursor complex Ru(TMP)(MeCN)$_2$ has also been characterized structurally [14,41,42].

Increasing substitution of halogens into the prophyrin ring has been demonstrated to enhance general oxidation activity of metalloporphyrins and has been rationalized in terms of (1) the electron-withdrawing effect increasing the electrophilic reactivity of the oxo intermediate toward the substrate; (2) the removal of electron density from the ring, making the metalloporphyrin less susceptible to attack by a high-valent oxo species (which leads to self-destruction of the catalyst); and (3), related to item 2, an increasing reduction potential within the [M(N) + e $\rightleftharpoons$ M(N - 1)] couples, where N is a relevant oxidation state for the metalloporphyrin complex M [43-45]. Thus, the ortho-substituted halogens of OCP provide favorable electronic effects, as well as the essential steric effect mentioned in Section I.

It should be noted that the dioxo derivatives Ru(OEP)(O)$_2$ (OEP = octaethylporphyrin) and Ru(TPP)(O)$_2$ (TPP = tetraphenylporphyrin dianion) have been made [19], and stoichiometric epoxidation of olefins is effected in solution. In alcohols, the OEP complex is converted to Ru(OEP)O(ROH), which is stabilized against μ-oxo formation by the presence of the alcohol [19]. This species, however, readily reacts with water present to form the catalytically inactive [Ru(OEP)(OH)]$_2$O complex [6,36].

B. Oxidation of Hydrocarbons

The complex Ru(TMP)(O)$_2$, 6a, has been shown under inert atmospheres to transfer in benzene solution both O atoms to olefins in stoichiometric reactions to give the corresponding epoxide. Under O_2 or air, the system becomes catalytic at ambient conditions and, although rather slow (with turnovers of up to 2 h^{-1} at 10 mM Ru using 0.5 M norbornene, cyclooctene, and β-methylstyrenes), shows high stereoselectivity: thus, cis- and trans-β-methylstyrene give the cis- and trans-epoxides, respectively [33]. The suggested catalytic cycle, outlined in equation (23.3), incorporates a key disproportionation of the Ru(IV) monooxo intermediate 10 (see Section C) to 6a and 9 [33], while a plausible route from 9 to 10 is given in equation (23.4) [14,39].

$$\underset{6a}{\underline{Ru}(O)_2} \xrightarrow{\text{alkene}} \underline{Ru}(O) + \text{epoxide} \qquad (23.3)$$

$$\underset{9}{\underline{Ru}} \qquad 1/2\ O_2 \qquad (\underline{Ru} = Ru(TMP))$$

$$\underset{9}{\underline{Ru}(II)} \xrightarrow{O_2} \underline{Ru}O_2 \xrightarrow{Ru(II)} \underline{Ru}(III)O_2\underline{Ru}(III) \longrightarrow 2\underline{Ru}(IV)O$$
$$10$$

$$(23.4)$$

The same TMP system has been used to effect aerobic epoxidation of steroids containing double bonds in the steroid nucleus, and sometimes in the attached side chains as well [46,47]. High conversions to epoxides have been achieved with high β-stereoselectivity for some cholesterol derivatives (e.g., in eq. 23.5, 90% of 11 with R = MeCO is converted to epoxide containing 95% of the β-isomer) [46]. With accompanying unsaturation in the R substituent or in the cyclopentane side chain, regioselective attack within the steroid nucleus at the 5,6-position generally occurs [47].

$$(23.5)$$

$$+ \quad \underset{\sim\sim}{6a} \longrightarrow \text{epoxide}$$

Rate data for oxidation of a series of ring-substituted styrenes by 6a support a mechanism involving attack by an electrophilic metal-oxo moiety [39]. A side-on approach of the alkene, which allows favorable interactions between its filled π-orbitals and the metal-oxygen π^* orbitals, is supported by structural data for Ru(OCP)(CO)(styrene oxide) [13], although mechanisms via intermediates of several other types (metallaoxetane, carbon radical, carbocation, ion pair, and charge transfer) have been proposed for first-row metalloporphyrin monooxo species [9,48].

It should be noted that alkene complexes of the type Ru(TMP) (η^2-alkene) are readily formed via 9; the ethylene complex is readily isolated and is stable to 10^{-3} torr at 20°C, but the cyclohexene complex readily loses the coordinated alkene [49]. Under catalytic oxidation conditions, the dioxo is formed preferentially, and there seems to be little doubt that epoxidation occurs via this species. There is analogy here to catalytic hydrogenation of alkenes, where

both so-called hydride and unsaturate routes have been delineated
[50]. The routes involve prior formation of either a metal-hydride
or a metal-alkene intermediate, respectively; in most cases,
hydrido(alkene) intermediates are eventually formed, although steps
involving hydrogen atom transfer from a metal to noncoordinated
alkene (akin to the net alkene epoxidation) have been established
in a few systems [50].

Toluene solutions of the dioxo complexes (6) are unreactive
toward saturated hydrocarbons such as cyclooctane under thermal
conditions up to 70°C [51] but, of interest, some O-insertion into
saturated C—H bonds has been achieved under photolytic conditions
in the presence of base [39] and with the electrochemically generated,
one-electron oxidized, cation-radical species $Ru(TMP^+)(O)_2$ [39].

C. Oxidation of Tertiary Phosphines and Thioethers

Solutions of $Ru(TMP)(O)_2$ react anaerobically at about 20°C in stages
with PPh_3 according to equation (23.6).

$$\underline{Ru}(O)_2 \xrightarrow[-OPPh_3]{PPh_3} \underline{Ru}(O) \xrightarrow[-OPPh_3]{2PPh_3} \underline{Ru}(PPh_3) \qquad (23.6)$$

$$\text{6a} \qquad\qquad\qquad \text{10} \qquad\qquad\qquad \text{12}$$

Equilibrium titration data imply that 6a is a more potent oxo transfer
agent than 10 [34], this latter species being characterized in solution
[34], while five-coordinate phosphine complexes such as 12 have
been isolated [52]. Species 10 has been observed also en route
from 8 to 6a (eq. 23.2) [34], consistent with the pathways suggested
in equations (23.3) and (23.4). In the presence of excess phosphine
under O_2, the phosphine oxide is produced catalytically [51]; O atom
transfer via 6a can be envisaged (as for alkene epoxidation), although
contributions from a pathway initiated via an outer sphere O_2 oxidation of six-coordinate bis(phosphine) species [29-31] formed in situ
cannot be ruled out.

Detailed kinetic studies by UV/visible spectroscopy have been
carried out on the stoichiometric oxidation of thioethers to sulfoxides
using 6a or 6b (the OCP analogue) in benzene, and interpreted
in terms of equation (23.7) [14,32], $\underline{Ru}$ = Ru(TMP) or Ru(OCP).

$$\underline{Ru}(O)_2 \xrightarrow[k]{SR_2} [\underline{Ru}(O)(OSR_2)] \xrightarrow[\text{fast}]{SR_2} \underline{Ru}(OSR_2)_2 \qquad (23.7)$$

$$\text{6} \qquad\qquad\qquad\qquad\qquad\qquad \text{13}$$

The k rate constants for the TMP systems at 20°C are 0.0075,
0.012, and 0.11 $M^{-1} s^{-1}$, respectively, for Et_2S, $n\text{-}Bu_2S$, and

Me(n-decyl)S, the differences perhaps being reflected more in $\Delta S^{\ddagger}$ than $\Delta H^{\ddagger}$ values [14]. The $\Delta S^{\ddagger}$ values are negative, as expected for an O atom transfer induced by $\nu(Ru=0)$ vibrational motion [53], and this transfer might be more efficient with bulkier substrates Alkylaryl and diaryl sulfides did not react with 6a, possibly because of their decreased nucleophilicity, although steric factors cannot be ruled out. For the oxidation of Et_2S by $Ru(OCP)(O)_2$, k is 0.072 $M^{-1}s^{-1}$ at 20°C, showing that the chlorine substituents favor O atom transfer, presumably by increasing the electrophilicity of the oxo ligand (see Section II.A).

Under O_2, benzene solutions of 6 (at 2-6 mM) effect catalytic oxidation of Et_2S, first to the sulfoxide, and then to the sulfone; the TMP system gives a total of only about 15 turnovers (at 65°C) before complete degradation of the TMP ligand of 6a, while the OCP ligand is not degraded in a catalytic system, even after several hours at 100°C (see Section II.A) [14,32]. The catalytic cycle is thought to occur via equation (23.7), followed by regeneration of 6 when 13, containing labile O-bonded sulfoxide ligands [54], reacts with O_2. Slow conversion of 13 to $\underline{Ru(S(O)Et_2)_2}$, containing the more substitution-inert, S-bonded sulfoxides, also leads to a loss of catalytic activity [14,32]. This ligand isomerization process is much slower for the OCP system; a decrease of electron density at a Ru(II) center would reduce the extent of Ru → S π-back bonding, hence the likelihood of isomerization to the S-bonded form. Catalysis via $Ru(porp)(R_2S)_2$ species in an outer-sphere, O_2 oxidation process [30] seems unlikely because such species are not detected during the catalysis, and also because they are unreactive toward O_2 under the catalytic conditions [32].

The selectivity for oxidation of dialkyl sulfides contrasts with that for an FeCl(TPP)/PhIO system that operates via a supposed "ClFe(TPP)O" intermediate (see Section I); this system effects catalytic formation of sulfoxides from dialkyl, alkylaryl, and diaryl sulfides [55]. If the differences in selectivity between the Ru^{VI}-dioxo system and the formally Fe^{V}-monooxo system are electronic, then the ligand trans to the oxo ligand (transferred as an oxygen atom) probably plays a critical role, as it does in the biologically important oxoiron porphyrin system [5-7]. The kinetic data for reaction (23.7) show, somewhat surprisingly, that $(Et_2SO)Ru(TMP)O$ is a more effective O atom donor than $Ru(TMP)(O)_2$, while the titration data for reaction (23.6) imply that $Ru(TMP)(O)_2$ is more potent than five-coordinate Ru(TMP)O; again the trans axial ligand appears to be critical in governing activity (as seen also in the O atom transfer system to be discussed in Section III, and in the azide porphyrin systems of Ellis and Lyons, Section IV).

The porphyrin dioxo systems show some analogies to the O atom transfer from $[Ru(bipy)_2(py)O]^{2+}$ to Me_2S studied by Meyer's group [53]. The activation parameters are comparable, and isomerization from O- to S-bonded sulfoxide is again observed. The "polypyridine" oxo species can be regenerated electrochemically, but not via O_2 [53].

D. Oxidation of Phenols and Alcohols

Preliminary kinetic and spectroscopic studies [14] have been interpreted in terms of the chemistry outlined in equation (23.8), in which $Ru(TMP)(O)_2$ in benzene reacts overall with 2 mole equivalents of phenol to form, in the absence of O_2, a diamagnetic bis(para-hydroquinone) complex of Ru(II) (14). The selective attack at the para- (vs. ortho-) nucleophilic position of the phenol is presumably governed by steric factors, the TMP ligand allowing only an end-on approach. The second-order rate constant, k', at 20°C is 0.069 $M^{-1}s^{-1}$, in the higher range of values determined for the thioether substrates (see Section II.C).

$$Ru(O)_2 \xrightarrow[k']{PhOH} [Ru(O)(HO\text{—}\langle O \rangle\text{—}OH)] \xrightarrow[fast]{PhOH} Ru(HO\text{—}\langle O \rangle\text{—}OH)_2$$

6a   14

The mechanism of the net oxygen insertion into the C—H bond could be similar to that suggested for the reaction between phenol and $[Ru(bipy)_2(py)O]^{2+}$; this gives both o- and p-hydroquinone derivatives by sequential electron transfer (into the Ru=O moiety from the aromatic ring) and H atom transfer processes [56]. An alternative plausible mechanism involves H atom transfer from the para position of phenol to give Ru(TMP)(O)(OH), followed by inner sphere capture of the intermediate organic radical before it can rearrange or separate in solution, rather than direct insertion into the C—H bond; this latter route corresponds to the "oxygen rebound" mechanism postulated for hydroxylation of saturated hydrocarbons by P450 systems [5,57].

Under O_2, species 14 is rapidly converted to the paramagnetic (S = 1) Ru(IV) complex (15) containing the anion of the hydroquinone, perhaps according to equation (23.9) [14], although involvement of 6a in this oxidation step is possible.

$$14 + \tfrac{1}{2} O_2 \longrightarrow H_2O + Ru\ (O\text{—}\langle O \rangle\text{—}OH)_2,\ 15 \qquad (23.9)$$

The net reaction of 6a with 2 mole equivalents of phenol under O_2 is thus a stoichiometric one, giving 15 and water; catalysis via

6a to give the p-hydroquinone (an industrially important reaction [58]) has yet to be realized.

Benzene solutions of 6a under O_2 react slowly at 20°C with 2-propanol [14]. Preliminary spectroscopic data suggest a non-catalytic conversion of the alcohol to acetone and water, with concomitant loss of 6a; the rate of loss of 6a is some 500 times slower than observed in the phenol reaction under corresponding conditions. The Ru product appears from ^{1}H NMR data to be a paramagnetic, S = 1, Ru^{IV} species (cf. 15) [14], but the axial ligands have not been identified [59].

III. O ATOM TRANSFER REACTIONS CATALYZED BY RUTHENIUM(III) COMPLEXES

In efforts to mimic the shunt pathway 1 → 5 of Figure 23.1, the Ru(III) precursors Ru(porp)Br(PR$_3$) (porp = OEP, TMP; R = Ph or n-Bu) were used in conjunction with PhIO for catalytic epoxidation of olefins and hydroxylation of cyclohexane [6,60]. In the OEP system, an isolated, green catalytic intermediate was tentatively formulated as containing O=Ru^{IV} (porp^{+})Br based on spectroscopic data [6,60], but contamination by phosphine oxide (formed by oxidation of the axial phosphine) prevented isolation of the pure compound. The TMP system was more effective because this porphyrin ligand prevents formation of catalytically inactive μ-oxo species (see Section II.A), and a similarly formulated catalytic species (also green) was assumed. In retrospect, formation of some of the reddish-brown Ru(TMP)(O)$_2$ species during this O atom transfer catalysis cannot be ruled out, but the differences in color qualitatively suggest that the iodosylbenzene systems are quite different from O_2 systems. It should be noted that the corresponding Fe(IV)-oxoporphyrin cation-radical species are also green [61].

More appealing Ru(III) precursors of the type Ru(porp)X (X = Cl,Br) have recently been synthesized from the corresponding Ru(IV) dihalide complexes [62], by the procedure outline in equation (23.10) [63].

$$\underline{Ru}X_2 \xrightarrow{NH_3} \underline{Ru}X(NH_3) \xrightarrow{H^+} \underline{Ru}X \; ; \; \underline{Ru} = Ru(porp) \quad (23.10)$$

Surprisingly, use of Ru(OEP)Br with PhIO under the conditions used earlier with the Ru(OEP)Br(PPh$_3$) complex [6,60] is not effective for oxidation of cyclohexene [63]. However, addition of OPPh$_3$ to Ru(OEP)Br does regenerate the activity of the Ru(OEP)Br(PPh$_3$) system (in which OPPh$_3$ is generated in situ) [64]. The implication is that phosphine oxide plays a role in the catalysis, possibly by binding as an axial ligand (trans to an oxo group?) and/or preventing formation of inactive μ-oxo species. Once again, the critical nature of the axial ligand is alluded to.

IV. O_2 OXIDATIONS CATALYZED BY FIRST-ROW TRANSITION METAL PORPHYRINS

There have been reports on the use of metalloporphyrins and O_2 in the presence of a coreductant for catalytic epoxidation of olefins in monooxygenase-type activity [7,8] (see Section I), but it is only very recently that the use of O_2 (or air) alone has been used successfully with metalloporphyrins to generate what is apparently dioxygenase-type catalysis (no coreductants are needed, and no H_2O is formed; see eq. 23.1). In a series of short communications [43,44,65], Ellis and Lyons first reported on the use of benzene solutions of five-coordinate Cr(III), Mn(III), and Fe(III) porphyrins at 27–80°C and 100 psig O_2 for catalytic oxidation of isobutane to t-BuOH (with up to 97% selectivity vs. minor amounts of acetone and CO_2 cleavage products).

Within the tetraphenylporphyrin complexes M(TPP)X, activity (up to 264 turnovers over 6 h) was observed, but only when the axial anionic ligand X was N_3^-, although the role of the ligand is unclear; the azide complexes do decompose to catalytically inactive nitride species [65]. Replacement of the pyrrole hydrogens by Br, and/or replacement of the phenyl hydrogens with Cl and/or F, led to markedly increased activity (up to 2245 turnovers), and in the halogenated systems the azide ligand was not so critical; indeed, the optimum catalyst was the tetrakis(pentafluorophenyl) complex, Fe(TPPF)OH, 16, and of interest the corresponding μ-oxo species [Fe(TPPF)]$_2$O, 17, was almost as active [43,44]. Quite remarkable is that use of 16 in neat isobutane, at room temperature with N_2-diluted air, gave 12,150 turnovers producing, t-BuOH in 95% selectivity, with no catalyst deactivation over the 5-day reaction time [43]. Of interest, Ru porphyrin complexes are not active catalysts for the isobutane oxidations [66].

Use of 16 and 17 in benzene or acetonitrile also effects catalytically the more difficult oxidation of propane, at 100–150°C under 1000 psig of air, to give mixtures of i-PrOH and acetone; the maximum turnover achieved was 870, before oxidative degradation of the fluorinated ligand was apparent [43,44].

A suggested pathway for the alkane oxidations is shown in Figure 23.2. As in Figure 23.1 for the monooxygenase system, an initial one-electron reduction yields Fe(II), 18, which now reacts with O_2 to yield 20 via 19 (cf. eq. 23.4), chemistry that has been

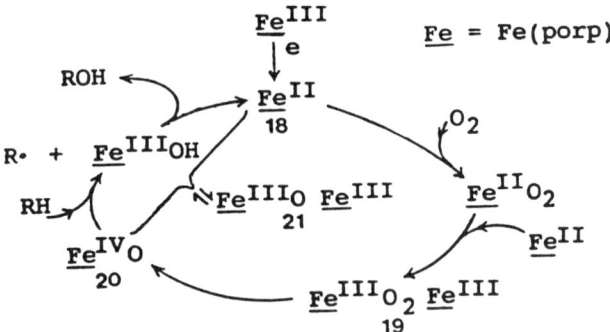

Figure 23.2   Pathway for O_2 oxidation of alkanes.

demonstrated for TPP and TMP systems at *low temperatures* [67].
At room temperature conditions, the usual O_2 oxidation products
of nonsterically hindered Fe(II)-porphyrins are the μ-oxo species
21, formed via reaction of 20 with 18. Ellis and Lyons speculate
[43] that a sufficient concentration of the active Fe^{IV}=O species
20 might be maintained in the halogenated TPP systems, via the
equilibrium shown. The source of the required initial electron has
not been identified, but suggested possibilities were adventitious
impurity or the alkane. In situ UV-visible spectral data indicated
the presence of 18 and 21 in solution. The catalyst 20 operates
at an oxidation level one below that of the monooxygenase systems
(5 in Fig. 23.1).

V.   CONCLUDING REMARKS

Biomimetic studies using Ru porphyrin complexes, in attempts to
model oxygenase systems that function via Fe porphyrin moieties,
have led to systems that effect (under mild conditions) selective
O_2 oxidations of a range of substrates, which include olefins, saturated hydrocarbons, thioethers, phenols, and alcohols. The systems
operate via well-characterized Ru(VI) dioxo species and are amenable
to detailed kinetic and mechanistic studies. This Ru porphyrin chemistry, coupled with recent reports of less well-defined dioxygenase-type
activity from some first-row transition metal porphyrin complexes
in O_2 oxidation of isobutane and propane, suggests that the scene
is set for some major developments in catalyzed, selective O_2 oxidations.

A drawback for possible commercial use of protein-free metalloporphyrins generally is their self-destruction by oxidative degradation;

extensive halogen substitution into the porphyrin ligand can circumvent this problem, but such ligands are expensive. Nevertheless, the fundamental chemistry learned from the metalloporphyrin systems should aid in the development of the catalytic oxidation chemistry of the more commercially attractive heteropolyacid systems (considered as oxidatively resistant inorganic analogues of metalloporphyrins) and zeolites [18].

ACKNOWLEDGMENTS

I thank sincerely the students (C. Alexander, A. Pacheco, M. Ke, N. Rajapakse, C. Sishta) and postdoctorals (M. J. Camenzind, T. W. Leung), the last five of these being jointly supervised by D. Dolphin, and the X-ray crystallographic groups at the University of British Columbia (J. Trotter, S. J. Rettig) and at Northwestern University (J. A. Ibers, J. W. Sparapany), who all contributed to the studies reported here. The Natural Sciences and Engineering Research Council of Canada (B.R.J., D.D.) and the U.S. National Institutes of Health (grant AM17989 to D.D.) provided financial support, and Johnson Matthey Ltd. the loan of the ruthenium.

REFERENCES

1. Collins, T. J., ed., *Report of the International Workshop on Activation of Dioxygen Species and Homogeneous Catalytic Oxidations*, Galzignano, Italy, 1984.
2. Ingraham, L. L., and Meyer, D. L., *Biochemistry of Dioxygen*, Plenum Press, New York, 1985.
3. Martell, A. E., and Sawyer, D. T., eds., *Oxygen Complexes and Oxygen Activation by Transition Metals*, Plenum Press, New York, 1988.
4. Parshall, G. W., *Homogeneous Catalysis*, Wiley, New York, 1980, Chap. 10.
5. Groves, J. T., and McMurry, T. J., in *Cytochrome P-450: Structure, Mechanism, and Biochemistry* (P. Ortiz de Montellano, ed.), Plenum Press, New York, 1986, Chap. 1.
6. James, B. R., in *Fundamental Research in Homogeneous Catalysis*, Vol. 1 (A. E. Shilov, ed.), Gordon & Breach, New York, 1986, p. 309.
7. Mansuy, D., *Pure Appl. Chem.*, *59*, 759 (1987).
8. Tabushi, I., *Coord. Chem. Rev.*, *86*, 1 (1988).
9. Bruice, T. C., *Aldrichim. Acta*, *21*, 87 (1988).
10. Poulos, T. L., Finzel, B. C., Gunzalus, I. C., Wagner, G. C., and Kraut, J., *J. Biol. Chem.*, *260*, 16122 (1985).

11. Mashiko, T., Reed, C. A., Haller, K. J., and Scheidt, W. R., *Inorg. Chem.*, *23*, 3192 (1984).
12. Holm, R., *Chem. Rev.*, *87*, 1401 (1987).
13. Groves, J. T., Han, Y., and Engen, D. V., *J. Chem. Soc.*, *Chem. Commun.*, 436 (1990).
14. Rajapaske, N., James, B. R., and Dolphin, D., *Stud. Surface Sci. Catal.*, *55*, 109 (1990).
15. Preprints, *Symposium on Oxygen Activation in Catalysis*, Division of Petroleum Chemistry, American Chemical Society, Boston, 1990.
16. (a) Meunier, B., *Bull. Soc. Chim. Fr.*, 578 (1986). (b) Amatsu, H., Miyamoto, T. K., and Sasaki, Y., *Bull. Chem. Soc. Jpn.*, *61*, 3193 (1988). (c) Balasubramanian, P. N., Lindsay-Smith, J. R., Davies, M. J., Kaaret, T. G., and Bruice, T. C., *J. Am. Chem. Soc.*, *111*, 1477 (1989).
17. Suslick, K. S., and Watson, R. A., in ref. 15, p. 169.
18. Sheldon, R., *Stud. Surface Sci. Catal.*, *55*, 1 (1990).
19. Che, C.-M., Ho, C., Lee, W.-O., and Lau, T.-C., in ref. 15, p. 179; Leung, W.-H., and Che, C.-M., *J. Am. Chem. Soc.*, *111*, 8812 (1989).
20. Taqui Khan, M. M., and Shukla, R. S., *J. Mol. Catal.*, *58*, 405 (1990).
21. Bresson, M., and Morvillo, A., *Stud. Surface Sci. Catal.*, *55*, 119 (1990); *Inorg. Chem.*, 950 (1989).
22. Faraj, M., and Hill, C. L., *J. Chem. Soc.*, *Chem. Commun.*, 1487 (1987).
23. Griffith, W. P., and Ley, S. V., *Aldrichim. Acta*, *23*, 13 (1990).
24. Coon, M. J., and White, R. E., in *Metal Ion Activation of Dioxygen* (T. G. Spiro, ed.), Wiley, New York, 1980, p. 73.
25. Gamage, S. N., and James, B. R., *J. Chem. Soc.*, *Chem. Commun.*, 1624 (1989).
26. Barton, D. H. R., *Aldrichim. Acta*, *23*, 3 (1990).
27. James, B. R., Addison, A. W., Cairns, M., Dolphin, D., Farrell, N. P., Paulson, D. R., and Walker, S., in *Fundamental Research in Homogeneous Catalysis*, Vol. 3 (M. Tsutsui, ed.), Plenum Press, New York, 1979, p. 751.
28. Paulson, D. R., Addison, A. W., Dolphin, D., and James, B. R., *J. Biol. Chem.*, *254*, 7002 (1979).
29. James, B. R., Pacheco, A., Rettig, S. J., Thorburn, I. S., Ball, R. G., and Ibers, J. A., *J. Mol. Catal.*, *44*, 147 (1987).
30. James, B. R., Pacheco, A., Rettig, S. J., and Ibers, J. A., *Inorg. Chem.*, *27*, 2414 (1988).
31. James, B. R., Mikkelsen, S. R., Leung, T. W., Williams, G. M., and Wong, R., *Inorg. Chim. Acta*, *85*, 209 (1984).
32. Rajapakse, N., James, B. R., and Dolphin, D., *Catal. Lett.*, *2*, 219 (1989).

33. Groves, J. T., and Quinn, R., *J. Am. Chem. Soc.*, 107, 5790 (1985).
34. Groves, J. T., and Ahn, K.-H., *Inorg. Chem.*, 26, 3831 (1987).
35. Rillema, D. P., Nagle, J. K., Barringer, L. F., and Meyer, T. J., *J. Am. Chem. Soc.*, 103, 56 (1981).
36. Collman, J. P., Barnes, C. E., Brothers, P. J., Collins, T. J., Ozawa, T., Galluci, J. C., and Ibers, J. A., *J. Am. Chem. Soc.*, 106, 5151 (1984).
37. Groves, J. T., and Quinn, R., *Inorg. Chem.*, 23, 3844 (1984).
38. Camenzind, M. J., James, B. R., and Dolphin, D., unpublished results, October 1984.
39. Groves, J. T., 199th ACS National Meeting, Boston, 1990, Abstract PETR 60.
40. Camenzind, M. J., James, B. R., and Dolphin, D., *J. Chem. Soc., Chem. Commun.*, 1137 (1986).
41. Camenzind, M. J., James, B. R., Dolphin, D., Sparapany, J. W., and Ibers, J. A., *Inorg. Chem.*, 27, 3054 (1988).
42. Camenzind, M. J., Rettig, S. J., James, B. R., and Dolphin, D., to be published.
43. Ellis, P. E., and Lyons, J. E., in ref. 15, p. 174; *Catal. Lett.*, 3, 389 (1989).
44. Ellis, P. E., and Lyons, J. E., *J. Chem. Soc., Chem. Commun.*, 1189, 1315 (1989).
45. Traylor, T. J., and Tsuchiya, S., *Inorg. Chem.*, 26, 1338 (1987).
46. Marchon, J. C., and Ramasseul, R., *J. Chem. Soc., Chem. Commun.*, 298 (1988); *J. Mol. Catal.*, 51, 29 (1989).
47. Tavares, M., Ramasseul, R., and Marchon, J.-C., *Catal. Lett.*, 4, 163 (1990).
48. Ostović, D., and Bruice, T. C., *J. Am. Chem. Soc.*, 111, 6511 (1989).
49. Rajapakse, N., James, B. R., and Dolphin, D., *Can. J. Chem.*, 68, 2274 (1990).
50. James, B. R., in *Comprehensive Organometallic Chemistry*, Vol. 8 (G. Wilkinson, F. G. A. Stone, and E. W. Abel, eds.), Pergamon Press, Oxford, 1982, Chap. 51.
51. Rajapakse, N., and James, B. R., unpublished observations.
52. Sishta, C., Camenzind, M. J., James, B. R., and Dolphin, D., *Inorg. Chem.*, 26, 1181 (1987).
53. Roecker, L., Dobson, J. C., Vining, W. J., and Meyer, T. J., *Inorg. Chem.*, 26, 779 (1987).
54. Davies, J. A., *Adv. Inorg. Chem. Radiochem.*, 24, 115 (1981).
55. Ando, W., Tajima, R., and Takata, T., *Tetrahedron Lett.*, 23, 1685 (1982).
56. Seok, W. K., Dobson, J. C., and Meyer, T. J., *Inorg. Chem.*, 27, 3 (1988).

57. Groves, J. T., *J. Chem. Educ.*, *62*, 928 (1985).
58. Romano, U., Esposito, A., Maspero, F., and Clerici, M. G., *Stud. Surface Sci. Catal.*, *55*, 33 (1990); B. Notari, *Stud. Surface Sci. Catal.*, *37*, 413 (1988).
59. Rajapakse, N., Ph.D. dissertation, University of British Columbia, Vancouver, 1990.
60. Leung, T., James, B. R., and Dolphin D., *Inorg. Chim. Acta*, *79(B7)*, 180 (1983).
61. Balch, A. L., Latos-Grazynski, L., and Renner, M. W., *J. Am. Chem. Soc.*, *107*, 2983 (1985).
62. Sishta, C., Ke, M., James, B. R., and Dolphin, D., *J. Chem. Soc., Chem. Commun.*, 787 (1986).
63. Sishta, C., Ph.D. dissertation, University of British Columbia, Vancouver, 1990.
64. Alexander, C., and James, B. R., work in progress.
65. Ellis, P. E., and Lyons, J. E., *J. Chem. Soc., Chem. Commun.*, 1187 (1989).
66. Lyons, J. E., private communication.
67. Balch, A. L., Chan, Y. W., Cheng, R. J., LaMar, G. N., Latos-Grazynski, L., and Renner, M. W., *J. Am. Chem. Soc.*, *106*, 7779 (1984).

24

The Use of Pd-Pt-Bi Containing Catalysts Supported on Activated Carbon for the Selective Oxidation of Glucose to Gluconic Acid

K. Deller and B. Despeyroux

Degussa AG, Hanau, Germany

I. INTRODUCTION

A. Catalysts Already Known for Gluconate Formation

To compete economically with the industrial fermentative synthesis of gluconic acid, catalysts with high activity, high selectivity, and high stability are desirable. Different catalysts are known already, most of them based on palladium on activated carbon. The use of bismuth leads to the selective formation of gluconic acid from glucose, especially under alkaline reaction conditions [1,2].

B. Behavior of Platinum on Activated Carbon Catalysts

Pd and Pt behave quite differently during the oxidation of glucose to gluconic acid. Under low catalyst concentration with Pt on activated carbon (catalyst-to-substrate ratio ≈ 1.5 wt %) the oxidation of glucose leads to lower yields of gluconic acid than are found using Pd on activated carbon catalysts. At high concentrations of platinum on activated carbon (catalyst-to-substrate ratio ≈ 20 wt %) the oxidation of glucose produces glucaric acid (through oxidation of the 1,6-position of the glucose molecule chain) at longer reaction times.

II. EXPERIMENTAL

A. Materials

The following Degussa catalysts were used:

CEF   196 RA/W 4% Pd, 1% Pt, 5% Bi
CF   196 RA/W 5% Pt, 5% Bi
CE   196 RA/W 5% Pd, 5% Bi
F   196 RA/W 5% Pt

Glucose as glucose monohydrate from Fluka ("p.a." quality) resp.
Riedel de Haen ("rein" quality) was used as received. Pure O_2 was
used.

B. Catalyst Preparation

An active carbon powder having a Brunauer–Emmett–Teller (BET)
surface area of 1000 m^2/g, with high macropore content, was prepared using a solution of Bi_2O_3 (dissolved in concentrated HCl)
and/or hexachloroplatinic acid and/or palladium(II) chloride, which
was added to an aqueous suspension of activated carbon. Coprecipitation by treatment with NaOH and final reduction leads to the
desired metallic phase. The catalyst suspension was filtered and
washed. The catalyst was used in the glucose oxidation without
further treatment.

C. Standard Reaction Conditions

The following standard reaction conditions were maintained.

Glucose	16 g ($\approx$ 17.6 g of glucose monohydrate) dissolved in 100 mL of water
Temperature	55°C
O_2 pressure	10 mbar
pH	10.0
NaOH solution	10 wt % for neutralization
Stirrer rate	1800 rpm
Catalyst conc.	0.24 g (1.5 wt % based on glucose)

III. RESULTS AND DISCUSSION

A. Catalyst System Pd-Pt-Bi on Activated Carbon

Table 24.1 gives the formation of all reaction products as a function
of time for the trimetallic catalyst Pd-Pt-Bi on activated carbon
[3].
    The influence of the temperature and pH of the reaction as
well as the metal concentrations used in the trimetallic catalyst
was investigated. It was found that a metal content of 4% Pd, 1%
Pt, and 5% Bi gives the best results. Activity values higher than

Table 24.1 Pd-Bi on Activated Carbon Catalyst: Glucose Oxidation Under Standard Reaction Conditions with Degussa Catalyst CE 196 RA/W 4% Pd, 1% Pt, 5% Bi

Reaction product	Amount formed (mol $\times 10^{-2}$) at:			
	18 min	20 min	25 min	30 min
Glucose	0.15	< 0.01	< 0.01	< 0.01
Gluconic acid	8.50	8.58	8.44	8.13
Fructose	0.08	0.13	0.13	0.13
Glucaric acid	0.03	0.05	0.09	0.32
Tartaric acid	< 0.01	< 0.01	0.05	0.07
Tartronic acid	< 0.01	< 0.01	0.09	0.17
Oxalic acid	< 0.01	0.01	0.06	0.14

Conversion (20 min):   100%
Selectivity (20 min):   98%
Activity (20 min):   4200 g[gluconic acid]/g[precious metal] $\times$ h

4000 g [gluconic acid]/g[precious metal] $\times$ h can be obtained. The optimized reaction conditions already given were used as a standard to compare different catalyst systems. Figure 24.1 shows the rate of formation of gluconic acid as a function of time using a trimetallic Pd-Pt-Bi on activated carbon catalyst under standard reaction conditions in comparison to other catalyst systems:

5% Pt on activated carbon (catalyst F 196 RA/W)
5% Pt, 5% Bi on activated carbon (catalyst CF 196 RA/W)
5% Pd, 5% Bi on activated carbon (catalyst CE 196 RA/W)

It is evident that the use of Pt can boost the activity of Pd-Bi on activated carbon catalyst without exerting an influence on the selectivity.

B. Catalyst Characterization of Catalyst CEF 196 RA/W
   4% Pd, 1% Pt, 5% Bi

The analysis of the metallic phase with energy dispersive X-ray (EDX) analysis shows that the catalyst particles are totally impregnated and the metal homogeneously dispersed throughout the catalyst particles. The metal dispersion on the catalyst surface is low (measured by CO adsorption) and is comparable to values of other Pd/Pt bimetallic catalysts without Bi.

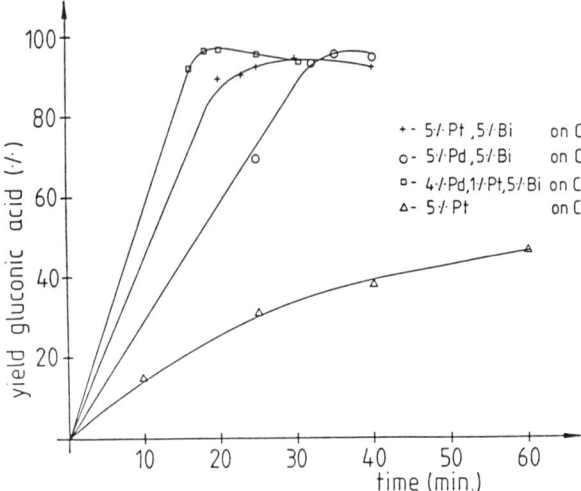

Figure 24.1 Yield of gluconic acid obtained for the different catalyst systems used in the oxidation of glucose under standard reaction conditions.

The crystallite size was measured by transmission electron microscopy and revealed well-crystallized, rod-shaped particles of Bi besides Pt-Pd agglomerates of about 2–5 nm size (comparable to crystallite size of Pd/Pt bimetallic catalysts without Bi). Spectroscopic (ESCA/SIMS) investigations demonstrated that under optimized preparation conditions the Pd phase is still mainly oxidized, whereas the Pt phase is mainly reduced. The Bi phase was found to be in the oxide form as Bi_2O_3 and $Bi_2O_2CO_3$. The last compound could be interpreted as an interaction of Bi with the support leading to the carbonate formation. Pure Bi on activated carbon catalyst (without precious metal) is totally inactive in this reaction. The presence of precious metal (Pd or Pt) is necessary.

REFERENCES

1. European Patent EP 233816 (Jan. 30, 1986); *Chem. Abstr.*, *108*(21), 187206k; European Patent EP 232202 (Jan. 30, 1986); *Chem. Abstr.*, *108*(21), 187205j.
2. European Patent EP 142725, to Kao Corporation (May 29, 1985); *Chem. Abstr.*, *103*(23), 196366m.

3. Part of this work was presented at the 1st International Symposium on New Developments in Selective Oxidation, Rimini, Italy, Sept. 18-22, 1989, and at the 5th Brazilian Catalysis Seminar, Guaruja City, Brazil, Sept. 13-15, 1989.

25

Enzymatic Catalysis in Organic Media: Prospects for the Chemical Industry

Jonathan S. Dordick, Damodar R. Patil, Sanghamitra Parida, Keungarp Ryu, and David G. Rethwisch

Department of Chemical and Biochemical Engineering, University of Iowa, Iowa City, Iowa

I. INTRODUCTION

Nature is extremely diverse in terms of the large number and many types of organic molecules required for life. This diversity is possible solely due to the wide catalytic scope of enzymes. It is the ability to harness the catalytic power of enzymes and use it for the synthesis of commercially important products that represents the core technology of applied biocatalysis. Enzymes are highly selective catalysts, which operate typically under mild reaction conditions (e.g., ambient temperatures and pressures, neutral solutions, etc.). Such properties have enabled enzymes to become valuable catalysts in the food, beverage, and diagnostics industries. These same properties make enzymes potentially attractive in synthetic chemistry, especially in the synthesis of pharmaceuticals, chiral intermediates, specialty polymers, and biochemicals.

The chemical industry, however, has been slow to employ enzymes. Perhaps the most significant reason for this has been the strict adherence by both chemists and biochemists to the conventional notion that enzymes function only in aqueous solutions. Indeed, it is stated in virtually every biochemistry textbook that enzymes are aqueous-based substances, requiring water for activity and that organic solvents, with few exceptions, only serve to destroy enzyme function. This is unfortunate because water, while an ideal solvent for the predominantly polar species required for life (e.g., amino and nucleic acids, carbohydrates, proteins, cofactors), is a poor solvent for nearly all applications in industrial chemistry. Most organic compounds of commercial interest are very sparingly soluble and often unstable in aqueous solutions. Hence, industrial chemistry is based on organic solvent and has mostly ignored the

Table 25.1   Potential Advantages of Employing Enzymes in
Organic Media

---

1. Increased solubility of nonpolar substrates.
2. Shifting of thermodynamic equilibria to favor synthesis over
   hydrolysis.
3. Suppression of water-dependent side reactions (e.g., hydrolysis of acid anhydrides and halides, polymerization of quinones).
4. Alteration in substrate specificity.
5. Immobilization is often unnecessary, since enzymes are insoluble
   in organic solvents.
6. Enzymes may be recovered by simple filtration or centrifugation.
7. If immobilization is required for optimal flow considerations,
   then simple adsorption onto nonporous surfaces (e.g., glass
   beads) is satisfactory. Enzymes are unable to desorb from these
   surfaces in nonaqueous media.
8. Ease of product recovery from low boiling, high vapor pressure
   solvents.
9. Enzymes exhibit enhanced thermostability.
10. Elimination of microbial contamination.
11. Potential of enzymes to be used directly in a chemical process.

Source: Ref. 3.

potential benefits of highly selective enzymic catalysts. This situation
is quickly changing. The application of enzymes in nonaqueous
media has enabled biocatalysts to compete successfully with traditional
chemical catalysts in a variety of synthetic processes. The notion
that enzymes are active only in water is fast becoming obsolete.

    Indeed, it is ironic that many enzymic processes in the cell
do not take place in a true aqueous environment and moreover would
not be optimal in such environments. Many enzymes or multienzyme
complexes, including lipases, esterases, dehydrogenases, and cyto-
chromes, function in natural hydrophobic environments, usually
in the presence of or immobilized to a membrane [1,2]. The water
activity under these conditions is significantly less than unity,
and the concentration of water is well below the 55 M value found
in aqueous solutions.

    From a biotechnological standpoint, there are numerous potential
advantages in employing enzymes in organic media (Table 25.1) [3,4].
These advantages are attractive in selective chemical synthesis,
and a number of approaches using enzymes in nonaqueous environments have been conceived. For example, water-miscible organic
solvents in relatively low concentrations (e.g., < 30% v/v) have

been used to increase the solubility of hydrophobic substrates [5]. Soluble enzymes have also been used in the aqueous components of both biphasic aqueous-organic systems [6-8] and inside reversed micelles dissolved in nonpolar solvents [9,10]. The enzyme in these three solvent designs, however, resides in a predominantly aqueous environment.

Several of the properties described in Table 25.1 are impossible to obtain unless the bulk water phase is eliminated. Such a situation exists using enzymes in monophasic organic solvents [3]. These systems are defined as those that lack a distinct aqueous phase with an insoluble enzyme catalyst (enzymes are insoluble in nearly all organic solvents) suspended in nearly anhydrous (no added water) media or water-miscible cosolvents employing the organic solvent as the predominant system component. Unlike the other solvent reaction systems, enzymes in monophasic solvents are not in direct contact with a bulk aqueous phase and, therefore, all the potential advantages listed in Table 25.1 have been realized. In this chapter we focus on the application of enzymes in organic media and the prospects for specific application in the chemical industry. Section II discusses a commercially important example of enzymes in organic synthesis, and Section III covers the use of enzymes in polymer synthesis and the control of such synthetic schemes.

II. ENZYMES IN ORGANIC SYNTHESIS

The use of enzymes as selective catalysts in organic synthesis is well established [11]. Enzymes are primarily used when high regio- or stereoselectivity is desired and respectable turnover numbers are needed. Many advances have been made in the areas of food additives (flavors, colorants, sweeteners, etc.), amino acid modifications, and pharmaceutical synthesis [12,13]. The use of enzymes in organic solvents has significantly extended the application of enzymes in this area, particularly when combined with an existing chemical process. One such application developed in our laboratory is the combined enzymochemical synthesis of chiral 1,2-diols. These diols are useful synthons for the production of optically active polymers, antibiotics and other pharmaceuticals, and food additives [14,15].

The preparation of 1,2-diols can be carried out chemically (e.g., base-catalyzed aldol condensation of formaldehyde followed by reduction; catalytic hydrogenation of hydroxyesters; alkene epoxidation followed by hydrolysis [16]). Nearly all commercial uses of the most common types of 1,2-diol (1,2-propanediol, 1,2-butanediol,

1,2-hexanediol, etc.) employ the racemic mixture. For specialty applications, however, when a single isomer is desired, chemical methods are inadequate. Several biological alternatives have been developed, including microbial fermentation and enzymatic resolution. For example, several obligate anaerobes are known to produce 1,2-propanediol in high yields from glucose [17]. Unfortunately, the breadth of products obtained is not large, and the reaction productivities are low as a result of the formation of dilute product solutions in water. Both factors limit the general commercial applicability of the microbial approach.

The enzymatic resolution of racemic 1,2-diols has been performed in nonaqueous media with lipases [18]. For example, porcine pancreatic lipase catalyzes the efficient acylation of 1,2-propanediol in ethyl acetate. Only the primary hydroxyl group is acylated via transesterification. The chirality introduced by the secondary hydroxyl group determines the isomer that can react with the enzyme. Once the chiral primary ester has been formed, simple base hydrolysis to the optically active 1,2-propanediol is achieved.

We have developed an alternative approach to the synthesis of optically active 1,2-diols that involves combined enzymatic and chemical processing. The two-step process employs racemic α-hydroxy acids as starting materials. The enzymatic step involves an initial acylation in organic media of the racemic acid to give an optically active α-hydroxyester. This is followed by a reduction with $LiAlH_4$ or $LiAl(OCH_3)_3H$ to give the optically active 1,2-diol. The approach is summarized in Figure 25.1.

A variety of commercially available lipases are known to catalyze the esterification of carboxylic acids in organic solvents in the

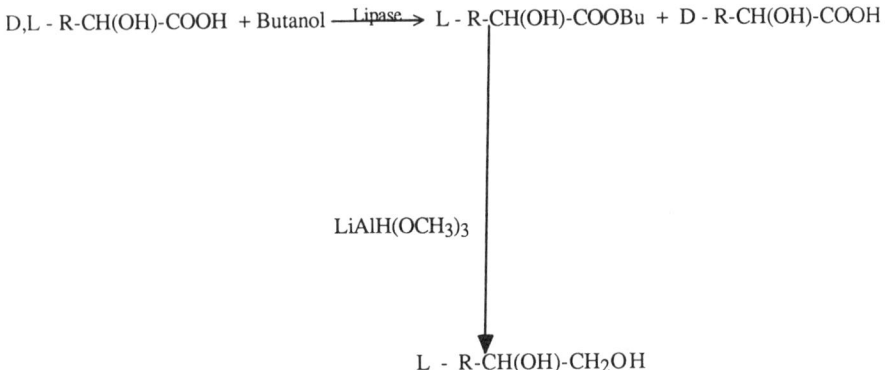

Figure 25.1 Synthesis of optically active 1,2-diols.

Table 25.2 Screen of Enzymes for α-Hydroxy Acid Esterification in Toluene[a]

Enzyme	Conversion after 36 h (%)
Lipase from *Candida cylindracea* (Sigma)	50
Lipase from *C. cylindracea* (Amano AY)	50
Lipase (Enzeco K 16824)	51
Lipozyme (Novo)	52
Lipase from *Pseudomonas* sp. (Amano P)	45
Fungal lipase (Amano FAP-15)	3
Lipase from *Mucor* sp. (Amano MAP-10)	34
Lipase from *Geotrichum* sp. (Amano GC-20)	4
Lipase from *Aspergillus niger* (Amano APF)	4
Subtilisin Carlsberg (Sigma)	0.3
Lipase from *Aspergillus niger* (Amano AP)	0.3
Lipase from *Penicillium* sp. (Amano G)	12
Protease from *Bacillus* sp. (Amano N)	0.7
Chymotrypsin (Sigma)	0.6
Cellulase from *T. viridae*	0.4
Amylase from *Aspergillus oryzae* (Amano Biozyme S)	0.5
Amylase from *B. subtilis* (Amano CONC)	0.2
Hemicellulase from *A. niger* (Amano)	0.6
Amylase from *Rhizopus oryzae* (Amano AG-975)	2
Amylase from *B. subtilis* (rapidase from Gist-Brocades)	1
Alkaline protease from *Bacillus* sp. (Amano proleather)	0.4

[a]Conditions: 0.1 M α-hydroxycaproic acid dissolved in 10 mL of toluene containing 0.6 M *n*-butanol and 0.2 g per milliliter of enzyme, shaken at 200 rpm at 30°C.

presence of a suitable alcohol [19]. As a model system, we chose the esterification of α-hydroxycaproic acid in toluene in the presence of *n*-butanol. Chemical reduction of the resulting optically active ester is expected to yield chiral 1,2-hexanediol. Our initial studies were aimed at screening a number of commercially available hydrolytic enzymes for activities of α-hydroxycaproic acid esterification in toluene. To that end, 0.1 M α-hydroxycaproic acid was dissolved in toluene containing 0.6 M *n*-butanol. The reactions were initiated by adding 0.2 g per milliliter of enzyme and the suspensions (enzymes are insoluble in nearly all organic solvents) shaken at 30°C at 200 rpm. Out of 21 enzymes tested, only 7 showed significant esterification

activity, and 2 lipases from *Candida cylindracea* were the most
active (Table 25.2).

The time course of α-hydroxycaproic acid esterification using
a *Candida* enzyme obtained from Sigma is shown in Figure 25.2.
Within 8 hours, 38% of the organic acid was esterified to the butyl
ester (same conditions as aforementioned). The resulting ester was
purified by extraction of the unreacted organic acid and evaporation
of the toluene and residual butanol. The isolated yield of ester
was 32% (2.2 g). Optical polarographic analysis of the ester gave
a $[\alpha]_D^{25}$ of -6.8(c0.22, $CHCl_3$).

The ester (2.2 g) was then dissolved in 0.5 L of diethyl ether
and 6.0 g of $LiAl(OCH_3)_3H$ was added. After 1 hour, the solution
was filtered and extracted with water to remove excess reductant
and the organic phase was dried, yielding 0.83 g (96% yield) *S*-
1,2-hexanediol $\{[\alpha]_D^{25} = -15.6$ (c0.5 ethanol}, with an enantiomeric
excess greater than 99%.

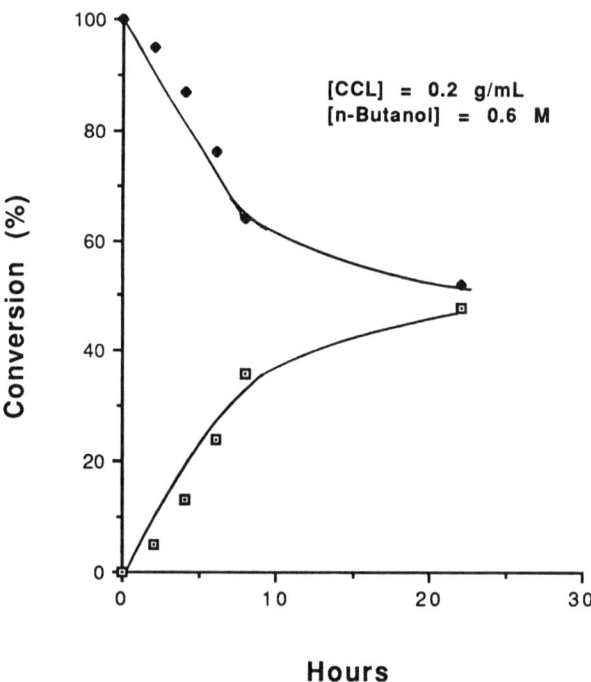

Figure 25.2 Time course of α-hydroxycaproic acid esterification
in toluene.

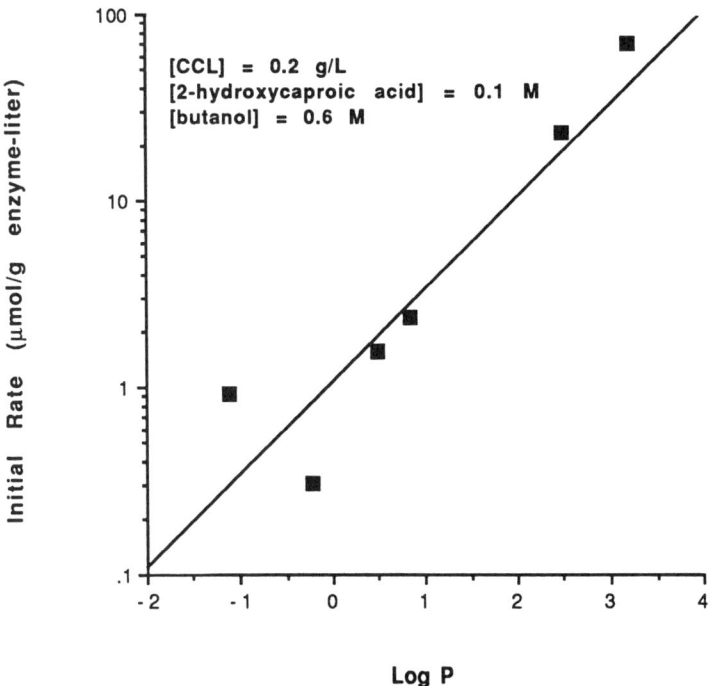

Figure 25.3   Effect of solvent hydrophobicity on lipase-catalyzed esterification of 2-hydroxycaprioic acid.

A variety of organic solvents were capable of sustaining lipase-catalyzed esterification of α-hydroxycaproic acid (Fig. 25.3). The highest enzyme activities were observed in the more hydrophobic solvents; this is not surprising insofar as hydrophobic solvents are less likely to distort the native enzyme structure [20]. A roughly logarithmic relationship exists between initial rates of enzymic catalysis and solvent hydrophobicity as represented by log P (a partitioning parameter of a given solvent between 1-octanol and water). Such a relationship allows for predictions of catalytic function in organic solvents and is useful as an approximating tool for lipase-catalyzed esterifications.

In the case of cyclohexane and toluene, reactions were terminated at approximately 45% conversion to optimize recovery of the optically pure ester. In addition to α-hydroxycaproic acid, several other α-hydroxy acids could be esterified by this approach, including lactic acid, 2-hydroxyisocaproic acid, 2-hydroxy-4-methylbutyric acid, 3-phenyllactic acid, and mandelic acid, resulting in the

Table 25.3 Combined Enzymochemical Synthesis of Optically Active 1,2-Diols[a]

Acid substrate	Reaction time (h)	Overall yield [% (g)]	Product	Enantiomeric excess (%)
Lactic acid	8	31 (1.2)	S-1,2-Propanediol	98
2-Hydroxyisocaproic acid	63	86 (0.6)	S-4-Methyl-1,2-pentanediol	89
2-Hydroxycaproic acid	8	65 (0.83)	S-1,2-Hexanediol	99
3-Phenyllactic acid	192	78 (0.9)	S-3-Phenyl-1,2-propanediol	75
3-Methyl-2-hydroxybutyric acid	408	35 (0.2)	S-3-Methyl-1,2-butanediol	17
Mandelic acid	168	40 (0.8)	S-2-Phenyl-1,2-ethanediol	40

[a]Conditions: 0.1 M hydroxyacid 6 M n-butanol, and 0.2 g per milliliter lipase from C. cylindracea (Sigma); reduction carried out as described in the text.

preparation of S-1,2-propanediol, S-4-methyl-1,2-pentanediol, S-3-methyl-1,2-butanediol, S-3-phenyl-1,2-propanediol, and S-2-phenyl-1,2-ethanediol, respectively (Table 25.3). n-Butanol could be replaced with n-hexanol and n-propanol without significant loss in activity.

A significant advantage of enzymatic catalysis in organic as opposed to aqueous solutions is the ability to carry out consecutive, "single-pot" syntheses. We have used the synthesis of optically active 1,2-diols to demonstrate this potential. Diethyl ether is able to sustain lipase catalysis as well as to act as a suitable solvent for chemical reduction of the resulting hydroxyester to the diol. To that end, 0.1 M α-hydroxycaproic acid was dissolved in 150 mL of diethyl ether containing 0.6 M n-butanol, and 0.33 g/mL lipase from *C. cylindracea* was added and the reaction stirred at 150 rpm at 25°C. After 5 days, the reaction was terminated at 45% conversion of the acid. The reaction mixture was filtered to remove enzyme, and the ether was washed three times with 2% sodium bicarbonate to extract out the unreacted acid. The ether phase was dried over magnesium sulfate and 3.0 g of $LiAl(OCH_3)_3H$ was added at 4°C. The chemical reduction was terminated after 1 hour, 200 mL of 5% sulfuric acid was added to destroy the unreacted reductant, and the ether phase was removed by extraction. The ether was evaporated on a rotary evaporator, removing the solvent and the unreacted butanol. The resulting product contained 0.79 g of 1,2-hexanediol (98% purity by gas chromatography), which amounted to an overall yield of 98%. The product was dissolved in ethanol to give an optical rotation of $[\alpha]_D^{25} = -15.0(c\ 0.22,\ \text{ethanol})$. This corresponds to an enantiomeric excess of 96% for the S-isomer.

The combined enzymatic/chemical synthesis of optically active 1,2-diols is just one example of the selective advantages inherent with enzymatic catalysis in organic solvents. One may envision a wide variety of reactions that are either unfeasible or impractical to perform alone, enzymatically or chemically, yet are attractive if carried out in a combined synthesis. Certainly enzymatic catalysis offers high reaction selectivity, while chemical catalysis offers simplicity and high catalytic rates. Together, they suggest an approach to overcome a technical constraint imposed by conventional chemical catalysis.

III. ENZYME-CATALYZED POLYMER SYNTHESIS

The synthesis of commercially useful polymers has generally been outside the realm of biocatalysis. Poor solubilities of both substrates and growing polymer chains in aqueous solutions result in low productivities and high processing costs. Chemical alternatives are cheaper and simpler. Selective (both regio- and stereo-) polymer synthesis is difficult, however, using conventional chemical approaches. We have used enzymes in organic media to overcome several catalytic constraints inherent in the synthesis of sugar-based polyesters and phenolic resins. The former materials have applications as biodegradable plastics, while the latter have been used as biocompatible replacements for phenol-formaldehyde plastics.

A. Enzyme-Catalyzed Synthesis of Sucrose-Based Polyesters

The specificity of enzymes in organic solvents is clearly evident in the regioselective acylations of multifunctional molecules. These compounds range from the relatively simple aliphatic glycols to complex oligosaccharides. Chemical syntheses universally lack regiospecificity, giving mixtures of products [21,22] and requiring expensive protection and deprotection steps for regiospecific synthesis [23,24]. This is never more apparent than in the synthesis of sugar esters. Chemical methods include base-catalyzed esterifications and acylations with anhydrides or acyl halides. Multifunctional acylations result that reduce the yield of the desired sugar ester. Separation of the desired ester from a mixture of mono-, di-, and oligoesters is expensive and tedious.

To overcome the difficulty associated with selective chemical acylations, Klibanov and coworkers developed a strategy for enzymic acylations in organic media. Using lipase from porcine pancreas, Therisod and Klibanov showed that a variety of monosaccharides (e.g., glucose, galactose, mannose, fructose) could be selectively acylated with trichloroethylesters of acetate, butyrate, and laurate in anhydrous pyridine [25]. Only primary hydroxyl functionalities of the sugars were acylated.

Dordick and coworkers used lipases in pyridine to carry out the selective acylation of a variety of di- and trisaccharides, such as lactose, sucrose, trehalose, and raffinose [26]. In the case of sucrose, high yields of sucrose 1,4'-diacetate were obtained in the presence of isopropenyl acetate using a lipase from a *Pseudomonas* sp. Finally, Riva and Klibanov used subtilisin in anhydrous dimethylformamide to selectively acylate sucrose, maltose, cellobiose, lactose, and several oliogosaccharides, all in primary positions [27].

It occurred to us that selective acylation of sugars could be advantageously employed in the synthesis of sugar-containing polyesters. For example, if the acyl donor were a diacid derivative, then selective links between sugar molecules could be obtained through ester bonds, and a long-chain sugar-fatty acyl polyester would result. It may be envisioned that such a polymer would be highly water-absorbent, due to the large number of free hydroxyl groups left underivatized by the selective enzymatic treatment. Furthermore, the polyester would be expected to be highly biodegradable, because both aerobic and anaerobic microorganisms would have little difficulty breaking ester bonds and metabolizing both sugar and fats. These polyesters may find significant use as diaper liners, packaging materials, and drug delivery polymers. Furthermore, enzymatic polymerization offers several potentially valuable advantages over chemical sugar-polyester synthesis, including the ability to retain labile functionalities that might be destroyed in conventional chemical processing, and the potential for synthesis of polymers with regular size and backbone structures. Clearly, an enzymatic solution to this common catalytic constraint may be warranted.

To identify enzymes capable of catalyzing the synthesis of sugar-containing polyesters, our initial strategy was to screen a variety of hydrolytic enzymes for activity in sucrose-butyrate synthesis in pyridine. In this manner, simple esters of sucrose could be obtained and structurally analyzed without the added complication of polymer formation. Trifluoroethylbutyrate was chosen as the butyrate donor. To that end, 15 enzymes were studied for sucrose-butyrate synthesis (Table 25.4). A typical reaction mixture contained 0.1 M sucrose dissolved in 2 mL of anhydrous pyridine containing 0.6 M trifluoroethylbutyrate. The reactions were initiated by the addition of 0.25 g per milliliter of enzyme (0.015 g/mL in the case of "proleather," an alkaline protease obtained from Amano) and the reactions shaken at 250 rpm and 45°C. Sucrose disappearance was monitored by high performance liquid chromatography (HPLC). The four most active enzymes were subjected to a 100 mL reaction scale (same concentrations of reactants and enzyme as before). In accordance with the time scale indicated in Table 25.5, the reactions were terminated and the solvent evaporated. The residual solids were chromatographed on silica gel (17:2:1 ratio of ethyl acetate to methanol to water) and the sucrose ester products separated. Clearly, the proleather produced the highest ratio of sucrose dibutyrate to monobutyrate. Such a situation is vital for synthesis of sucrose polyesters. ^{13}C-NMR analysis of the proleather mono- and diester products indicated that the sucrose is first acylated in the 1'-position followed by acylation at the 6-position (Fig. 25.4).

Proleather was the ideal choice to carry out polyester synthesis. In this case, bis(2,2,2-trifluoroethyl) adipate was chosen as the diacid derivative. Sucrose (0.1 M) was dissolved in 25 mL of anhydrous pyridine containing 0.1 M bis(2,2,2-trifluoroethyl) adipate. The reaction was initiated by the addition of 0.015 g per milliliter of proleather and the reaction magnetically stirred at 100 rpm at 45°C under a slight nitrogen stream. The ratio of sucrose to the diacid derivative was purposely chosen to be equimolar because it

Table 25.4   Screen of Enzymes for Sucrose-Butyrate Synthesis[a]

Enzyme	Sucrose conversion after 120 h (%)
Control (no enzyme)	0
Lipase from *Aspergillus* sp.	0
Aminoacylase	70
Lipozyme (Novo)	8
Fungal amylase (HT from Rohm)	34
Bacterial protease (Bioenzyme)	100
Amylase from *B. subtilis* (Rapidase from Gist-Brocades)	24
Rhizopus sp. lipase	0
Alkaline protease (Amano-Proleather)	96
Bacillus protease	65
Lipase from *Pseudomonas* sp. (Amano P)	0
Lipase from *Candida cylindracea* (Sigma)	7
Lipase from porcine pancreas (Sigma)	13
Yeast esterase (Sturge, Ltd.)	0
Crude subtilisin (Amano protease N)	83[b]
Lipase from *Penicillium* sp. (Amano G)	24

[a]Conditions: sucrose (0.1 M) dissolved in 2 mL of pyridine containing 0.6 M trifluoroethylbutyrate. Reaction initiated by addition of 0.25 g per milliliter of enzyme and shaken at 250 rpm at 45°C.
[b]In dimethylformamide.

Table 25.5   Enzymatic Synthesis of Sucrose Butyrates[a]

Enzyme	Conversion (%)	Total isolated yield [g (%)]	Yield (g)	
			1'-Ester	6,1'-Diester
Proleather	99 (8 days)	0.5 (43)	0.12	0.38
Bioenzyme	100 (8 days)	0.57 (52)	0.30	0.27
Bacillus protease	62 (21 days)	0.39 (37)	0.31	0.08
Subtilisin in DMF	62 (25 days)	0.91	0.66	0.25

[a]Conditions: sucrose (0.1 M) dissolved in 25 mL of pyridine (except with subtilisin) containing 0.25 g per milliliter of enzyme and 0.6 M trifluoroethylbutyrate, magnetically stirred at 150 rpm at 45°C.

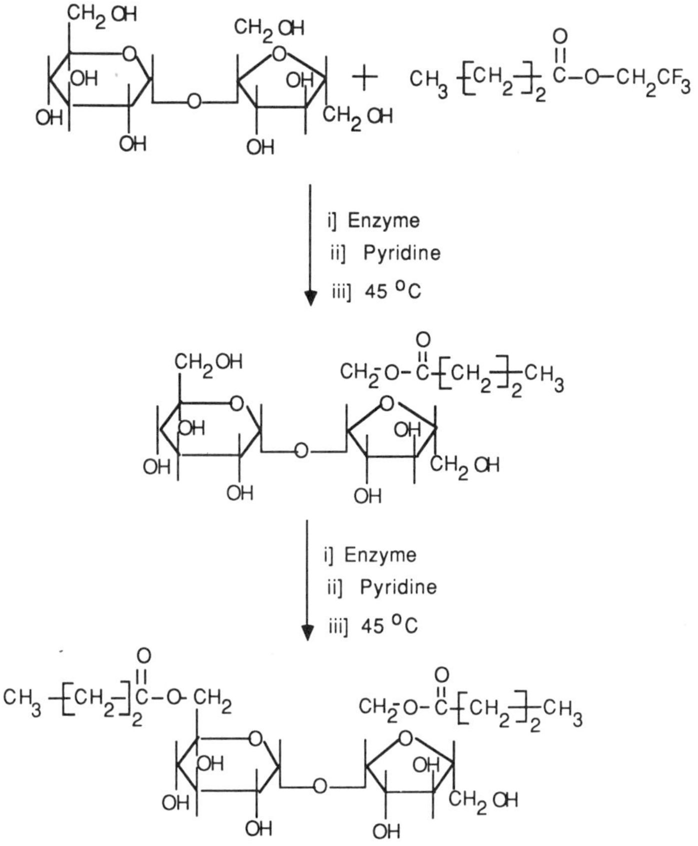

Figure 25.4 Sequential acylation of sucrose with trifluoroethylbutyrate in pyridine catalyzed by proleather.

was expected that two hydroxyls on sucrose would readily react with the diacid functionalities (proleather did not catalyze the synthesis of sucrose tributyrates in the aforementioned experiment).

The progress of the reaction was followed by gel permeation chromatography (GPC) HPLC. The reaction was terminated after 28 days (80% conversion of the sucrose), the enzyme removed by filtration, and the pyridine and bis(2,2,2-trifluoroethyl) adipate removed by rotary evaporation. The products of the reaction were completely soluble in water as well as having high solubilities in polar organic solvents including methanol, ethanol, pyridine, dimethylformamide, and dimethyl sulfoxide. Figure 25.5 depicts the

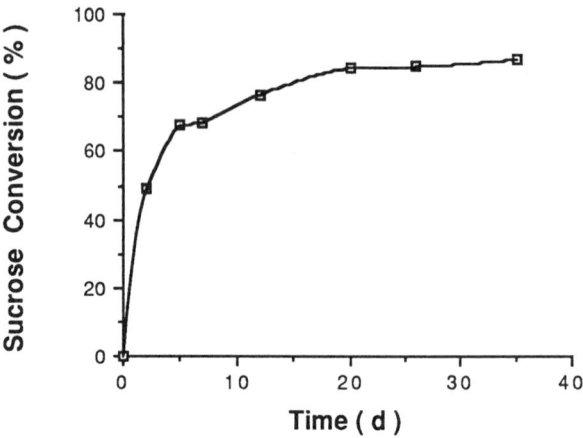

Figure 25.5   Synthesis of polysucrose adipate: sucrose conversion versus time.

conversion of sucrose during the reaction. While the reaction was slow, GPC data showed the formation of higher molecular weight species as reaction time increased (Fig. 25.6). Clearly, molecules with molecular weights in excess of 10,000 were produced. The average molecular weight was determined following dialysis of the product (through a 1000 dalton dialysis bag to remove unreacted sucrose and low molecular weight mono- and diester products). The dialyzed product was shown to have a weight average molecular weight of 2110 and a number average molecular weight of 1555, therefore giving a polydispersity of 1.36. The polyester showed selective linkages between the adipic acid functionalities and the 6- and 1'-positions of the sucrose as determined by ^{13}C-NMR spectrometry (Fig. 25.7). From the NMR data, it is clear that a shift in the positions of the 6 and 1' carbons has occurred, indicative of acylation at those positions.

The synthesis of sucrose-based polyesters has shown clearly that enzymes are capable of acting as highly selective polymerization catalysts. Their selectivity is unmatched by any chemical catalyst and has been used to overcome a technical constraint that has limited the usefulness of chemical catalysts in the polysaccharide field.

B.   Peroxidase-Catalyzed Polymerization of Phenols in Organic Media

1.   Polymer Synthesis

Phenolic polymers, generally in the form of phenol-formaldehyde resins, are used in a variety of materials including laminates,

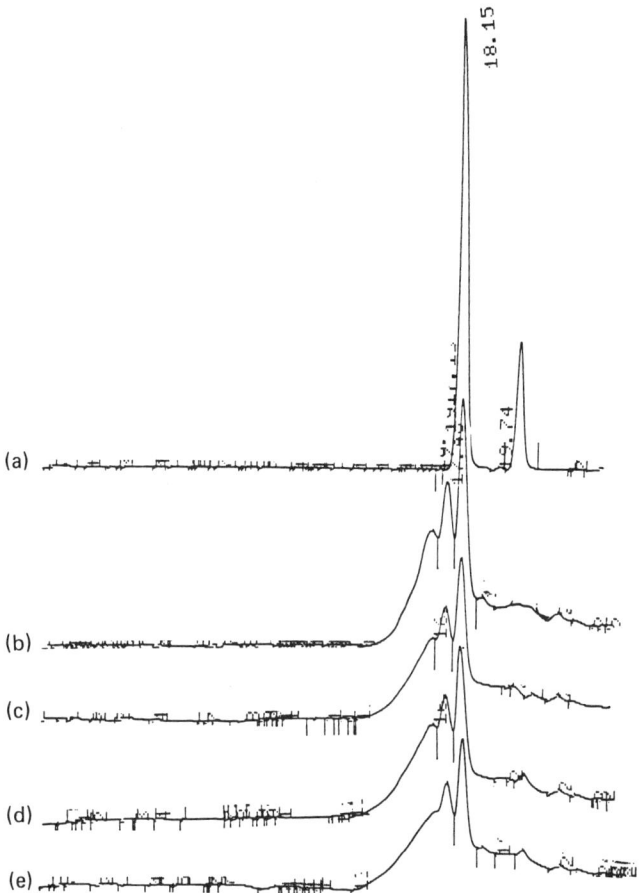

Figure 25.6   Gel permeation chromatograms of (a) sucrose reaction product after (b) 7, (c) 14, (d) 18, and (e) 25 days.

adhesives, particle boards, photographic developer resins, bonding agents, and soil conditioning agents [28]. A major problem with this synthesis is the dependence of polymerization on formaldehyde. Recent government regulations have limited the use of formaldehyde in the workforce as well as its presence in final products [29]. Alternatives to formaldehyde are needed. Conventional chemical approaches are uneconomical and do not alleviate the health concerns.

The application of peroxidases for phenolic polymerizations has been attempted in aqueous solutions, yet only low molecular weight coupling products are formed [30]. High molecular weight polymers cannot be synthesized because the growing polymer chains have

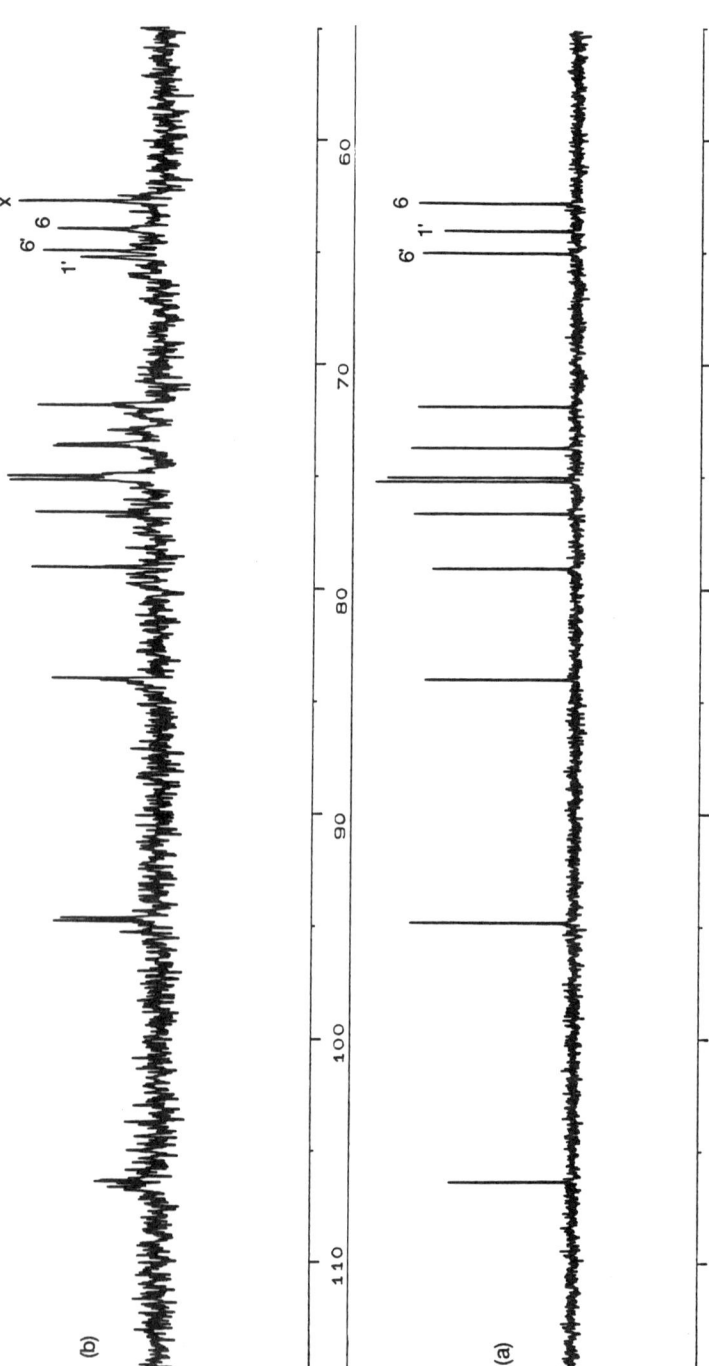

File: DP1T9.002   User.
Scans   280/-1   Pulse width   4.1   usec
Size   32 K   Recycle delay 1.312   sec
Spect. Freq.   90.56   MHz   01   75461.93
Spect. width   23809.52   Hz   02   6400.00
Total LB   0.25   Hz
Mon. Jan 29.   1990   10:27:43

Figure 25.7 ^{13}C–NMR spectra of (a) sucrose and (b) polysucrose adipate: X = enzyme peak.

poor solubility in water. Poor substrate solubility also lowers the productivity of phenolic polymerizations. We have examined the polymerization of phenols using horseradish peroxidase in nonaqueous media [31]. Using p-phenylphenol, a hydrophobic phenol of great commercial interest in the synthesis of oil-soluble phenol-formaldehyde resins [28], peroxidase was shown to catalyze polymerization in dioxane (containing 15% aqueous buffer) yielding molecular weights in excess of 25,000, nearly 50-fold higher than that obtained in aqueous solutions. Furthermore, the productivity of poly(phenyl-phenol) in dioxane is far higher than in water as a result of high solubility of the phenol in dioxane; a 1 M p-phenylphenol solution in 85% dioxane yielded 150 g per liter of poly(phenylphenol). This compares favorably to the maximum saturation of 1.5 mM p-phenylphenol in water. In addition to the increased size of the polyphenols, peroxidase catalysis in dioxane can be controlled selectively [31]. Polymer size is found to be highly sensitive to the water content in dioxane; below 30% v/v water, relatively small polymers are produced, whereas in 15% v/v water, molecular weights in excess of 25,000 are synthesized. Specific molecular weights can be obtained by varying the water content in dioxane (Fig. 25.8).

The polymerization of phenols catalyzed by peroxidase in organic media is a general phenomenon. A wide variety of electron donors are capable of acting as substrates (Table 25.6), as well as a number of water-miscible solvents capable of supporting the polymerization reaction. This reaction is being commercialized as a phenol-formaldehyde replacement [32]. To develop a general methodology for phenolic polymerizations, however, the kinetics of the phenol oxidation and of the actual polymerization reaction must be elucidated.

2. Computer Simulation of Polymer Synthesis Through Kinetic Analysis

Recently, Ryu and Dordick, in an attempt to quantify the effects of substrate and solvent on peroxidase catalysis, discovered that linear free energy relationships existed between substrate and solvent hydrophobicities and the catalytic efficiency of peroxidase [33]. These relationships were subsequently used to develop a predictive model of peroxidase catalysis in organic solvents as shown in equation (25.1), where π and log P represent the hydrophobicities of substrate (phenolic substituent) and solvent, respectively.

$$\frac{V_{max}}{K_m} = 680 \exp[-(0.69 \log P + 1.50) \pi] \qquad (25.1)$$

This equation can be used to predict the effects of solvent and substrate on the initiation of phenoxy radicals during the polymerization

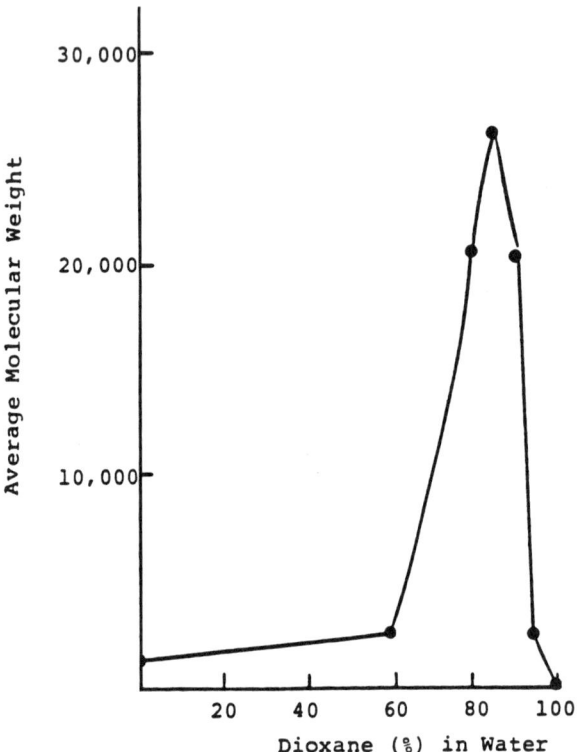

Figure 25.8 Effect of dioxane concentration on the average molecular weight of fractionated poly(p-phenylphenol).

of phenols catalyzed by peroxidase. In this manner, by studying the basic kinetics of peroxidase catalysis in organic media, information can be obtained about the properties and molecular weights of phenolic polymers produced via peroxidase catalysis.

The polymerization of phenols catalyzed by peroxidase follows a free radical mechanism as outlined in Figure 25.9. After initiation of phenoxy radicals from peroxidase-catalyzed oxidation of phenolic substrates, chain growth is initiated by radical transfer from a phenoxy radical to the growing polymer chain to produce a polymer radical. Chain termination of the polymer radical with a phenoxy radical leads to the growth of the polymer chain. Similar reactions continue until polymer solubility has been reduced and product precipitates out of solution, thereby terminating polymerization. In water, the degree of polymerization is low because the solubility

(1) Initiation Step:

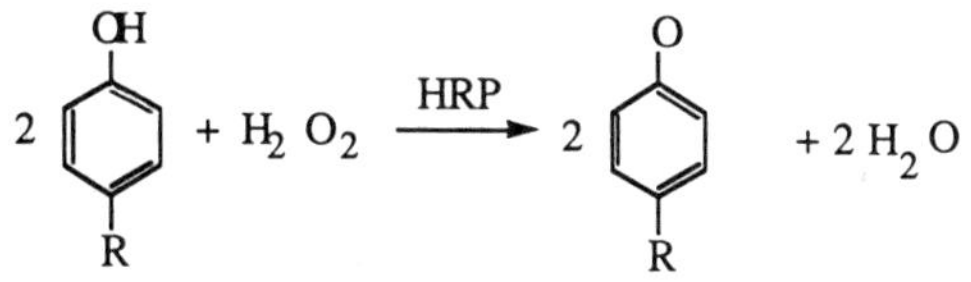

Phenol -----> Phenoxy radical

(2) Coupling Step:

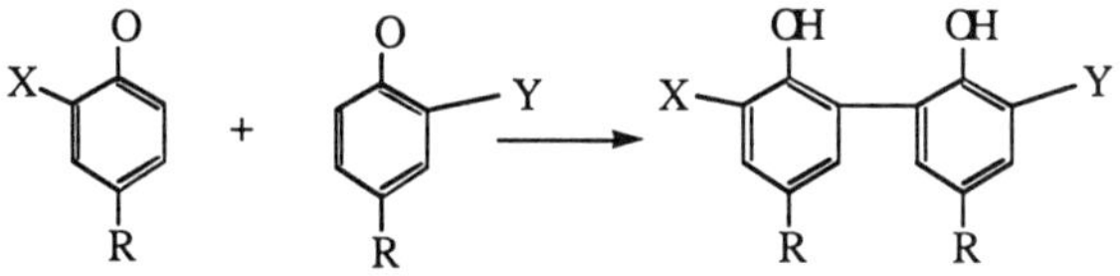

Phenoxy radicals ---> Higher molecular weight
phenol

(3) Radical Transfer Steps:

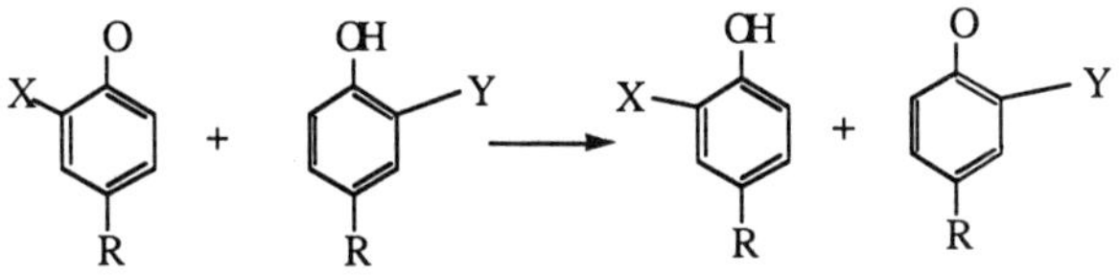

Phenoxy Radicals ---> Phenols

Figure 25.9 Mechanism of the enzymatic polymerization of the phenols.

Table 25.6 Spectrum of Compounds That Are Polymerized by
Horseradish Peroxidase in Dioxane[a]

Compound	M_w (daltons)
Phenol	1400
p-Methoxyphenol	2000
p-Cresol	1900
p-Chlorophenol	600
p-tert-Butylphenol	1900
p-Phenylphenol	26000
Aniline	1700
2,6-Dimethylphenol	500
4,4'-Biphenol	400
1-Naphthol	very high[b]
2-Naphthol	2000

[a]Containing 15% aqueous acetate buffer, 20 mM, pH 5.0,
20 mM phenolic substrate, 20 mM H_2O_2 (added slowly throughout reaction).
[b]Insoluble product.
Source: Ref. 31.

of the growing polymer chain in solution is poor. The presence
of organic solvents significantly enhances the polymer solubility,
hence the polymer size attainable.

As a model, bisphenol A, a commercially useful phenolic monomer
has been used. A Monte Carlo simulation has been run to predict
the progress of bisphenol A polymerization using peroxidase in
organic media. This approach uses a simple random number generator
to determine which sets of reactions will be performed in the organic
solvent leading to polymer synthesis. At early reaction times, when
the concentration of growing polymer is minimal, only initiation
of phenoxy radicals takes place. As the reaction proceeds, however,

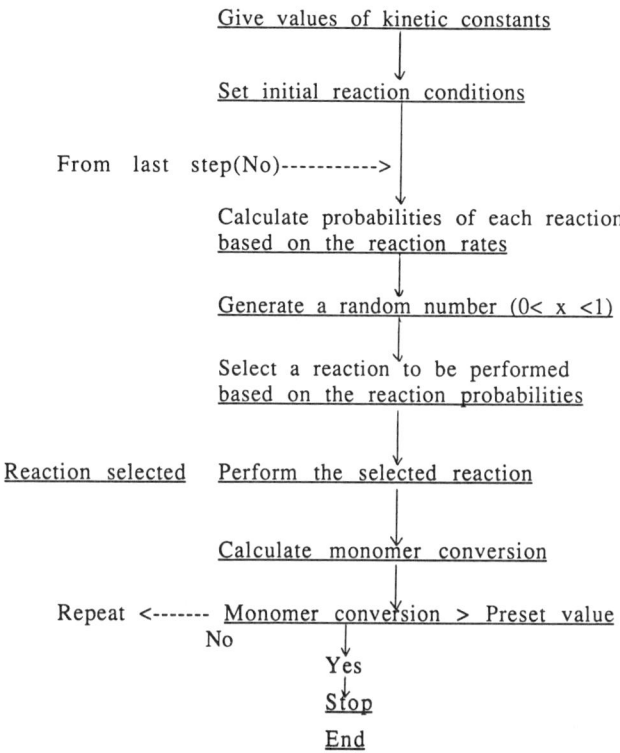

Figure 25.10 Monte Carlo simulation flow of the enzymatic polymerization of phenols.

polymerization reactions predominate, and the fraction of reactions governing polymer growth increases relative to radical initiation. A flowchart is given in Figure 25.10.

The reaction progress is well described by this simulation. Figure 25.11 plots the conversion of bisphenol A as a function of reaction time. Excellent agreement between experimental results and the predictive model is obtained. Similarly, excellent agreement between number average degree of polymerization and reaction conversion is obtained (Fig. 25.12), indicating that the Monte Carlo simulation technique can accurately predict the progress of phenolic polymerizations in organic media catalyzed by peroxidase once basic enzyme kinetics have been elucidated.

Dordick et al.

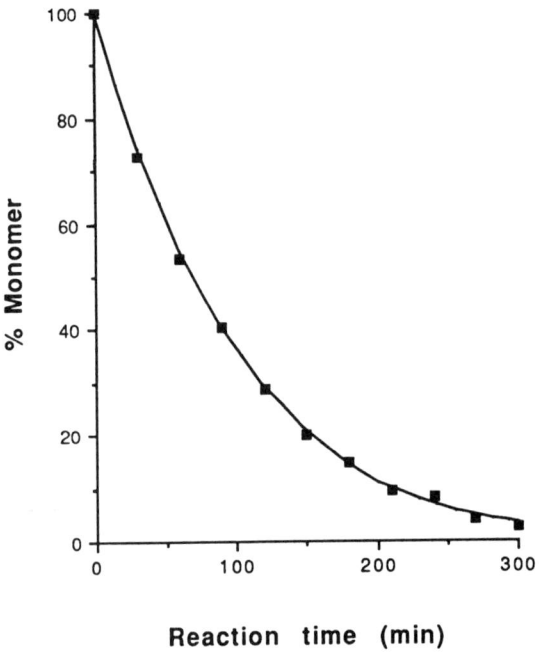

Figure 25.11   Conversion of bisphenol A versus reaction time.

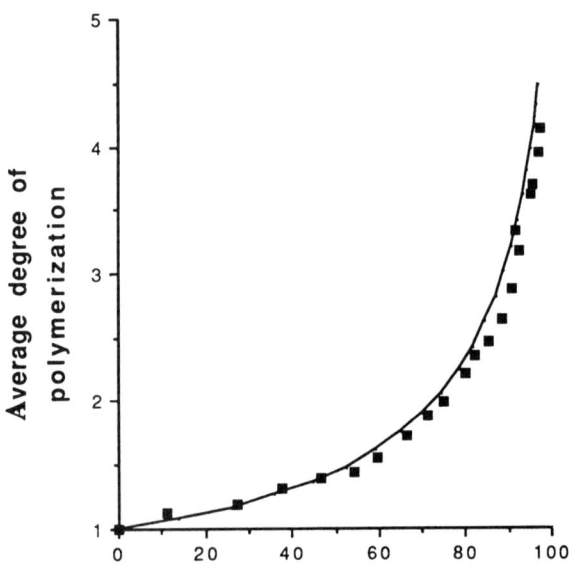

Figure 25.12   Effect of reaction conversion on molecular weight of phenolic polymer.

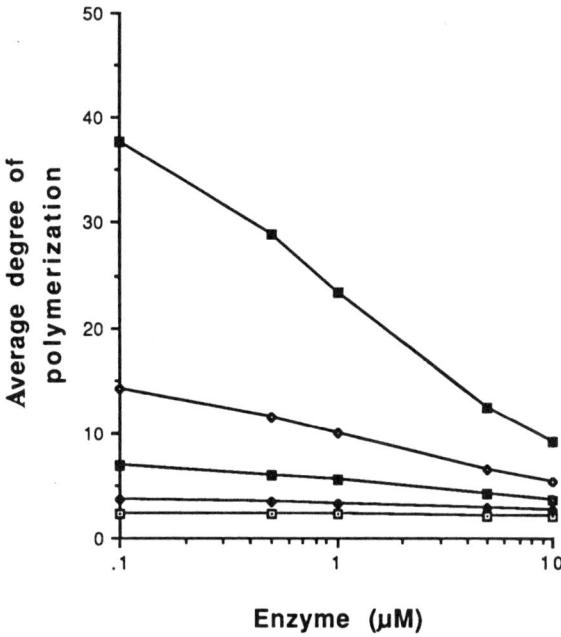

Figure 25.13 Effect of enzyme concentration on degree of polymerization. The degree of conversion ranged from 40-90%.

Simulation models using the Monte Carlo method also provide information of molecular weight distributions of products as a function of enzyme concentration (which is directly related to the rate of radical initiation). Figure 25.13 shows the changes in number average molecular weight units as a function of enzyme concentration from 0.1 to 10 μM at various levels of monomer conversion. To maximize polymer size, radical coupling rate constants were chosen to be low ($\approx 10^6$ $M^{-1}s^{-1}$) and radical transfer constants high ($\approx 10^3$ $M^{-1}s^{-1}$). [It should be noted that polymer size is expected to be maximized when rate constants of radical transfer are optimized with respect to radical coupling. The average range of values for radical coupling and radical transfer for phenols is 10^6 to 10^9 $M^{-1}s^{-1}$ and 1 to 10^3 $M^{-1}s^{-1}$, respectively.] Molecular weights of polymers were insensitive to enzyme concentration until 70% conversion was attained. At lower rates of radical generation, the Monte Carlo simulation predicts that polymer synthesis will predominate over radical generation. More polymeric species are available for polymer growth to take place at higher degrees of conversion, rather than at low degrees of conversion; hence the dependence on enzyme concentration at high degrees of conversion.

These results show that peroxidase-catalyzed phenolic polymerizations can be accurately predicted using a simulation approach. Information needed for such a simulation can come from phenol and solvent chemistries and initial rate data of peroxidase catalysis in organic solvents.

IV. CONCLUSIONS AND FUTURE PROSPECTS

The use of enzymes in organic solvents enables enzymologists, chemists, and chemical engineers to overcome common catalytic constraints inherent in conventional chemical catalysis. It may be envisioned that enzymatic catalysis in nonaqueous media will be used increasingly to solve common problems in the chemical industry. Where high selectivity is required, enzymes will be investigated as a valuable alternative. Furthermore, coupling enzymatic reactions with conventional chemical catalysis will enhance the selectivity of organic synthesis while maintaining the existing process parameters. Even in the area of polymer synthesis, commercially available enzymes will continue to make an impact, remaining as an alternative for the synthesis of specialty polymers.

The examples presented here are fairly simple, yet they provide potentially significant commercial possibilities. Many more applications of enzymes in organic media have been developed [3], and the coupling of enzymology with synthetic chemistry is fast maturing into a discipline of biocatalysis that will significantly expand the use of biotechnology in the chemical, polymer, and pharmaceutical industries.

ACKNOWLEDGMENTS

The authors acknowledge support from the National Science Foundation (J.S.D. is a Presidential Young Investigator), the Mead Corporation, the Sugar Association, and the U.S. Department of Agriculture.

REFERENCES

1. Borgstrom, B., and H. L. Brockman, eds., *Lipases*, Elsevier, Amsterdam, 1984.
2. Gunsales, I. C., Meeks, J., Lipscomb, J., Debrunner, P., and Munck, E., in *Molecular Mechanisms of Oxygen Activation* (O. Hayashi, ed.), Academic Press, New York, 1974, pp. 561-614.

3. Dordick, J. S., *Enzyme Microb. Technol.*, *11*, 194-211 (1989).
4. Klibanov, A. M., *Chemtech.*, *16*, 354-359 (1986).
5. Butler, L. G., *Enzyme Microb. Technol.*, *1*, 253-259 (1979).
6. Lilly, M. D., *J. Chem. Tech. Biotechnol.*, *32*, 162-169 (1982).
7. Carrea, G., *Trends Biotechnol.*, *2*, 102-106 (1984).
8. Halling, P. J., *Biotechnol. Adv.*, *5*, 47-84 (1987).
9. Martinek, K., Levashov, A. V., Klyachko, N., Khmelnitski, Y. L., and Berezin, I. V., *Eur. J. Biochem.*, *155*, 453-468 (1986).
10. Luisi, P. L., and Laane, C., *Trends Biotechnol.*, *4*, 153-161 (1986).
11. Porter, R., and Clark, S., eds., *Enzymes in Organic Synthesis*, London, 1985.
12. Jones, J. B., *Tetrahedron*, *42*, 3351-3403 (1986).
13. Dordick, J. S., *Appl. Biochem. Biotechnol.*, *22*, 361-373 (1989).
14. Fujisawa, T., Kojima, E., Itoh, T., and Sato, T., *Tetrahedron Lett.*, *26*, 6089-6092 (1985).
15. Spassky, N., and Sigwalt, P., *Eur. Polym. J.*, *7*, 7-16 (1971).
16. Schossig, J., in *Kirk-Othmer Encyclopedia of Chemical Technology*, Vol. A.1., Wiley-Interscience, New York, pp. 305-312.
17. Cameron, D. C., and Cooney, C. L., *Bio/technology*, *4*, 651-654 (1986).
18. Cesti, P., Zaks, A., and Klibanov, A. M., *Appl. Biochem. Biotechnol.*, *11*, 401-407 (1985).
19. Kirchner, G., Scollar, M. P., and Klibanov, A. M., *J. Am. Chem. Soc.*, *107*, 7072-7076.
20. Zaks, A., and Klibanov, A. M., *Proc. Natl. Acad. Sci. USA*, *82*, 3192-3196 (19).
21. Notheisz, F., Bartok, M., and Remport, V., *Acta Phys. Chem.*, *18*, 89-98 (1972).
22. Mosher, H. S., and Morrison, J. D., *Science*, *221*, 1013-1019 (1983).
23. Haines, A. H., *Adv. Carbohydr. Chem. Biochem.*, *39*, 13-70 (1981).
24. Desgupta, F., Hay, G. W., Szarek, W. A., and Schilling, W. L., *Carbohydr. Res.*, *114*, 153-157 (1983).
25. Therisod, M., and Klibanov, A. M., *J. Am. Chem. Soc.*, *108*, 5638-5640 (1986).
26. Dordick, J. S., Hacking, A. J., and Khan, R. A., U.S. Patent pending.
27. Riva, S., and Klibanov, A. M., *J. Am. Chem. Soc.*, *110*, 3291-3295 (1988).
28. Whitehouse, A. A. K., Pritchett, E. G. K., and Barnett, G., *Phenolic Resins*, Elsevier, New York, 1968, Chap. 2.
29. Marshall, E., *Science*, *237*, 381 (1987).

30. Schwartz, R. D., and Hutchinson, D. B., *Enzyme Microb. Technol.*, *3*, 361–363 (1981).
31. Dordick, J. S., Marletta, M. A., and Klibanov, A. M., *Biotechnol. Bioeng.*, *30*, 31–36 (1987).
32. Pokora, A. R., and Cyrus, W. L., U.S. Patent 4,647,952 (1987).
33. Ryu, K., and Dordick, J. S., *J. Am. Chem. Soc.*, *111*, 8026–8027 (1989).

26

Addition Reactions Catalyzed by Rhodium(II) Carboxylates

Michael P. Doyle, Kenneth G. High, and Carey L. Nesloney

Department of Chemistry, Trinity University, San Antonio, Texas

I. INTRODUCTION

Addition reactions catalyzed by transition metals and transition metal compounds offer a complex variety of pathways for product formation that are not possible with traditional free radical or ionic processes. Consequently, opportunities for regio- and stereocontrol in transition metal catalyzed addition reactions are substantial. Dependent on the catalyst (ML_n), activation of the unsaturated center (A=B) or of the reagent (X—Y) may occur (eqs. 26.1, 26.2) and selectivity is dependent on the sequence of mechanistic events

$$X-Y + L_nM-\overset{A}{\underset{B}{\|}} \longrightarrow \overset{A\diagup X}{\underset{B\diagdown Y}{|}} + ML_n \tag{26.1}$$

$$A=B + L_nM-\overset{X}{\underset{Y}{|}} \longrightarrow \overset{A\diagup X}{\underset{B\diagdown Y}{|}} + ML_n \tag{26.2}$$

that occur along the pathway for product formation.

Our interest in transition metal catalyzed addition reactions has occurred as a result of our growing realization of the great versatility of rhodium(II) carboxylates and carboxamides as catalysts for a broad variety of organic transformations. Rhodium(II) carboxylates, especially rhodium(II) acetate, have become the catalysts of choice for carbenoid transformations of diazo compounds [1-3], and recent advances in catalyst design using chiral carboxamides as ligands have even led to carbon–carbon bond forming reactions that occur with exceptionally high enantioselectivities (e.g., equation 26.3) [4].

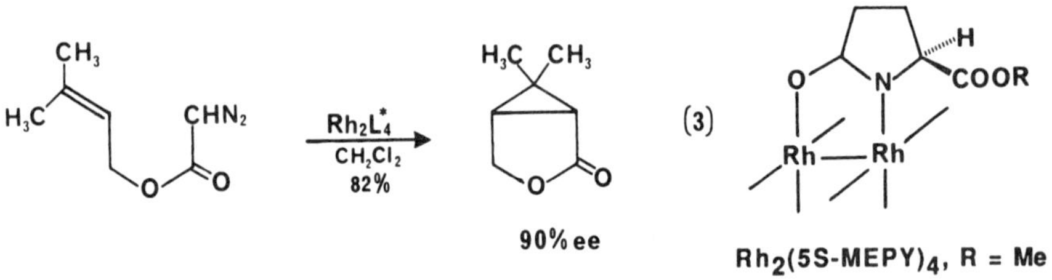

(26.3)

Less well known is the use of $Rh_2(OAc)_4$ as an efficient and selective hydrogenation catalyst for terminal olefins (eq. 26.4) [5].

$$CH_3(CH_2)_5CH=CH_2 + H_2 \xrightarrow[DMF]{Rh_2(OAc)_4} CH_3(CH_2)_6CH_3 \quad (26.4)$$

Rhodium(II) acetate, the parent of an extensive list of rhodium(II) carboxylates, is a binuclear compound with four bridging acetate ligands and having D_{4h} symmetry (1, $R = CH_3$) [6]. In the absence of coordinating ligands that include nitriles, alcohols, and ketones, $Rh_2(OAc)_4$ possesses one vacant coordination site per metal atom. Electron-withdrawing groups have a profound effect

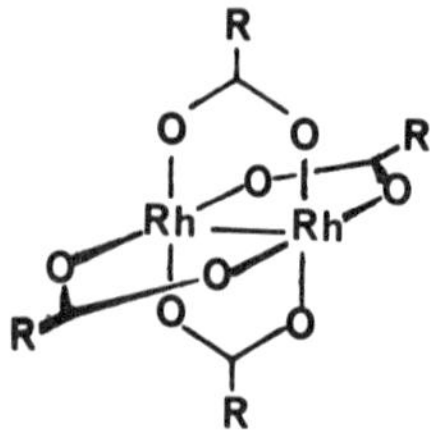

1

on the electrophilic reactivity of these transition metal compounds and on their selectivity in carbenoid reactions [7,8]. Whereas $Rh_2(OAc)_4$ does not give evidence for coordination with olefins in solution, rhodium(II) perfluorobutyrate, $Rh_2(pfb)_4$, forms olefin complexes (eq. 26.5) with equilibrium constants that range from 70

$$|| + Rh_2(pfb)_4 \rightleftharpoons Rh(pfb)_4Rh-|| \quad (26.5)$$

(with styrene) to 860 (with ethyl vinyl ether) [9]. However, α,β-unsaturated esters do not form detectable complexes with $Rh_2(pfb)_4$. This capability for olefin coordination and the promise of reactant activation suggested a broad applicability of rhodium(II) carboxylates and, potentially, carboxamides as catalysts for addition reactions. The use of these compounds for catalytic reactions of organosilanes is described in this chapter.

Catalytic hydrosilylation of unsaturated organic compounds, a process of substantial importance for the synthesis of silicon derivatives, is similar to, but less exothermic than, catalytic hydrogenation [10-13]. Numerous transition metal compounds, including many that are employed for hydrogenation and especially those of Pt, Ni, and Rh(I) [10-18], are effective catalysts for these addition reactions, but hexachloroplatinic acid (Speier's catalyst) [19] is generally recognized to be the catalyst of choice for hydrosilylation of compounds that possess a carbon-carbon multiple bond [20]. Even with trialkylsilanes, chloroplatinic acid catalyzed addition occurs at room temperature. With carbonyl compounds, hydrosilane reductions are effectively catalyzed by rhodium(I) compounds, which with chiral ligands can lead to alkoxysilanes having moderate to high enantiomeric excesses [21-23].

II. RESULTS AND DISCUSSION

A. Catalytic Hydrosilylation of 1-Alkynes

Chloroplatinic acid catalyzed hydrosilylation of alkynes proceeds by cis addition across the carbon-carbon triple bond to yield the corresponding vinyl silanes [24,25]. Limited regiocontrol in intermolecular organosilane addition to unsymmetrical internal alkynes has been achieved, although high regioselectivity is obtained in intramolecular hydrosilylation with silane derivatives of homopropargyl alcohols (e.g., eq. 26.6) [25].

$$(26.6)$$

Terminal alkynes also exhibit low selectivities in hydrosilylation reactions, and preferential addition of hydrogen to the 1-position resulting in the formation of 2-silylalkenes has been reported [24].

The stereochemical outcome of this transformation, however, is cis addition. However, with the use of a catalytic amount of $Rh_2(pfb)_4$, hydrosilylation of terminal alkynes takes an entirely different course. Controlled addition of triethylsilane to 1-hexyne, for example, results almost exclusively in the product of trans addition to the carbon-carbon triple bond (eq. 26.7).

$Rh_2(pfb)_4$	85	10	5
H_2PtCl_6	3	87	10
$(Ph_3P)_3RhCl$	20	79	1

(26.7)

Both chloroplatinic acid and Wilkinson's catalyst, $Rh(PPh_3)_3Cl$, exhibit a contrasting high preference for cis addition when this hydrosilylation reaction was performed under the same conditions. Among the three catalysts, chloroplatinic acid has the lowest regioselectivity for triethylsilane addition, although the extent of regiocontrol with all three catalysts is relatively high. Nearly identical results were obtained with 1-octyne, but in neither case, 1-octyne or 1-hexyne with $Rh_2(pfb)_4$, did the use of *tert*-butyldimethylsilane significantly change the stereoselectivity for addition. However, when 1-ethynyl-1-cyclohexanol was added to triethylsilane in the presence of $Rh_2(pfb)_4$, only the product from trans addition was formed (eq. 26.8), suggesting that steric or stereoelectronic factors may be important determinants of product stereochemistry.

(26.8)

In direct contrast to these results, if the mode of addition of reagents is reversed and 1-octyne, for example, is added to triethylsilane in the presence of $Rh_2(pfb)_4$, the products from simple addition of the silane are minor constituents of the reaction mixture, and the dominant products are isomeric allylsilanes (eq. 26.9).

	Rh$_2$(pfb)$_4$	H$_2$PtCl$_6$	(Ph$_3$P)$_3$RhCl
	8	16	49
	0	0	17
	2	4	32
	90 (t/c = 2.0)	80 (t/c = 8.9)	2
yield, %	77	84	58

(26.9)

A similar outcome is obtained with chloroplatinic acid, but with this catalyst the trans/cis ratio of isomeric allylsilanes is 8.9 instead of the 2.0 ratio obtained with Rh$_2$(pfb)$_4$. Use of Wilkinson's catalyst does not lead to allylsilane products. Similar results and product distributions are obtained with 1-hexyne, and propargyl ethers show an even greater facility for allylsilane formation (eq. 26.10), but only with Rh$_2$(pfb)$_4$ catalysis, since use of chloroplatinic acid results in low product yields.

$$RO-CH=CHCH_2SiEt_3 \qquad (26.10)$$

R = CH$_3$	72%	>98% (t/c = 2.4)
PhCH$_2$	77%	91% (t/c = 2.6)

The contrasting results obtained for hydrosilylation of alkynes with Rh$_2$(pfb)$_4$, H$_2$PtCl$_6$, and (Ph$_3$P)$_3$RhCl suggest their mechanistic differences. Cis addition of hydrogen and a trialkylsilyl group to a carbon-carbon triple bond is consistent with an oxidative addition mechanism [26], and this pathway is most likely for hydrosilylation reactions catalyzed by (Ph$_3$P)$_3$RhCl. With Rh$_2$(pfb)$_4$, however, the dominant reaction stereochemistry is trans addition and, given the availability of only one coordination site per rhodium in the dirhodium complex, the oxidative addition mechanism probably does

not apply. Allylsilane formation is consistent with a carbocation
mechanism in which 1,2-hydride migration from the first-formed
vinyl cation to a silyl-substituted allyl cation may account for the
generation of this product (Fig. 26.1). In dichloromethane and
in the absence of a better coordinating ligand, $Rh_2(pfb)_4$ forms
a complex with triethylsilane, and it is this coordinated organosilane
that is activated for electrophilic addition of the trialkylsilyl group
to the alkyne. The metal hydride intermediate serves as the hydride
source for capture of the allyl cation intermediate in this scheme.
A similar mechanism may be operative with the active catalyst derived
from chloroplatinic acid in allylsilane formation. However, a "free"
allyl cation may not actually exist in either catalytic reaction, since
the greatly different geometrical isomer ratios of the allylsilanes
suggest an intimate involvement of the catalytic species in product
formation.

When triethylsilane is added to the alkyne in the presence of
$Rh_2(pfb)_4$, allylsilane is not produced, and the product from trans
addition of triethylsilane across the carbon-carbon triple bond is
formed. Here we believe that the coordinated alkyne reacts with
the organosilane to form a reaction intermediate in which the vinyl
cation is stabilized by a bridging oxygen of the perfluorobutyrate
(Fig. 26.2). Because triethylsilyl group addition, which itself must
be aided by the dirhodium catalyst, occurs, this bulky substituent
is trans to the incipient cationic center. The rhodium hydride is
now in close proximity to the electrophilic vinyl carbon center,

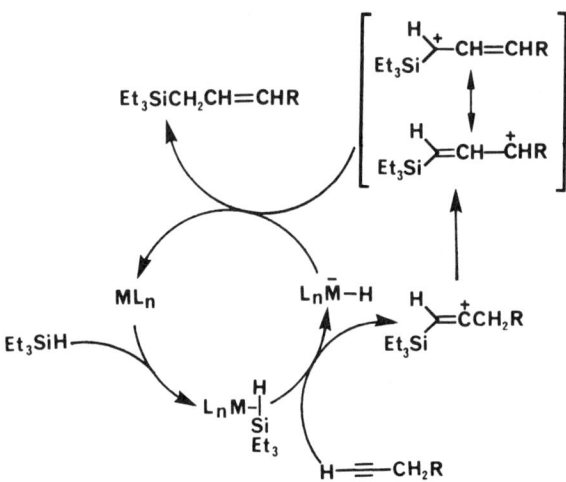

Figure 26.1

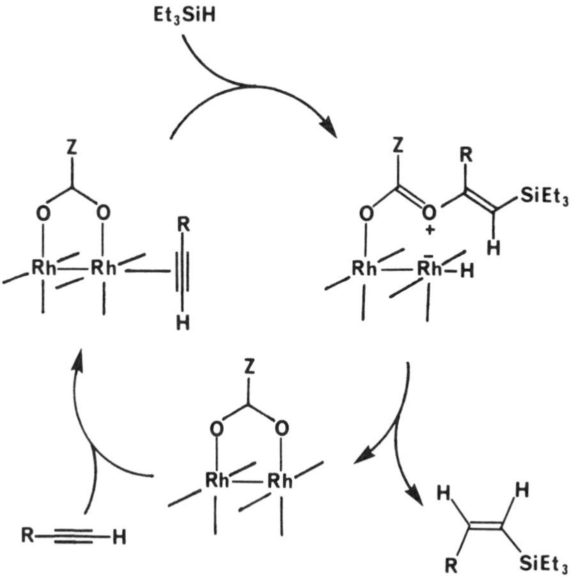

Figure 26.2

and thus collapse of this intermediate to products can be expected to result in the observed stereochemistry. Lappert et al. had reported the use of rhodium(II) acetate for hydrosilylation of alkynes [27], but the unique transformations observed with the use of $Rh_2(pfb)_4$ were not evident.

B. Catalytic Hydrosilylation of Alkenes

The greatest attention afforded hydrosilylation reactions has been with alkenes [10,11]. Although $Rh_2(pfb)_4$ is an effective and highly regioselective catalyst for organosilane addition to alkenes, we will not describe its reaction characteristics here. Instead, we focus on a transformation for which $Rh_2(pfb)_4$ appears to be uniquely suitable: substitution reactions resulting from addition/elimination.

Treatment of styrene with triethylsilane in refluxing chloroform containing a catalytic amount of $Rh_2(pfb)_4$ resulted in the exclusive formation of trans-β-triethylsilylstyrene (eq. 26.11).

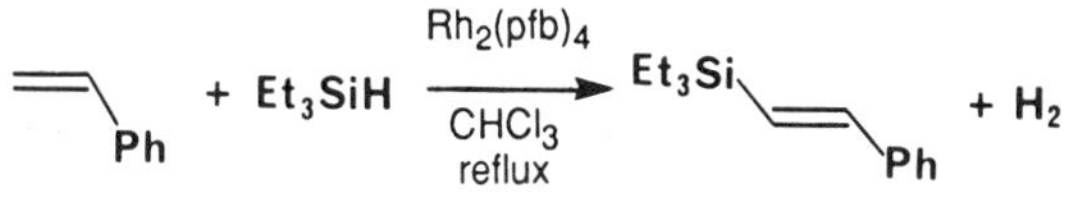

(26.11)

This product, formally derived from electrophilic triethylsilyl addition
to styrene followed by proton elimination, has been reported to be
the major, but not exclusive, product in reactions catalyzed by
$(Ph_3P)_3RhCl$ [28]. Subsequent efforts suggested that this process
could not be generalized [29].

We have investigated this formal organosilane substitution reaction
with acrylate esters and, as shown in equation (26.12), the selectivity
for formation of the vinylsilane is subject to steric influences.

R' = Et	Et_3SiH	71%	29%
Et	t-$BuMe_2SiH$	82%	18%
Et	$(MeO)_3SiH$	50%	50%
t-Bu	Et_3SiH	84%	16%

(26.12)

As in addition reactions of triethylsilane with alkynes, the mode
of addition controls product formation. Addition of the acrylate
ester to the trialkylsilane results exclusively in the addition product.
When the mode of addition is reversed, however, the results described in equation (26.12) are obtained. Curiously, only the products from addition to the beta position are observed; no α-substituted
product was obtained. Further studies are required to determine
the generality of this transformation and to unravel its mechanism.

C. Silane Alcoholysis

Although silane alcoholysis is certainly not an addition reaction,
a great deal of what we have learned about the catalysis of organosilane reactions by $Rh_2(pfb)_4$ has resulted from ivnestigations of
this transformation (eq. 26.13). A wide variety of catalysts are
reported to be effective for the alcoholysis of organosilanes [30-34],

$$R'OH + R_3SiH \xrightarrow{Rh_2(pfb)_4} R'OSiR_3 + H_2 \qquad (26.13)$$

but none of them offer the versatility afforded by the use of
$Rh_2(pfb)_4$ [35]. Reactions catalyzed by $Rh_2(pfb)_4$ are rapid at room
temperature. In reactions with triethylsilane, relative reactivities
followed the order: 1° ROH > 2° ROH >> 3° ROH. Even greater
differentiation could be achieved with *tert*-butyldimethylsilane, so

that diols such as 1-3 were silylated only at the 1° carbinol. Yields of greater than 90% for triethylsilyl ethers derived from alcohols ranging from cholesterol and glycidol to phenol and borneol were obtained, and with olefinic alcohols silyl ether formation occurred without competition from silane addition to the carbon-carbon double bond.

With chloroplatinic acid catalysis, methanolysis of α-naphthyl-phenylmethylsilane occurs with racemization, although heterogeneous catalysts give exclusive or predominant inversion [36]. In contrast, $(Ph_3P)_3$RhCl-catalyzed methanolysis of this optically active silane proceeds with predominant retention of configuration when this reaction is performed in benzene, but predominant inversion of configuration is observed in methanol [37]. Rhodium(II) perfluoro-butyrate catalyzed methanolysis of (S)-(−)-α-naphthylphenylmethyl-silane in dichloromethane at 25°C yielded the corresponding methoxysilane with complete inversion of configuration at silicon (eq. 26.14),

$$(26.14)$$

whereas performing the same reaction in methanol as the solvent gave racemic methoxysilane. Control experiments established that the enantiomeric methoxysilane underwent racemization in methanol that contained $Rh_2(pfb)_4$.

The coordination of triethylsilane with $Rh_2(pfb)_4$, observed by both UV/visible and NMR spectral analyses [35], and the inversion of configuration at silicon from methanolysis of optically active α-naphthylphenylmethylsilane suggest the mechanism for silane alcoholysis that is described in Figure 26.3. The rhodium(II) catalyst activates the silane for backside nucleophilic attack by the alcohol at silicon, resulting in the formation of a dirhodium hydride

$$R'OH + Rh_2L_4 \rightleftharpoons Rh(L)_4Rh(R'OH)$$

$$R_3SiH + Rh_2(L)_4 \rightleftharpoons Rh(L)_4Rh\text{-}(H\text{-}SiR_3)$$

$$Rh(L)_4Rh\text{-}(H\text{-}SiR_3) + R'OH \longrightarrow [Rh(L)_4Rh\text{-}H]^- + R_3Si\overset{+}{\underset{H}{O}}R'$$

$$[Rh(L)_4Rh\text{-}H]^- + R_3Si\overset{+}{\underset{H}{O}}R' \longrightarrow Rh_2(L)_4 + H_2 + R_3SiOR'$$

Figure 26.3

complex and the protonated silyl ether. Subsequent protonation of the hydride releases the dirhodium catalyst and forms the observed products, molecular hydrogen and the alkoxysilane. Alcohol inhibits this transformation by coordination with the active site of the rhodium(II) catalyst.

D.   Transfer Hydrogenation of Alkenes

The effectiveness of $Rh_2(OAc)_4$ as a hydrogenation catalyst that is selective for monosubstituted alkenes has been reported [5]. Since hydrogen is generated during $Rh_2(pfb)_4$-catalyzed silane alcoholysis (eq. 26.13), we were intrigued by the potential of this reaction for selective hydrogenation of alkenes. Indeed, treatment of styrene with an equivalent amount of triethylsilane in the presence of ethanol and a catalytic amount of $Rh_2(pfb)_4$ caused quantitative reduction of styrene to ethylbenzene (eq. 26.15, R = H).

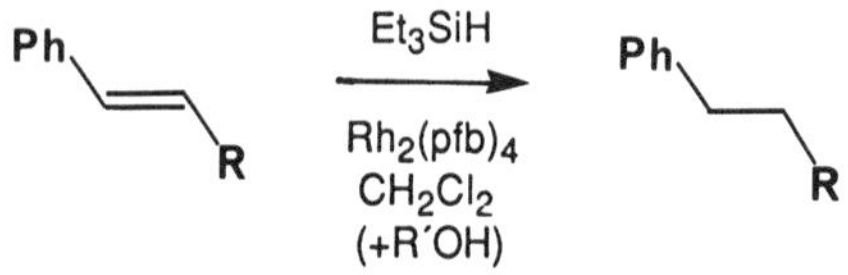

$$\underset{R}{Ph\diagdown} \xrightarrow[\substack{Rh_2(pfb)_4 \\ CH_2Cl_2 \\ (+R'OH)}]{Et_3SiH} \underset{R}{Ph\diagdown} \qquad\qquad (26.15)$$

Similar treatment of ethyl cinnamate (R = COOEt) and cinnamyl alcohol (R = CH_2OH), without added alcohol also resulted in quantitative reduction of the carbon-carbon double bond. However, 1-octene was not hydrogenated under the same conditions. Indeed the alkenes that are better coordinating ligands for $Rh_2(pfb)_4$ than is styrene [9] exhibit little tendency for hydrogenation by this method, which suggests that the alkene, acting as an inhibitor, plays a critical role in defining the selectivity for hydrogenation.

III. CONCLUSIONS

Rhodium(II) perfluorobutyrate is a uniquely effective and selective catalyst for the hydrosilylation of alkynes and alkenes, for silane alcoholysis, and for selective hydrogenation. The predominant trans addition observed for hydrosilylation of alkynes differentiates this catalyst, both synthetically and mechanistically, from those employed earlier [e.g., hexachloroplatinic acid and $(Ph_3P)_3RhCl$]. Rearrangement reactions yielding allylsilanes, although characteristic of chloroplatinic acid catalyzed hydrosilylation of alkynes, are more generally suitable as a synthetic methodology with $Rh_2(pfb)_4$ catalysis. Reactant coordination with the dirhodium catalyst appears to play a critical role in determining both the nature of the transformation and its selectivity.

ACKNOWLEDGMENTS

The authors are grateful to the Robert A. Welch Foundation and to the National Science Foundation for their support of this research, and express our appreciation to the Johnson Matthey Company for their loan of rhodium(III) chloride. We also acknowledge the contributions to these investigations of Gene Devora, Sara Dastgheib-Hosseini, Jonathan Lin, and Roland J. Pieters.

REFERENCES

1. (a) Doyle, M. P., *Acc. Chem. Res.*, *19*, 348 (1986). (b) Doyle, M. P., *Chem. Rev.*, *86*, 919 (1986).
2. Maas, G., *Top. Current Chem.*, *137*, 75 (1987).
3. Doyle, M. P., in *Catalysis of Organic Reactions* (R. L. Augustine, ed.), Dekker, New York, 1985, Chap. 4.
4. Doyle, M. P., unpublished results.
5. Hui, B. C. Y., Teo, W. K., and Rempel, G. L., *Inorg. Chem.*, *12*, 757 (1973).
6. (a) Felthouse, T. R., *Prog. Inorg. Chem.*, *29*, 73 (1982). (b) Cotton, F. A., and Walton, R. A., *Multiple Bonds Between Metal Atoms*, Wiley, New York, 1982, p. 311. (c) Boyar, E. B., and Robinson, S. D., *Coord. Chem. Rev.*, *50*, 109 (1983).
7. Doyle, M. P., Bagheri, V., Wandless, T. J., Harn, N. K., Brinker, D. A., Eagle, C. T., and Loh, K.-L., *J. Am. Chem. Soc.*, *112*, 1906 (1990).
8. Doyle, M. P., Bagheri, V., Pearson, M. M., and Edwards, J. D., *Tetrahedron Lett.*, *30*, 7001 (1989).

9. Doyle, M. P., Mahapatro, S. N., Caughey, A. C., Chinn, M. S., Colsman, M. R., Harn, N. K., and Redwine, A. E., *Inorg. Chem.*, *26*, 3070 (1987).

10. Speier, J. L., *Adv. Organomet. Chem.*, *17*, 407 (1979).

11. Lukevics, E., *Russ. Chem. Rev.*, *46*, 264 (1977).

12. (a) Chalk, A. J., and Harrod, J. F., *J. Am. Chem. Soc.*, *87*, 16 (1965). (b) Chalk, A. J., *J. Organomet. Chem.*, *21*, 207 (1970).

13. (a) Cundy, C. S., Kingston, B. M., and Lappert, M. F., *Adv. Organomet. Chem.*, *11*, 253 (1973). (b) Lappert, M. F., Nile, T. A., and Takahashi, S., *J. Organomet. Chem.*, *72*, 425 (1974).

14. Prignano, A. L., and Trogler, W. C., *J. Am. Chem. Soc.*, *109*, 3586 (1987).

15. (a) Brunner, H., *J. Organomet. Chem.*, *300*, 39 (1986). (b) Brunner, H., *Synthesis*, 645 (1988).

16. Green, M., Howard, J. A. N., Proud, J., Spencer, J. L., Stone, F. G. A., and Tsipsi, C. A., *J. Chem. Soc., Chem. Commun.*, 671 (1976).

17. (a) Tamao, K., Kobayashi, K., and Ito, Y., *J. Am. Chem. Soc.*, *111*, 6478 (1989). (b) Tamao, K., Nakagawa, Y., Arai, H., Higuchi, N., and Ito, Y., *J. Am. Chem. Soc.*, *110*, 3712 (1988).

18. (a) Haszeldine, R. N., Parish, R. V., and Taylor, R., *J. Chem. Soc. (D)*, 2311 (1974). (b) Dickers, H. M., Haszeldine, R. N., Mather, A. P., and Parish, R. V., *J. Organomet. Chem.*, *161*, 91 (1978).

19. (a) Speier, J. L., Webster, J. A., and Barnes, G. H., *J. Am. Chem. Soc.*, *79*, 974 (1957). (b) Saam, J. C., and Speier, J. L., *J. Am. Chem. Soc.*, *80*, 4104 (1958).

20. (a) Onopchenko, A., and Sabourin, E. T., *J. Org. Chem.*, *52*, 4118 (1987). (b) Eddy, V. J., and Hallgren, J. E., *J. Org. Chem.*, *52*, 1903 (1987).

21. Corriu, R. J. P., and Moreau, J. J. E., *J. Organomet. Chem.*, *91*, C27 (1975); *85*, 19 (1975).

22. Hayashi, T., Yamamoto, Y., Kasuga, K., Omizu, H., and Kumada, M., *J. Organomet. Chem.*, *113*, 127 (1976).

23. Brunner, H., and Kurzinger, A., *J. Organomet. Chem.*, *346*, 413 (1988).

24. Pukhnarevich, V. B., Kopylova, L. I., Trofimov, B. A., and Voronkov, M. G., *Zh. Obshch. Khim.*, *45*, 89 (1975); *45*, 2638 (1975).

25. (a) Tamao, K., Miyaki, N., Kiso, Y., and Kumada, M., *J. Am. Chem. Soc.*, *97*, 5603 (1975). (b) Tamao, K., Maeda, K., Tanaka, T., and Ito, Y., *Tetrahedron Lett.*, *29*, 6955 (1988).

26. Collman, J. P., Hegedus, L. S., Norton, J. R., and Finke, R. G., *Principles and Applications of Organotransition Metal Chemistry*, University Science Books, Mill Valley, CA, 1987, p. 564.

27. Cornish, A. J., Lappert, M. F., Filatovs, G. L., and Nile, T. A., *J. Organomet. Chem.*, *172*, 153 (1979).

28. Onopchenko, A., Sabourin, E. T., and Beach, D. L., *J. Org. Chem.*, *48*, 5101 (1983).

29. Onopchenko, A., Sabourin, E. T., and Beach, D. L., *J. Org. Chem.*, *49*, 3389 (1984).

30. (a) Haszeldine, R. N., Parish, R. V., and Riley, B. F., *J. Chem. Soc., Dalton Trans.*, 705 (1980). (b) Dwyer, J., Hilal, H. S., and Parish, R. V., *J. Organomet. Chem.*, *228*, 191 (1982).

31. Luo, X.-L., and Crabtree, R. H., *J. Am. Chem. Soc.*, *111*, 2527 (1989).

32. Corriu, R. J. P., and Moreau, J. J. E., *J. Organomet. Chem.*, *127*, 7 (1977).

33. Yamamoto, K., and Takemae, M., *Bull. Chem. Soc. Jpn.*, *62*, 2111 (1989).

34. Ojima, I., and Kogure, T., *Organometallics*, *1*, 1390 (1982).

35. Doyle, M. P., High, K. G., Bagheri, V., Pieters, R. J., Lewis, P. J., and Pearson, M. M., *J. Org. Chem.*, *55*, 6082 (1990).

36. Sommer, L. H., and Lyons, J. E., *J. Am. Chem. Soc.*, *91*, 7061 (1969).

37. Corriu, R. J. P., and Moreau, J. J. E., *J. Organomet. Chem.*, *114*, 135 (1976).

27

Involvement of $Co_4(CO)_8(\mu_2\text{-}CO)_2(\mu_4\text{-}PPh)_2$ Catalysts in Olefin Hydroformylation

Charles U. Pittman, Jr., and Hikmat Hilal
Department of Chemistry, Mississippi State University,
Mississippi State, Mississippi

Ming-Jaw Don and Michael G. Richmond
Department of Chemistry, Center for Organometallic Research
and Education, University of North Texas, Denton, Texas

I. INTRODUCTION

Homogeneous hydroformylation reactions of terminal olefins (eq. 27.1) have been conducted using a variety of single metal and mixed metal atom clusters as catalyst precursors. Catalysis using carbonyl clusters of ruthenium [1-4], iron [5-8], cobalt [9-12], and others [12] have been reported.

$$R-CH=CH_2 + CO/H_2 \xrightarrow{\text{catalyst}} R-CH_2CH_2CHO + R-CH(CH_3)-CHO$$

$$(27.1)$$

Among the cobalt clusters studied was $Co_4(CO)_8(\mu_2\text{-}CO)_2(\mu_4\text{-}PPh)_2$, 1 [9-11]. This cluster is especially stable due to the presence of the two μ_4-bridging PhP groups. Thus several cobalt-cobalt bonds can be broken without excising a Co atom. This cluster appeared to be a likely candidate for true cluster catalysis, since a total of five bonds to cobalt would need to be broken to lead to a species of lower nuclearity. The structurally related bimetallic cluster $Fe_2Co_2(CO)_{10}(\mu_2\text{-}CO)(\mu_4\text{-}PPh)_2$, 2, also fits this reasoning. Thus, both 1 and 2 were in employed in this study.

Previous reports have presented evidence in favor of Co_4 cluster catalysis with 1 [9-11]. More than a 95% yield of the original cluster 1 has been recovered unchanged in the hydroformylation of 1-pentene after 30-60 hours, using temperatures of 130-150°C and CO/H_2 pressures of 600-1200 psig [9]. Product distribution studies have also been used in the search of further evidence in favor of cluster

1

2

catalysis [9-11]. If 1 were to fragment, it would most likely yield
cobalt species $HCo(CO)_4$ (3) and/or $PhP[Co(CO)_4]_2$ (4), which could
then serve as catalysts or catalyst precursors. Experiments performed
using 1 and $Co_2(CO)_8$ (a precursor to 3) separately as the starting
catalysts, found that at 600 psi, 150°C, and a 1:1 ratio of H_2 to
CO, the aldehyde/alcohol ratio was much lower using 3 as the catalyst
than it was with the use of 1. Furthermore, addition of PPh_3 to 1
resulted in an increase of selectivity toward aldehyde production
(without significant alcohol formation), whereas in case of 3, addition
of PPh_3 is known to enhance the production of the alcohols [13,14].
This evidence was used to exclude $HCo(CO)_4$ as the catalyst, or
catalyst precursor, in reactions in which 1 was the cobalt species
charged.

In this work, three techniques were employed to examine the
Although these observations are in favor of cluster catalysis
by 1, one cannot yet unequivocally exclude noncluster catalysis
or catalysis by a species of lower nuclearity. Other lower nuclearity
transient species may be formed and may be responsible for all
or some of the catalytic events. Such species might exist in an
equilibrium such that return to ambient conditions regenerates the
parent cluster. Two or more concurrent reaction pathways with
different catalytic species may also operate. So far, no reports
have appeared in the literature that either unequivocally confirm
or unequivocally exclude catalysis by cluster 1 in hydroformylation
reactions.

In this work, three techniques were employed to examine the
ability of 1, or a related Co_4 species, to catalyze the hydroformylation reaction of 1-octene at different temperatures. First, the kinetic
criterion proposed by Laine [15] was used. In this technique, plots
of turnover frequency (TF) versus the total cluster concentration
are obtained. A decrease in TF with increasing catalyst concentration
indicates that cluster fragmentation occurs to give a catalytically
active species of lower nuclearity [15]. On the other hand, an

increase in TF with increasing total cluster concentration is consistent
with catalysis by a cluster catalyst of higher nuclearity [15]. A
constant TF with increasing cluster concentration is indicative of
catalysis by the starting cluster or species of equivalent nuclearity
[15]. The second technique involved studies of the product distribution (selectivity), both with and without added phosphine present,
when cluster 1 was the catalyst. These selectivities were compared
to the corresponding studies performed using $Co_2(CO)_8$ as the
catalyst precursor. The third technique employs cylindrical internal
reflectance-Fourier transform infrared (CIR-FTIR) spectroscopy
to look directly at the cobalt species under reaction conditions.
This chapter presents the observations we obtained using each
technique.

II. EXPERIMENTAL

1-Octene and the mixed xylenes were purchased from Aldrich and
used as received after further degassing by N_2 streams. The purity
of 1-octene was 99.8% by gas-liquid chromatography (GLC). Synthesis
gas having a CO/H_2 ratio of 1:1 was obtained by mixing the two
gases at equal pressures in a gas reservoir prior to use. The cobalt
cluster, 1, was prepared as described in the literature [6]. Catalytic
hydroformylation reactions were conducted in 300 mL Parr stainless
steel autoclave reactors. In some reactions these reactors were
fitted with precision stainless steel inserts to give an internal cylindrical reaction volume of approximately 20 mL. The reactor contained
a variable speed stirrer, a sampling tube connected via valving
to allow sampling under the reaction pressures and temperatures,
and inlet ports. A thermocouple well permitted measurements of
the temperature of the reacting solution. A CO/H_2 reservoir was
used with a line constant pressure regulator between the reservoir
and the reactor to keep the pressure constant throughout the course
of the reactions. In a typical experiment, the reactor was charged
with the parent cluster, then assembled and flushed with CO. The
olefin (1-octene, 30 mL = 190.5 mmol), the solvent (mixed xylenes,
20 mL), and the GC internal standard (dried toluene, 10 mL) were
then syringed into the reactor under a stream of CO gas via a
side valve. The reactor was closed and pressure tested (at 900
psi CO without stirring at room temperature for 1 min) by submerging
in water bath; then the pressure of CO was reduced to 50 psi. The
reactor was then heated to the desired internal temperature (throughout a period of 40 min) using a thermocouple-controlled close fitting
heating sleeve. Thus, no hydrogen was present until after the
reaction temperature had been reached. Then the reactor was

pressurized with CO/H_2 (800 psi) from the reservoir to maintain the 1:1 ratio. Stirring was immediately begun. The progress of the reaction was monitored by taking aliquots at different times via the sampling tube loop. Each aliquot was diluted with xylenes (1/25 volume), and passed over silica gel to eliminate metallic species. Aliquots were quantitatively analyzed using gas chromatography (see earlier) employing a temperature program with an initial temperature of 50°C rising at a ramp rate of 10°C/min to a final temperature 160°C, which was then held for 4 minutes.

Kinetics were followed by monitoring the pressure drop in the gas reservoir with a digital pressure transducer and by sampling as a function of time followed by GLC analysis. The rates measured by the uptake of CO/H_2 were compared to those obtained by following the formation of products and disappearance of olefin by internally standardized GLC techniques. The latter method was found to be more accurate and more reproducible. Therefore, the kinetic data in this chapter were obtained via the GLC method. Similar techniques were employed using 2 except that 1-pentene was the olefin in those studies.

III. RESULTS AND DISCUSSION

Hydroformylation of 1-octene using 1 as the catalyst precursor has been conducted in xylenes at different temperatures. At 130°C and below the reaction was very slow, and reasonable kinetic studies could not be made. At 140 and 150°C, the reaction proceeds at reasonable rates giving linear and branched aldehydes at ratios of 1 to 1.5. Alcohols and other side products were formed and analyzed by gas chromatography-mass spectroscopy (GC/MS: Table 27.1). At 150°C, no induction periods were found except at catalyst concentrations below 0.4×10^{-3} M (Fig. 27.1). Induction periods are more obvious with reactions conducted at 140°C (Fig. 27.2).

Pretreatment of the catalytic solutions at 140°C, for 90 minutes or longer under hydrogen or nitrogen, affected the hydroformylation rate. Pretreatment with hydrogen increased the rate but not by a large amount. Hydroformylation did not occur after nitrogen pretreatment. Aliquots taken at different times were turbid, and the clear purple color of the solution of 1 was gone. After 2-3 hours, a dark metallic residue was formed on the reactor walls which was not catalytically active. Thus, 1 decomposes at 140°C under nitrogen.

Pretreatment with CO at 140°C does not affect the activity of the catalyst within experimental error, and CO protects the cluster from decomposition during the reactor heat-up period. In the absence

Table 27.1 Product Selectivities in 1-Octene Hydroformylations Catalyzed by $Co_4(CO)_{10}(\mu_4-PPh)_2$[a] and $Co_2(CO)_8$ with and Without Added PPh_3[b]

Entry	Catalyst name	Catalyst (mmol)	Temperature (°C)	Aldehyde n/b selectivity[c]			Alcohol/Aldehyde at:		
				5 h	10 h	20 h	5 h	10 h	20 h
1	$Co_2(CO)_8$	0.1	140		0.82	0.84	0.024	0.040	0.105
2	$Co_4(CO)_{10}(PPh_3)_2$	0.1	140		1.2	1.3	0	0	0.030
3	$Co_2(CO)_8/PPh_3$	0.1/0.1	140		1	1	0.05	0.09	0.16
4	$Co_4(CO)_{10}(PPh)_2/PPh_3$	0.1/0.1	140		1.1	1.1	0	0	0
5	$Co_4(CO)_{10}(PPh_3)_2/Co_2(CO)_8$	0.04/0.01	140		1.1	2	0.01	0.020	0.043
6	$Co_4(CO)_{10}(PPh_3)_2/Co_2(CO)_8$	0.05/0.05	140	1	1	1.5	0.06	0.06	0.10
7	$Co_2(CO)_8$	0.1	150			0.9		0.16	0.20
8	$Co_4(CO)_{10}(PPh_3)_2$	0.1	150	1.1	1.0	1.0	0.01	0.017	0.10
9	$Co_2(CO)_8/PPh_3$	0.1/0.1	150	1.0	1.0	0.9		0.47	0.70
10	$Co_4(CO)_{10}(PPh_3)_2/PPh_3$	0.1/0.1	150	1.1		1.0	0	0.01	0.04
11	$Co_4(CO)_{10}(PPh_3)_2/PPh_3$	0.1/0.4	150	1.1		1.0	0	0	0.01

[a]$Co_4(CO)_{10}(\mu_4-PPh)_2$ is cluster 1.
[b]All reactions were carried out in mixed xylene solvent (20 mL) and toluene (10 mL) using 1-octene (30 mL, 190.5 mmol) and CO/H_2 (1:1, 800 psig).
[c]n/b = ratio of normal to branched aldehydes.

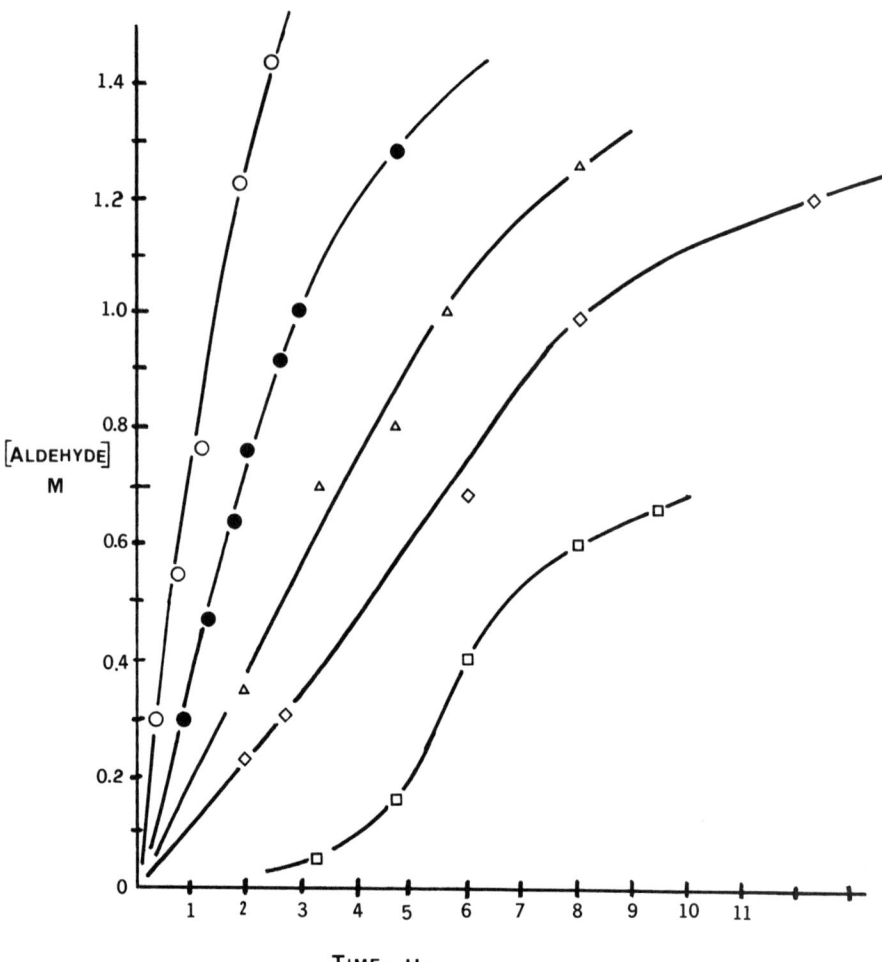

Figure 27.1 Total aldehyde concentration (molarity) versus time
(h) at 150°C in the hydroformylation of 1-octene (30 mL) at CO/H_2
(1:1, 800 psig) in xylene (20 mL) and toluene (10 mL) using
$Co_4(CO)_8(\mu_2\text{-}CO)_2(\mu_4\text{-}PPh)_2$ as the catalyst precursor at the following concentrations. (○) 2.416×10^{-3} M, (●) 2.083×10^{-3} M,
(△) 0.833×10^{-3} M, (◇) 0.416×10^{-3} M, and (□) 0.201×10^{-3} M.

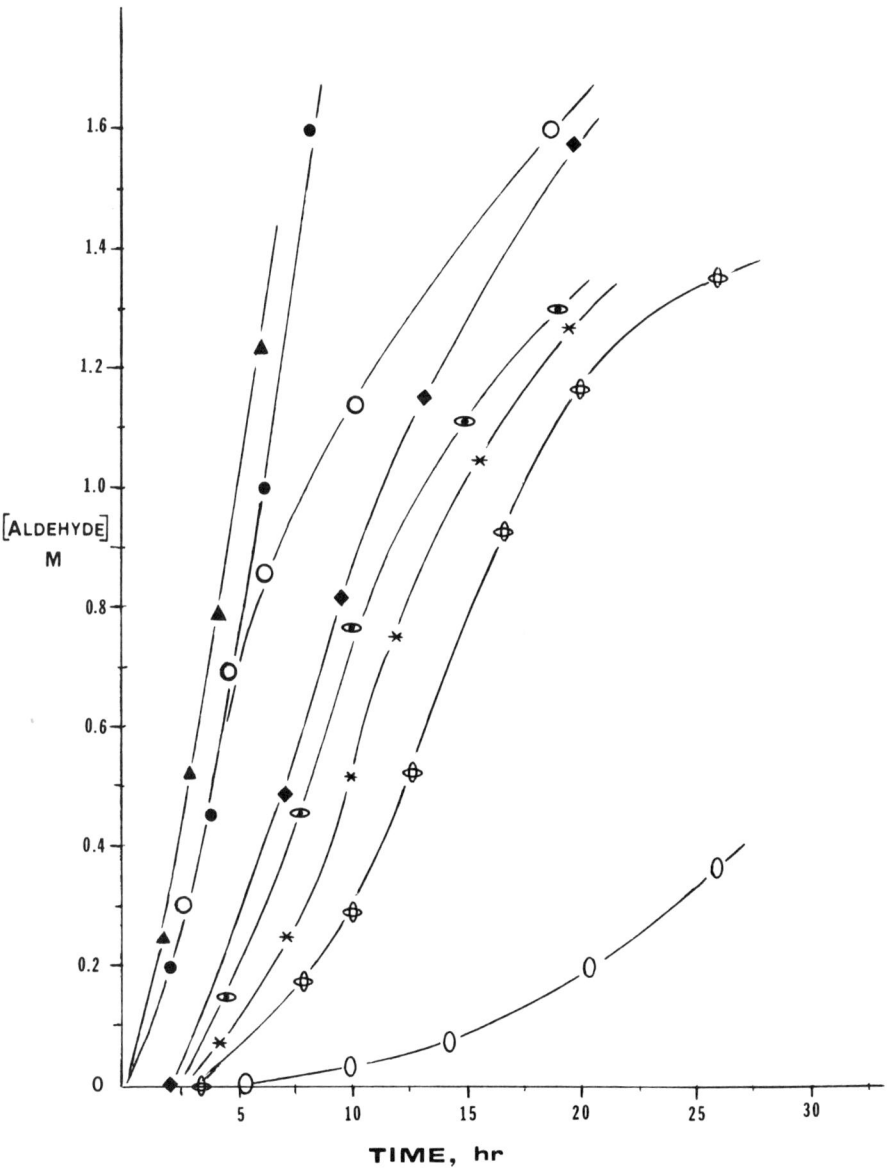

Figure 27.2   Total aldehyde concentration (molarity) versus time
(h) at 140°C in the hydroformylation of 1-octene (30 mL) at CO/H_2
(1:1, 800 psig) in xylene (20 mL) and toluene (10 mL) using
$Co_4(CO)_8(\mu_2\text{-}CO)_2(\mu_4\text{-}PPh)_2$ as the catalyst precursor at the following concentrations: (▲) 2.0×10^{-3} M, (●) 1.68×10^{-3} M, (○)
0.84×10^{-3} M, (◆) 0.71×10^{-3} M, (⬭) 0.60×10^{-3} M, (∗)
0.42×10^{-3} M, (⊕) 0.31×10^{-3} M, and (𝟶) 0.21×10^{-3} M.

of added CO, the cluster decomposition presumably occurs via CO
ligand loss as a first step. Similar observations have been reported
for other metal carbonyl catalytic systems [17]. CO pressure prevents cluster decomposition at 140°C. These conditions are not
yet sufficient to cause fragmentation to such lower nuclearity species
as $Co_2(CO)_4$. However, at 150°C under 800 psi H_2/CO, slow decomposition of 1 does occur, as evidenced in the CIR-FTIR and
turnover frequency experiments.

A. Turnover Frequency Measurements

The reaction was conducted over more than a tenfold range of cluster
concentrations. Measurements of the turnover frequency versus
cluster concentration were made at both 150 and 140°C. Plots of
TF versus cluster concentration at 150°C (and 800 psi $H_2/CO = 1$)
are shown after reaction times of 10 and 4 hours and based on
initial rates, 0 hour (Fig. 27.3). In each case, at 150°C, the TF
decreases sharply with increasing initial cluster concentration. This
suggests that fragmentation of the parent cluster occurs to give
a species of lower nuclearity as one major contributor to the observed
catalysis [15]. Catalysis could be occurring through one or more
catalytically active species of lower nuclearity in addition to some
catalysis proceeding through 1. This is consistent with observations
of this reaction using CIR-FTIR method, where 15-20% cluster fragmentation to $^-Co(CO)_4$ and/or other unidentified species was observed
at 150°C after 6 hours of exposure to CO, CO/H_2, or H_2 pressures
of 250 psi.

One can make the assumption that the only active catalysts
are based on 1 and $HCo(CO)_4$. If cluster 1 slowly fragments into
$HCo(CO)_4$ under reaction conditions (150°C, 800 psi $H_2/CO = 1$,
1-octene present), then $HCo(CO)_4$ should make a progressively
greater percentage contribution to the observed catalysis of hydroformylation as its concentration increases and as the concentration
of 1 decreases. Thus, early in the reaction, most of the catalysis
would occur through 1. One would then expect the curve of TF
versus cluster concentration to fall more slowly when the TF measure-
ment was based on *initial rates* than when TF was based on overall
conversion after several hours. This is consistent with the observed
results (compare the slope of the TF plot at time = 0 with time = 4
or 10 h in Fig. 27.3).

At 140°C, the TF *did not rapidly decrease with increasing cluster
concentration*. The erratic occurrence of induction periods prevented
investigators from obtaining highly accurate plots of highly accurate
TF versus catalyst concentration based on initial rate measurements
at low cluster concentrations. However, data taken at 140°C after

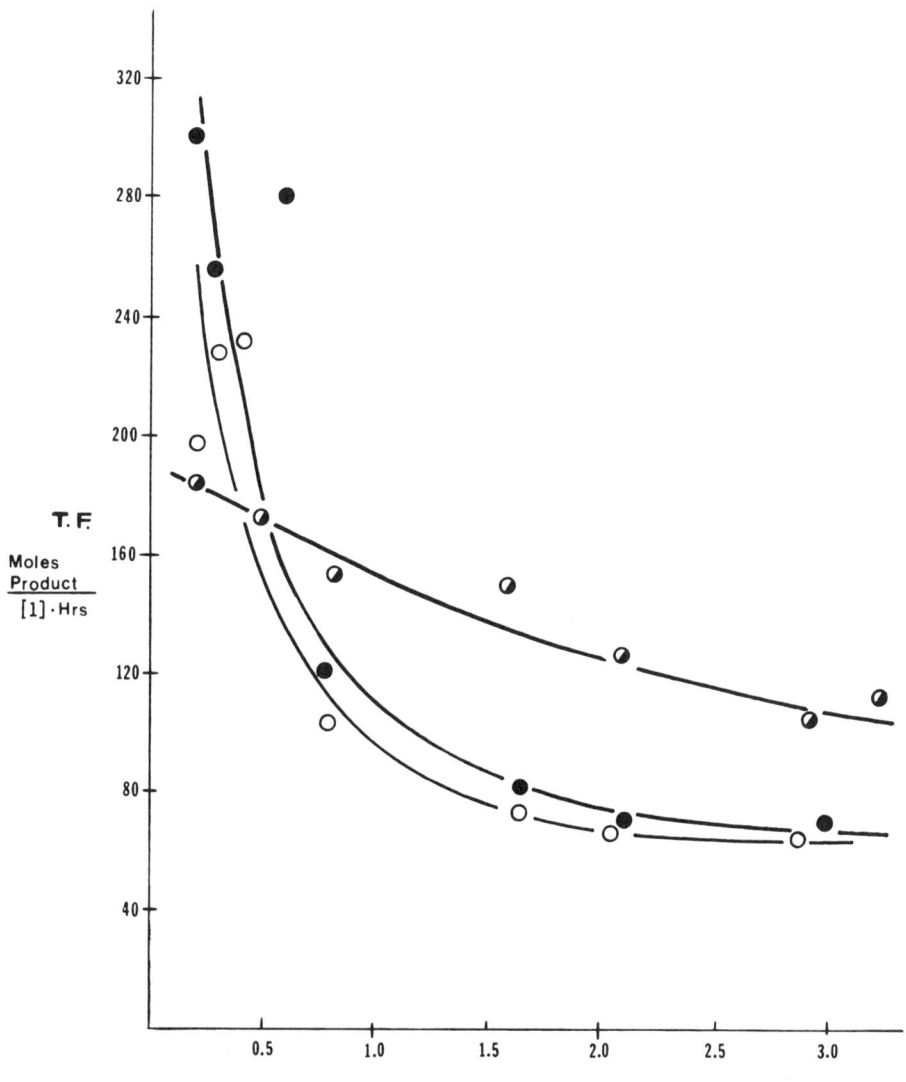

Figure 27.3 Plot of the turnover frequency versus $Co_4(CO)_8(\mu_2-CO)_2(\mu_4-PPh)_2$ molar concentration in 1-octene hydroformylations at 150°C under CO/H_2 (1:1, 800 psig) at 10 (○) and 4 (●) hours initial rate at time 0 (◕).

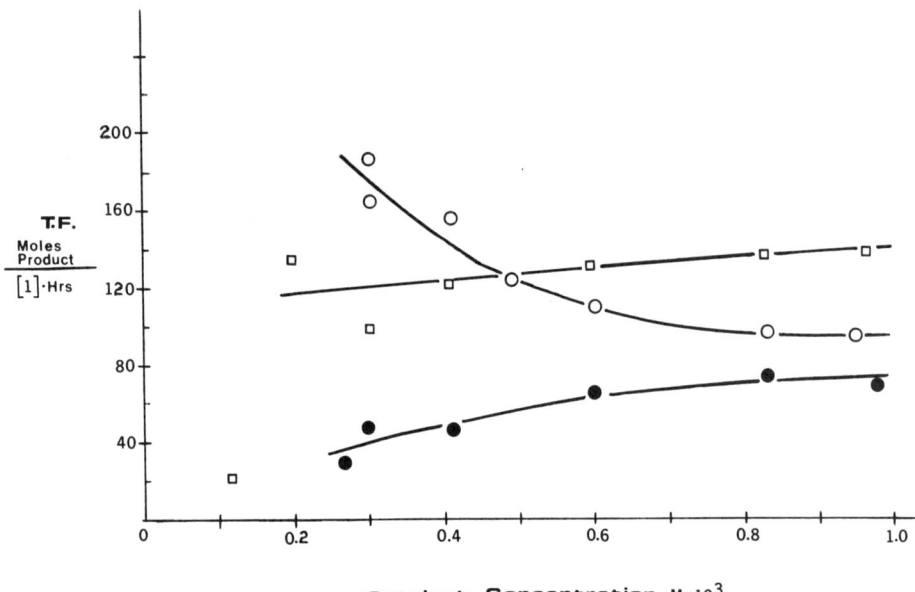

Figure 27.4   Plot of the turnover frequency versus $Co_4(CO)_8(\mu_2-CO)_2(\mu_4-PPh)_2$ molar concentration in 1-octene hydroformylation at 140°C under CO/H_2 (1:1, 800 psig) at 20 (o), 10 (□), and 5 (●) hours.

reaction periods of 5, 10, and 20 hours give the TF plots shown in Figure 27.4. There is no clear indication that TF decreases with increasing concentration of the cluster at either 5 or 10 hours within the error limits of the method. The data summarized in Figure 27.4 at 140°C and 800 psi $H_2/CO = 1$ are consistent with catalysis by an active species with Co_4 nuclearity and not consistent with catalysis caused by fragments of lower nuclearity even after 10 hours. The downward slope of the TF plot after 20 hours, however, shows that the catalytic contribution from $HCo(CO)_4$ is steadily growing with time.

   This interpretation is also consistent with CIR-FTIR spectra obtained at 130 and 140°C in benzene. No indication of cluster fragmentation was observed in the CIR-FTIR spectra under conditions of hydroformylation of 1-pentene. If cleavage of 1 to $HCo(CO)_4$ occurs slowly at 150°C under reaction conditions, such cleavage would be much slower at 140°C. If $HCo(CO)_4$ is produced only slowly in catalytically significant amounts at 140°C, then decreasing plots of TF versus cluster concentration should begin to appear

only at long reaction times. Indeed, after reaction times of 20 hours
(Fig. 27.4), there is evidence (downward slope of the TF plot)
that a significant catalysis contribution might be manifested from
a lower nuclearity species that results from cluster fragmentation.
This is certainly less pronounced than in the case of 150°C reaction,
since the downward slope of this TF plot is much smaller than at
150°C.

B. Selectivity Studies

The product distributions for 1-octene hydroformylations were compared using both 1 and $Co_2(CO)_8$ as the catalyst precursor at both
140 and 150°C (Table 27.1). Also, the effect of added triphenylphosphine on the product selectivities was studied with both catalyst
precursors. If fragmentation of 1 were to occur, likely fragments
would include $Co_2(CO)_8$ and $PhP[Co(CO)_4]_2$. Under reaction conditions, the $Co_2(CO)_8$ would give catalytically active $HCo(CO)_4$ [7]
according to:

$$H_2 + Co_2(CO)_8 \rightleftharpoons 2HCo(CO)_4$$

If 1 and $Co_2(CO)_8$ [e.g., $HCo(CO)_4$] were to have different catalytic
selectivities, the fragmentation of 1 into $HCo(CO)_4$ over time would
result in a shift in the product distribution to that expected for
$HCo(CO)_4$. Studies to investigate this point are summarized in
Table 27.1.

At 140°C the normal-to-branched aldehyde ratio is lower for
$Co_2(CO)_8$ (entry 1 in Table 27.1) than for 1 (entry 2). Most notably,
$Co_2(CO)_8$ produces more alcohol by-product than does catalysis
by 1. This is also true at 150°C (entries 7 and 8) although after
20 hours the reactions catalyzed by 1 show significant amounts
of alcohol. At long times the product distribution produced by 1
becomes more like that produced by $Co_2(CO)_8$. This is consistent
with decomposition of some 1 into $HCo(CO)_4$ over long periods at
150°C (as was shown by the CIR-FTIR work).

The effect of adding PPh_3 to the two catalyst precursors differed
at 140 and 150°C. The addition of PPh_3 to $Co_2(CO)_8$ at 140°C causes
an increase in alcohol formation (cf. entries 3 and 1). In contrast,
the addition of PPh_3 to 1 completely retards alcohol production
(cf. entries 4 and 2). At 150°C the same trends are seen. Adding
PPh_3 to $Co_2(CO)_8$ increases alcohol production (entry 9 vs. 7),
whereas adding PPh_3 to 1 retards alcohol formation (entries 11
and 10 vs. 8). Taken together, it is clear that at 140°C the catalytic
action from 1 cannot be due to $Co_2(CO)_8$ [e.g., $HCo(CO)_4$]. The
two catalyst precursors cannot give a common active species. At

150°C there is evidence for significant catalysis by $HCo(CO)_4$ produced from 1 after 20 hours, but added PPh_3 reduces this tendency.

Catalytic reactions were also carried out using mixtures of 1 and $Co_2(CO)_8$ at 140°C (entries 5 and 6). As the mole ratio is varied from only 1 to only $Co_2(CO)_8$ (e.g., entries 2, 5, 6, 1 in that order) one observes a steady increase in alcohol formation.

It is well known that addition of PPh_3 to $Co_2(CO)_8$ enhances the formation of alcohols in catalytic hydroformations [13,14]. The ability of PPh_3 to suppress alcohol formation using cluster 1 clearly establishes that fragmentation of 1 to $HCo(CO)_4$ is not important at 140°C and becomes significant only at longer time periods at 150°C. These findings are completely consistent with the TF trends and with the CIR-FTIR results. Even if some fragmentation of 1 occurs after long reaction times at 140°C (as expected from Laine's criterion applied to the 20 h TF plot), major involvement by $HCo(CO)_4$ appears to be ruled out by the product distribution studies. These conclusions are in agreement with an earlier CIR-FTIR study of 1-pentene hydroformylation catalyzed by cluster 1 at 130°C [8]. In that study no cluster fragmentation to $HCo(CO)_4$ was observed during the reaction.

Both the TF plots and the product distribution studies show that kinetically significant amounts of $HCo(CO)_4$ are formed at 150°C with the first hour and contribute to the catalysis. However, the addition of triphenylphosphine must stabilize 1 and reduce its fragmentation rate to $HCo(CO)_4$. If 1 continued to fragment to $HCo(CO)_4$ at the same rate as it did without added phosphine, the production of alcohol would have to become faster, since it is established that $HCo(CO)_4$ produces more alcohol in the presence of triphenylphosphine. This stabilization of 1 toward fragmentation is a curious result, since only cluster 1 is reisolated from hydroformylation runs at 130, 140, and 150°C in which 4 equivalents of triphenylphosphine had been added. Furthermore, when the authentic bis-triphenylphosphine adduct of 1 [e.g., cis-$Co_4(CO)_6(\mu_2$-$CO)_2(PPh_3)_2(\mu_4$-$PPh)_2$] is introduced as the catalyst precursor, only 1 is isolated after the reaction. Richmond's demonstration that 15-20% cluster decomposition occurs after 6 hours at 150°C under CO pressures [8] suggests that under such forcing conditions (150°C), the cluster fragments to achieve higher saturation of cobalt atoms with CO ligands. Thus, $^-Co(CO)_4$ is formed under CO pressure, and $HCo(CO)_4$ under H_2/CO pressures.

C.   Hydroformylation Activity of $Co_4(CO)_8(\mu_2$-$CO)_2(\mu_4$-$PPh)_2$ as Examined by Cylindrical Internal Reflectance Spectroscopy

IR and UV/visible spectra were obtained for the reaction mixtures, before and after reaction. These spectra did not show any significant change in the nature of the cluster at 140°C. However, though consistent with our discussion of cluster catalysis, these spectroscopic data alone cannot be accepted as confirming cluster catalysis at 140°C. Partial fragmentation to transient (undetectable) catalytic species, which recombine to form the parent cluster, may be involved. Used with turnover frequency criteria and product distribution studies, the case for cluster catalysis at 140°C is strong. To gain further insight, cylindrical internal reflectance (CIR) spectroscopy was used to study the hydroformylation of 1-pentene using $Co_4(CO)_8(\mu_2\text{-}CO)_2(\mu_4\text{-}PPh)_2$. The CIR method [18-21] was used to probe for cluster fragmentation under steady state catalytic conditions. This spectroscopic technique allows for the in situ examination of the working catalyst solution under steady state conditions as demonstrated by the work of Moser [22-26], Darensbourg [27-29], and Richmond [30,31]. Here $Co_4(CO)_8(\mu_2\text{-}CO)_2PPh)_2$ (~ 0.27 mmol) in benzene (8 mL) and 1-pentene (9 mmol) was examined as a function of synthesis gas pressure and temperature.

Figure 27.5 shows the CIR-FTIR spectrum of $Co_4(CO)_8(\mu_2\text{-}CO)_2(\mu_4\text{-}PPh)_2$ in benzene solution at room temperature under 600 psi of H_2/CO (1:1). The carbonyl bands observed at 2041 (vs), 2030 (s), 2014 (s), and 1869 (m) cm^{-1} are in good agreement with the IR frequencies recorded for the cluster in the normal transmission mode [10,32]. The parent cluster is observed to be stable for extended periods of time under these conditions, suggesting that mixtures of H_2/CO do not immediately promote cluster fragmentation. Figure 27.6 shows the CIR-FTIR spectrum of $Co_2(CO)_8$ recorded under analogous conditions. The observed carbonyl stretching bands at 2070 (vs), 2042 (vs), 2021 (vs), and 1851 (s) cm^{-1} are identical to those reported in the literature [33] and serve to illustrate the ability of this spectroscopic technique to identify this possible cluster fragmentation product if it forms during the hydroformylation reaction (see below).

The effect of H_2/CO pressure on the hydroformylation reaction was initially studied at 130°C (constant temperature). Initial H_2/CO (1:1) loading pressures chosen were 600, 800, and 1000 psi. CIR-FTIR spectra were then recorded as a function of time once the reactor had reached 130°C. All CIR-FTIR spectra revealed only the presence of the starting cluster [$v(CO)$:2040, 2030, 2015, 1871 cm^{-1}] and served to establish that the cluster does not fragment immediately to species of lower nuclearity under high pressures of H_2/CO at 130°C. Figure 27.7 shows the CIR-FTIR spectra of cluster 1 at 1000 psi; the spectrum in Figure 27.7A was recorded under autogenous hydroformylation conditions and is very similar to the corresponding initial spectrum shown in Figure 27.5 in the metal carbonyl region (2100-1800 cm^{-1}). It is evident that extensive

Co$_4$(CO)$_{10}$(PPh)$_2$

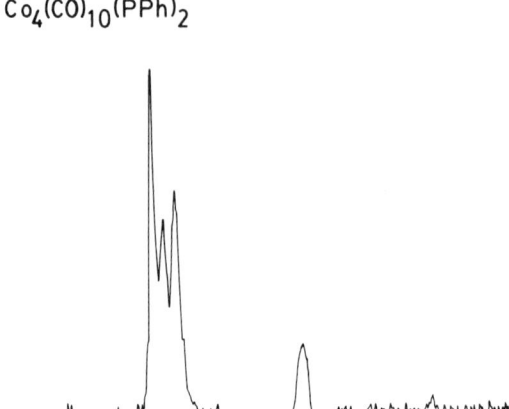

Figure 27.5   CIR–FTIR spectrum of Co$_4$(CO)$_{10}$ (μ_4-PPh)$_2$ in benzene with H$_2$/CO (600 psi; 1:1) at room temperature.

Co$_2$(CO)$_8$

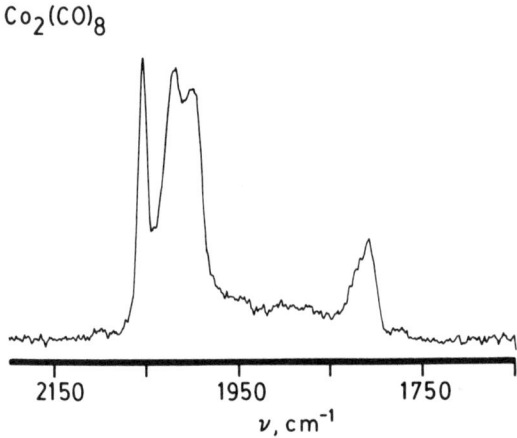

Figure 27.6   CIR–FTIR spectrum of Co$_2$(CO)$_8$ in benzene with H$_2$/CO (600 psi; 1:1) at room temperature.

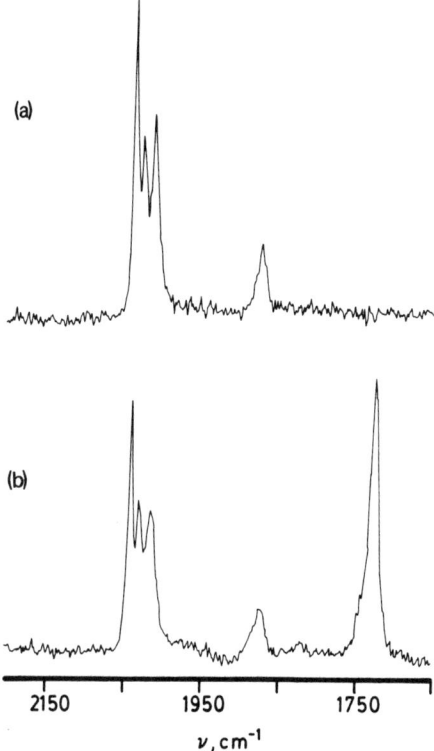

Figure 27.7 CIR-FTIR spectra of $Co_4(CO)_{10}(\mu_4-PPh)_2$/1-pentene in benzene with H_2/CO (1000 psi; 1:1) at 130°C after (A) 1 hour and (B) about 58 hours.

cluster fragmentation has not occurred because the absorbance values of the cluster's carbonyl groups have not appreciably changed during the course of the reaction. CIR experiments conducted at lower H_2/CO pressures are unexceptional compared to the 1000 psi reaction and require no description.

The appearance of the product aldehydes is readily observed by the 1729 cm^{-1} carbonyl stretching band, which also allows us to quantify the extent of the aldehyde production. Approximately 30-40% of aldehyde is present in each spectrum after 20 hours, based on an independent CIR calibration study of the product aldehydes (note, these spectra are not shown). Finally, Figure 27.7B shows the CIR-FTIR spectrum from the 1000 psi H_2/CO experiment after about 58 hours at 130°C. Even with prolonged heating and

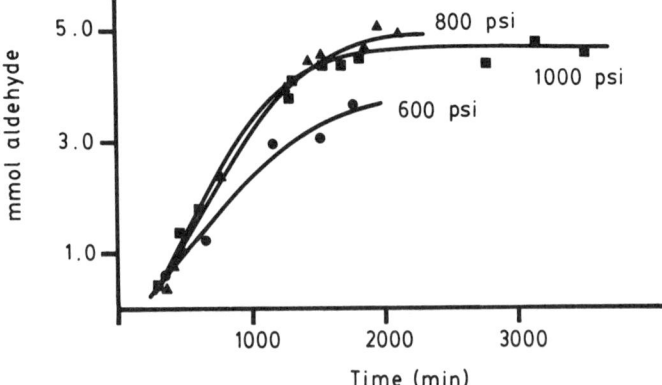

Figure 27.8   1-Pentene hydroformylation results (CIR-FTIR) using
$Co_4(CO)_{10}(\mu_4\text{-PPh})_2$ at 130°C as a function of the initial H_2/CO
(1:1) loading pressure: 600 (●), 800 (▲), and 1000 (■) psi.

hydroformylation activity, it is clear that the parent cluster does
not undergo irreversible fragmentation and/or decomposition, because
the absorbance values for the cluster's carbonyl stretching bands
are equal, within experimental error, to the CIR spectrum recorded
before warm-up. Furthermore, when the same spectrum is cooled
to room temperature and $Co_4(CO)_8(\mu_2\text{-CO})_2(\mu_4\text{-PPh})_2$ is subtracted,
only the product aldehyde band is observed. The resulting difference
spectrum reinforces the absence of other metal carbonyl containing
species (see below).

Figure 27.8 shows the pressure dependence of aldehyde formation
in the hydroformylation reaction at 130°C as monitored by CIR spectroscopy. The plot was constructed by plotting the millimoles of
aldehydes present (determined by the absorbance value associated
with the 1729 cm^{-1} band) as a function of time. Regardless of the
initial H_2/CO pressure, there appears to be a roughly 5 hour
induction period in each catalytic run. This is significant because
it may represent the time it takes for the cluster to transform into
a catalytically active species. Cluster fragmentation at this point
to give $HCo(CO)_4$ (or some derived species) is not observable at
130°C, given the similarity in the CIR spectra recorded before,
during, and after each hydroformylation reaction. In addition, spectral subtraction examination reveals no new metal carbonyl complexes
immediately following observable catalytic activity.

The effect of additives such as azobisisobutyronitrile AIBN,
hexanal, and hexyl alcohol on the hydroformylation reaction was also
examined at 130°C and 600 psi of H_2/CO (1:1). If these reactions pro-

ceed via a radical chain mechanism, it is conceivable that both AIBN and hexanal could accelerate the reaction by providing a high initial concentration of initiator radicals. AIBN (5 mol % based on cluster) did not appear to affect the hydroformylation as the CIR spectra recorded as a function of time mirrored the 130°C and 600 psi H_2/CO experiment already described (see Fig. 27.8). The presence of hexanal (1.3 mmol) clearly retarded the hydroformylation because the final amount of aldehydes was calculated to be about 1.4 mmol, based on the absorbance of the 1729 cm^{-1} carbonyl band ($\sim$ 16% conversion based on 1-pentene after subtraction of the initially charged hexanal). The observed aldehyde inhibition may reflect the fact that the aldehyde competes more effectively for a vacant coordination site on the cluster than 1-pentene, thereby preventing 1-pentene consumption. Inhibition has also been observed by CIR spectroscopy when $Co_4(CO)_8(\mu_2\text{-}CO)_2(\mu_4\text{-}PPh)_2$, $P(OMe)_3$ (2 mol equiv), and 1-pentene are treated with 600 psi H_2/CO at 130°C [34]. No hydroformylation activity was observed after 10 hours, presumably the result of rapid and reversible $P(OMe)_3$ coordination, which prohibits alkene coordination [35]. Added hexyl alcohol ($\approx$ 1.2 mmol) had no measurable effect on the reaction based on a reaction profile that was similar to that observed in Figure 27.8.

The reaction at 140°C and 600 psi of H_2/CO (1:1) was next examined. Figure 27.9A shows the CIR spectrum recorded after 1 hour at 140°C while Figure 27.9B shows the CIR spectrum about 20 hours into the hydroformylation reaction. The spectra are identical in the metal carbonyl region, indicating that the parent cluster is either the only species or the predominate species in solution. A plot of the aldehydes present as a function of time is shown in Figure 27.11, below. The initial rate of formation of aldehydes is greater than the rate found in the 130°C experiment, and the aldehyde production is observed to level off after about 33% 1-pentene conversion. After 35 hours, the reaction was terminated and allowed to cool to room temperature. The CIR spectrum showed both $Co_4(CO)_8(\mu_2\text{-}CO)_2(\mu_4\text{-}PPh)_2$ and aldehydes present; however, when $Co_4(CO)_8\text{-}(\mu_2\text{-}CO)_2(\mu_4\text{-}PPh)_2$ was subtracted out of the room temperature spectrum, a trace amount of $HCo(CO)_4$ could be discerned based on the weak νCO at 2038 cm^{-1} [36,37]. It is important to note that during the early phases of the reaction, the IR spectral subtraction revealed no $HCo(CO)_4$, reinforcing the involvement of some form of the parent cluster as the active hydroformylation catalyst.

Raising the temperature to 150°C shortened the induction period to less than 30 minutes (no induction periods were observed in most batch runs at 150°C: see Fig. 27.1). This led to an increasing production of aldehydes over the first 4.5 hours of the reaction. After this period, the total amount of aldehydes was observed to

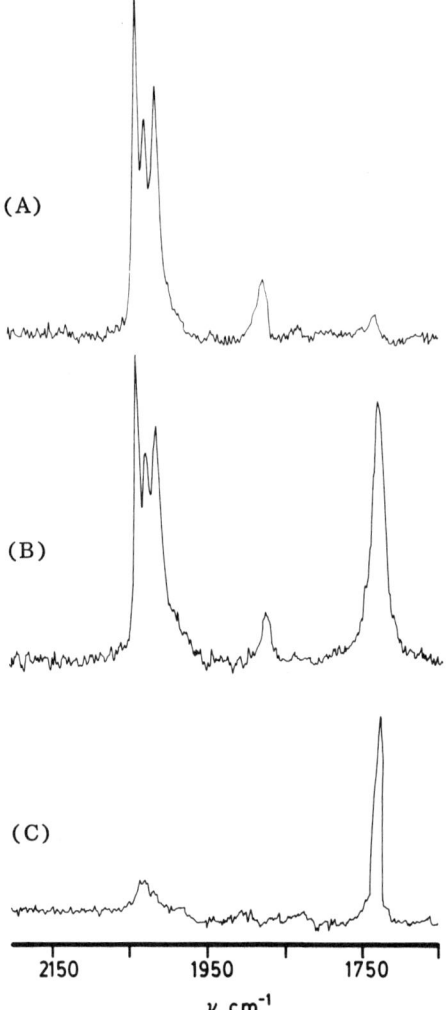

(A)

(B)

(C)

2150   1950   1750

ν, cm^{-1}

Figure 27.9 CIR–FTIR spectra of $Co_4(CO)_{10}$ (μ_4-PPh)$_2$/1-pentene in benzene with H_2/CO (600 psi; 1:1) at 140°C after (A) 2 hours and (B) about 20 hours; (C) represents the room temperature spectrum after the parent cluster had been subtracted from the reaction solution.

slowly decrease, as shown in Figure 27.10. The CIR-FTIR spectra
recorded over the first few hours are very similar to the other
experiments (see, e.g., Figs. 27.7, 27.9). After about 10 hours,
however, unequivocal evidence is available for the support of cluster
fragmentation to $HCo(CO)_4$. Figure 27.10B shows the CIR spectrum
after 11 hours at 150°C, where the terminal metal carbonyl bands
exhibit an asymmetry not previously observed. In addition, the
μ_2-carbonyl band is greatly diminished, which suggests that major
cluster alteration has taken place. Figure 27.10C, recorded 18 hours
into the reaction, clearly shows signs of fragmentation of the parent
cluster, based on the diminished terminal carbonyl bands at 2030
and 2015 cm^{-1} coupled with an increase in the high frequency band
at 2040 cm^{-1}. These changes are even more pronounced in the CIR
spectrum recorded after 33 hours as shown in Figure 27.10D and
the spectrum obtained upon cooling (Fig. 27.10E).

Conclusive proof for the presence of $HCo(CO)_4$ in the 150°C
reaction derives from both the CIR-FTIR spectra and the presence
of hexanol and 2-methylpentanol. Increasing amounts of $HCo(CO)_4$
are seen after 11 hours of reaction (starting with Fig. 27.10B)
based on the appearance of the 2038 cm^{-1} carbonyl band that is
present in the difference spectrum after $Co_4(CO)_8(\mu_2-CO)_2(\mu_4-PPh)_2$ is subtracted out [38]. Unfortunately, we are unable at this
point to quantify the amount of $HCo(CO)_4$ present in any given
spectrum and can use only the amount of $Co_4(CO)_8(\mu_2-CO)_2(\mu_4-PPh)_2$ subtracted out to provide a rough estimate of the amount
of remaining parent cluster. For example, we estimate that less
than 5% of the parent cluster has decomposed after 11 hours
(Fig. 27.10B) based on the fact that 95% of the parent cluster can
be subtracted out of the CIR spectrum. $HCo(CO)_4$ is clearly visible
in the difference spectrum of the final reaction solution, which is
shown in Figure 27.10F. Approximately 70% $Co_4(CO)_8(\mu_2-CO)_2(\mu_4-PPh)_2$ has been removed from the CIR-FTIR spectrum shown in
Figure 27.10E, indicating that 30% of the initially charged parent
cluster has decomposed in this higher temperature reaction. Our
observations concerning $Co_4(CO)_8(\mu_2-CO)_2(\mu_4-PPh)_2$ are not entirely
unexpected; indeed, King et al. [37] have also observed decomposition of this cluster to give $HCo(CO)_4$ at high temperature (> 185°C).

The aldehydes' concentration within the CIR reactor cell also
provides supporting evidence for the presence of $HCo(CO)_4$.
Figure 27.11 shows that at 130°C the concentration of aldehydes
steadily increases, while at 140°C the concentration levels off, more
or less, to a conversion value of approximately 37%. However, the
reaction conducted at 150°C shows a consumption of the aldehydes
after 8 hours based on the gradual decrease in the absorbance
of the 1729 cm^{-1} carbonyl band. We believe that as $HCo(CO)_4$ forms

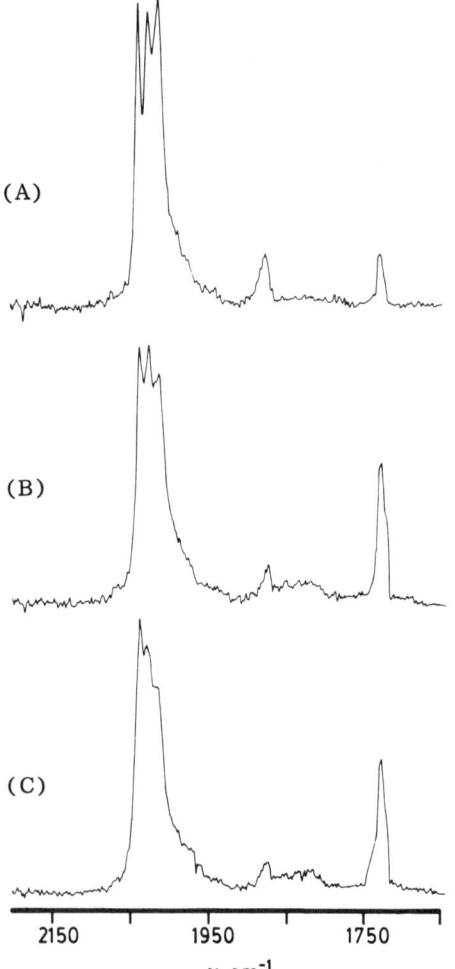

(A)

(B)

(C)

2150   1950   1750

ν, cm^{-1}

Figure 27.10   CIR–FTIR spectra of $Co_4(CO)_{10}(\mu_4\text{-PPh})_2$/1-pentene in benzene with H_2/CO (600 psi; 1:1) at 150°C after (A) 1, (B) 11, (C) 18, and (D) 33 hours. Spectrum (E) was recorded at room temperature after the hydroformylation reaction had been terminated (33 h). Spectrum (F) represents the room temperature spectrum after the subtraction of the parent cluster from the reaction solution.

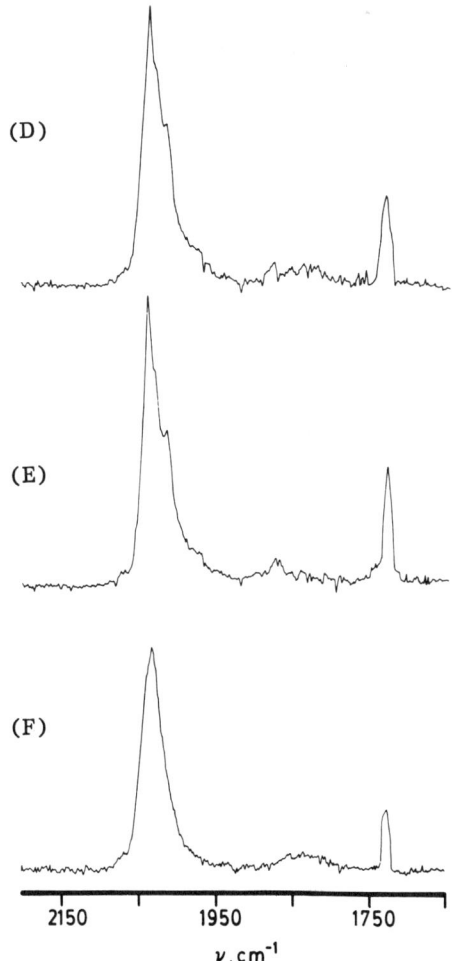

Figure 27.10 (Continued)

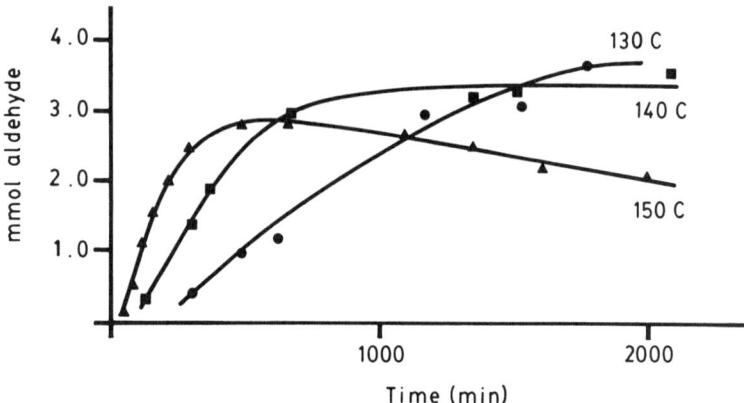

Figure 27.11 1-Pentene hydroformylation results (CIR-FTIR) using $Co_4(CO)_{10}(\mu_4\text{-}PPh)_2$ in benzene with H_2/CO (600 psi; 1:1) as a function of temperature: 130 ($\bullet$), 140 ($\blacksquare$), and 150°C ($\blacktriangle$).

it begins to hydrogenate the hexanal and 2-methylpentanal to the corresponding alcohols. This reaction is known to be facile for $HCo(CO)_4$ [39]. While the CIR technique does not lend itself to the ready quantitation of these alcohols, independent GLC analyses do indeed confirm the onset of aldehyde-alcohol hydrogenation with the appearance of $HCo(CO)_4$ in the CIR spectra.

D. Hydroformylation Activity of $Fe_2Co_2(CO)_{11}(\mu_4\text{-}PPh)_2$ as Examined by Cylindrical Internal Reflectance Spectroscopy and Batch Reactions

The catalysis of organic reactions by mixed metal clusters is an attractive topic due to the possibility of modulated chemical reactivity as a result of adjacent dissimilar metals [40-42]. Polar metal-metal bonds formed by a combination of early and late transition metals have been shown to undergo reaction with CO [43-45]. Combined with the interest in the cluster surface analogy [46,47] is the possibility that heteroatom clusters may serve as paradigms for bimetallic and alloy catalysts of known stoichiometry and composition [48,49]. The mixed metal cluster $Fe_2Co_2(CO)_{11}(\mu_4\text{-}PPh)_2$, 2 is a close structural relative of 1 [50]. Thus, it was prepared and tested in batch reactions and *in situ* CIR-FTIR studies of 1-pentene hydroformylation [30].

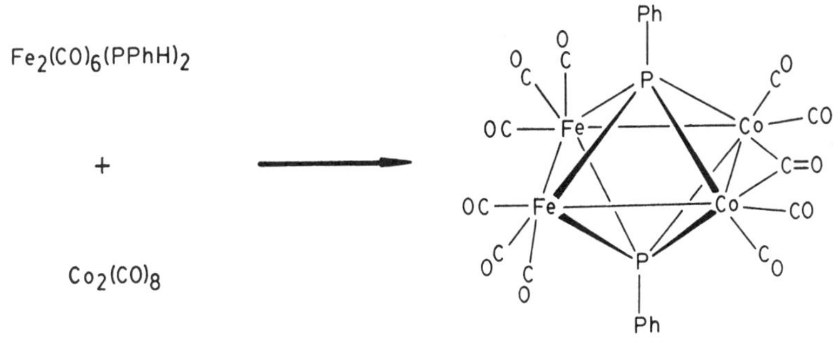

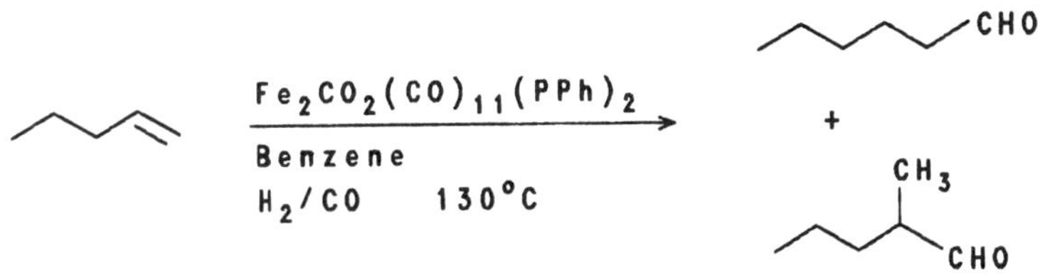

$Fe_2(CO)_6(PPhH)_2$

+

$Co_2(CO)_8$

$\longrightarrow$

$$ \xrightarrow[\substack{\text{Benzene} \\ H_2/CO \quad 130°C}]{Fe_2Co_2(CO)_{11}(PPh)_2} $$

Slow hydroformylation of 1-pentene to hexanal and 2-methyl-pentanal was observed under batch reactor conditions using 2. Table 27.2 illustrates representative results, showing the effect of H_2/CO pressure on the ratio of normal to branched (n/b) alde-hydes. Both FTIR and HPLC analyses of the final reaction solutions revealed 2 as the only organometallic species present (> 90%), strongly suggesting that intact 2 (or some Fe_2Co_2 nuclearity species) functions as the active catalyst.

This reaction was next examined by the *in situ* CIR-FTIR tech-nique, the results of which are shown in Figure 27.12. 1-Pentene (9 mmol), benzene (8 mL), and 2 (≈ 0.27 mmol) were charged to

Table 27.2   Hydroformylation of 1-Pentene Catalyzed by $Fe_2Co_2(CO)_{11}(\mu_4-PPh)_2$ at 130°C[a]

Entry	Pressure (psi)[b]	Aldehyde (% yield)	Alcohol (% yield)	Time (h)	n/b selectivity[c]
1	400	50		168	3.2
2	600	73		170	1.7
3	800	89	Trace	150	1.7

[a]Yields based on 1-pentene consumed as determined by GLC analysis.
[b]Initial H_2/CO pressure (1:1).
[c]Ratio of hexanal to 2-methylpentanal.

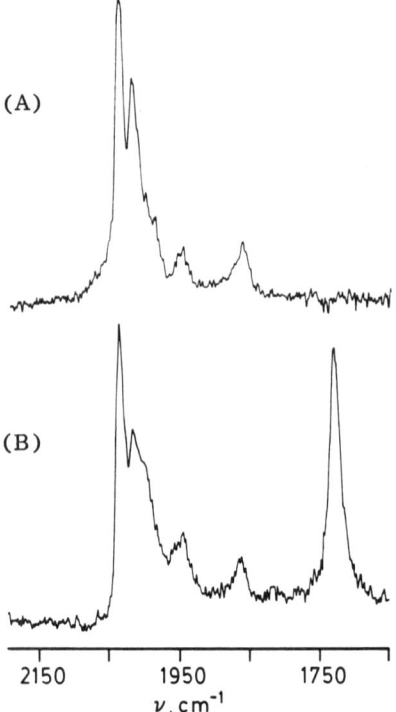

(A)

(B)

2150   1950   1750
ν, cm^{-1}

Figure 27.12   CIR-FTIR spectra of Fe$_2$Co$_2$(CO)$_{11}$(μ_4-PPh)$_2$/1-pentene in benzene with H$_2$/CO (600 psi; 1:1) at (A) room temperature and (B) 130°C and 30 hours into the hydroformylation reaction.

the CIR reactor and pressurized to 600 psi with H$_2$/CO (1:1). CIR-FTIR analysis both at ambient temperature and at 130°C over long periods revealed metal carbonyl stretching bands at 2039 (vs), 2020 (s), 1948 (m), and 1862 (m) cm^{-1}, in agreement with the absorbances reported for 2 [34]. Apart from the temperature broadening of the CIR-FTIR spectra, the ambient and 130°C spectra were identical except for the aldehyde product (1729 cm^{-1}), which is observed at the latter temperature. The intensity of the absorptions due to 2 remains constant even after long periods at 130°C under hydroformylation conditions. Furthermore, solution stability studies of 2 in the absence of olefin under CO, H$_2$/CO, or H$_2$ at 150°C showed negligible fragmentation associated with 2. Thus, cluster 2, or a Fe$_2$Co$_2$ nuclearity species derived from 2, functions as the active catalyst.

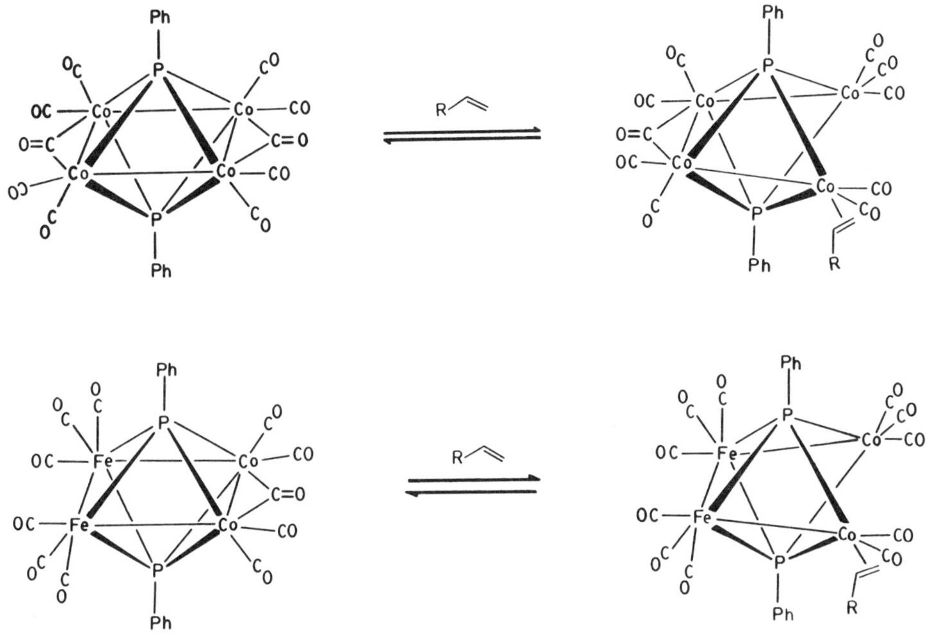

closo   nido

Figure 27.13

While the existing data does not permit the presentation of
a detailed mechanism, the CIR-FTIR measurements reveal that both
$Co_4(CO)_8(\mu_2-CO)_2(\mu_4-PPh)_2$ and $Fe_2Co_2(CO)_{11}(\mu_4-PPh)_2$ persist
during catalysis at 130°C, with the former cluster functioning as
the active catalyst at 140°C. Although the mixed metal cluster 2
was not examined at higher temperatures, the stability of 2 upon
exposure to CO, H_2/CO, and H_2 at high temperature underscores
the possibility of cluster-mediated catalysis. One possible mode
for substrate activation consistent with the foregoing results involves
the 1-pentene-induced closo → nido polyhedral expansion depicted
in Figure 27.13. Such ligand-induced transformation has been
observed in structurally analogous cobalt clusters [51-56], as predicted by polyhedral skeletal electron pair (PSEP) theory [57-62].
Once formed, the nido cluster 3 could generate aldehydes by the
accepted carbon monoxide insertion and acyl-metal hydrogenolysis
steps. Implicit in the formation of 3 is the formal transposition
of the μ_2-bridging carbonyl to a terminal carbonyl ligand, a process

that has been shown to be facile in 1 [63] and related derivatives [64,65]. Unfortunately, the direct detection of 3 has not been successful, suggesting that the formation of 3 could be rate determining. The hydroformylation reaction using $Fe_2Co_2(CO)_{11}(\mu_4-PPh)_2$ is currently being examined using Laine's kinetic criteria [15] in an effort to determine the nature of the active catalyst.

ACKNOWLEDGMENTS

This work was supported in part by the National Science Foundation through grant no. RII-8902064 (CUP), the State of Mississippi, and Mississippi State University (CUP). The Council for International Exchange of Scholars and the Fulbright Foundation are thanked for the award of a Senior Fulbright Fellowship for the period of August 1988–June 1989 at Mississippi State (H.H.). Mr. P. Zhu is thanked for conducting several experiments and Dr. E. Alley for assistance with the GC-MS. M.G.R. thanks the Robert A. Welch Foundation and the UNT faculty research program for financial support and Barnes Spectra-Tech for their gift of an external CIR sampling bench.

REFERENCES

1. Suss-Fink, G., and Herrmann, G., *J. Chem. Soc., Chem. Commun.*, 735 (1985).
2. Suss-Fink, G., *J. Organomet. Chem.*, *193*, C20 (1980).
3. Suss-Fink, G., *J. Mol. Catal.*, *42*, 361 (1987).
4. Schmidt, G. F., Reiner, J., and Suss-Fink, G., *J. Organomet. Chem.*, *355*, 379 (1988); Knifton, J. F., *J. Mol. Catal.*, *47*, 99 (1988).
5. Marrakchi, H., Effa, J. B. N., Haimeur, M., Leiko, J., and Aune, J. P., *J. Mol. Catal.*, *30*, 101 (1985).
6. Richmond, M. G., Halabi, M. A., and Pittman, C. U., Jr., *J. Mol. Catal.*, *22*, 367 (1984).
7. Pittman, C. U., Jr., Richmond, M. G., Wilemon, G. M., and Halabi, M. A., in *Catalysis of Organic Reactions* (J. R. Kosak, ed.), Dekker, New York, 1984, p. 101.
8. Ryan, R. C., Pittman, C. U., Jr., and O'Connor, J. P., *J. Amer. Chem. Soc.*, *99*, 1986 (1977).
9. Pittman, C. U., Jr., Ryan, R. H., Wilson, D., Wilemon, G. M., and Halabi, M. A., in *Symposium on Metal Clusters in Catalysis*, American Chemical Society, San Francisco meeting, Aug. 22–29, 1980, pp. 714–723.
10. Dahl, L. F., Ryan, R. C., O'Connor, J. P., and Pittman, C. U., Jr., *J. Organomet. Chem.*, *193*, 247 (1980).

11. Pittman, C. U., Jr., Wilemon, G. M., Wilson, W. D., and Ryan, R. C., *Angew. Chem.*, *Int. Ed. Engl.*, *19*, 478 (1980).
12. Gates, B. C., Guczi, L., and Knozinger, H., eds., *Metal Clusters in Catalysis*, Elsevier, Amsterdam, 1986.
13. Slaugh, L. H., and Mullineaux, R. D., *J. Organomet. Chem.*, *13*, 469 (1968).
14. Pregaglia, G. F., Andreeta, A., Ferrari, G. F., and Ugo, R., *J. Organomet. Chem.*, *30*, 38 (1987).
15. Laine, R. M., *J. Mol. Catal.*, *14*, 137 (1982).
16. Richmond, M. G., and Kochi, J. K., *Inorg. Chem.*, *25*, 656 (1986).
17. Parshall, G. W., *Homogeneous Catalysis*, Wiley, New York, 1980.
18. Rein, A. J., and Wilks, P. A., *Am. Lab.*, *14*, 153 (1982).
19. Harrick, N. J., *Internal Reflection Spectroscopy*, Wiley-Interscience, New York, 1967.
20. Kortum, G., *Reflectance Spectroscopy*, Springer-Verlag, New York, 1969.
21. Mirabella, F. M., Jr., and Harrick, N. J., *Internal Reflection Spectroscopy: Review and Supplement*, Harrick Scientific Corporation, Ossining, NY, 1985.
22. Moser, W. R., Chossen, J. E., Wang, A. W., and Krouse, S. A., *J. Catal.*, *95*, 21 (1985).
23. Moser, W. R., Papile, C. J., Brannon, D. A., Duwell, R. A., and Weininger, S. J., *J. Mol. Catal.*, *41*, 271 (1987).
24. Moser, W. R., Papile, C. J., and Weininger, S. J., *J. Mol. Catal.*, *41*, 293 (1987).
25. Moser, W. R., Wang, A. W., and Kildahl, N. K., *J. Am. Chem. Soc.*, *110*, 2816 (1988).
26. Moser, W. R., *Scan Time*, No. 3 (1984).
27. Tooley, P., Ovalles, C., Kao, S. C., Darensbourg, D. J., and Darensbourg, M. Y., *J. Am. Chem. Soc.*, *108*, 5465 (1986).
28. Darensbourg, D. J., and Ovalles, C., *Scan Time*, No. 4 (1984).
29. Darensbourg, D. J., and Ovalles, C., *J. Am. Chem. Soc.*, *106*, 3750 (1984).
30. Richmond, M. G., *J. Mol. Catal.*, *54*, 199 (1989).
31. Schulman, C. L., Richmond, M. G., Watson, W. H., and Nagl, A., *J. Organomet. Chem.*, *368*, 367 (1989).
32. Ryan, R. C., and Dahl, L. F., *J. Am. Chem. Soc.*, *97*, 6904 (1975).
33. Van Boven, M., Alemdaroglu, N., and Penninger, J. M. L., *J. Organomet. Chem.*, *84*, 65 (1975).
34. Don, M. J., and Richmond, M. G., unpublished results.
35. Richmond, M. G., and Kochi, J. K., *Inorg. Chem.*, *25*, 1334 (1986).

36. Whyman, R., *J. Organomet. Chem.*, *81*, 97 (1974).
37. King, R. B., King, A. D., Jr., and Tanaka, K., *J. Mol. Catal.*, *10*, 75 (1981).
38. Tannenbaum, R., Dietler, U. K., and Bor, G., *Inorg. Chim. Acta*, *154*, 109 (1988).
39. Heck, R. F., and Breslow, D. S., *J. Am. Chem. Soc.*, *83*, 4023 (1961).
40. Vahrenkamp, H., *Adv. Organomet. Chem.*, *22*, 169 (1983).
41. Geoffroy, G. L., *Acc. Chem. Res.*, *13*, 469 (1980).
42. Bullock, R. M., and Casey, C. P., *Acc. Chem. Res.*, *20*, 167 (1987).
43. Longato, B., Norton, J. R., Huffman, J. C., Marsella, J. A., and Caulton, K. G. *J. Am. Chem. Soc.*, *103*, 209 (1981); *104*, 6360 (1982).
44. Ferguson, G. S., and Wolczanski, P. T. *Organometallics*, *4*, 1601 (1985).
45. Roland, E., and Vahrenkamp, H., *Organometallics*, *2*, 183 (1984).
46. Knozinger, H., Gates, B. C., and Guczi, L., eds., *Metal Clusters in Catalysis*, Elsevier, New York, 1986.
47. Muetterties, E. L., Rhodin, T. N., Band, E., Brucker, C. F., and Pretzer, W. R., *Chem. Rev.*, *79*, 91 (1979).
48. Sinfelt, J. H., *Bimetallic Catalysts—Discoveries, Concepts, and Applications*, Wiley, New York, 1983.
49. (a) Sinfelt, J. H., *Acc. Chem. Res.*, *10*, 15 (1977).
    (b) Ichikawa, M., *J. Catal.*, *54*, 67 (1979).
50. (a) Vahrenkamp, H., and Wucherer, E. J., *Angew. Chem., Int. Ed. Engl.*, *20*, 680 (1981). (b) Vahrenkamp, H., Wucherer, E. J., and Wolters, D., *Chem. Ber.*, *116*, 1219 (1983).
51. (a) Richmond, M. G., and Kochi, J. K., *Inorg. Chem.*, *26*, 541 (1987). (b) Richmond, M. G., Korp, J. D., and Kochi, J. K., *J. Chem. Soc., Chem. Commun.*, 1102 (1985).
52. Bogan, L., Lesch, D. A., and Rauchfuss, T. B., *J. Organomet. Chem.*, *250*, 429 (1983).
53. Planalp, R. P., and Vahrenkamp, H., *Organometallics*, *6*, 492 (1987).
54. Huttner, G., Schneider, J., Muller, H. D., Mohr, G., von Seyerl, J., and Wohlfahrt, L., *Angew. Chem., Int. Ed. Engl.*, *18*, 76 (1979).
55. Knoll, K., Hutter, G., Zsolnai, L., Jibril, I., and Wasiucionek, M., *J. Organomet. Chem.*, *294*, 91 (1985).
56. Schneider, J., Minelli, M., and Huttner, G., *J. Organomet. Chem.*, *294*, 75 (1985).
57. Wade, K., *Adv. Inorg. Chem. Radiochem.*, *18*, 1 (1976).
58. Wade, K., in *Transition Metal Clusters* (B. F. G. Johnson, ed.), Wiley, New York, 1980, Chap. 3.

59. Mingos, D. M. P., *Nature (London) Phys. Sci.*, *236*, 99 (1972).
60. Halet, J. F., Hoffmann, R., and Saillard, J. Y., *Inorg. Chem.*, *24*, 1695 (1985).
61. Teo, B. K., *Inorg. Chem.*, *23*, 1251 (1984).
62. Teo, B. K., Longoni, G., and Chung, F. R. K., *Inorg. Chem.*, *23*, 1257 (1984).
63. Richmond, M. G., and Kochi, J. K. *Organometallics*, *6*, 777 (1987).
64. Richmond, M. G., and Kochi, J. K., *J. Organomet. Chem.*, *323*, 219 (1987).
65. Richmond, M. G., and Kochi, J. K., *Organometallics*, *6*, 254 (1987).

28

Structure Sensitivity and Mechanism of Isobutyraldehyde Decarbonylation over Pd/SiO₂ and Pt/SiO₂ Catalysts

Ruozhi Song and D. Ostgard
Department of Chemistry and Biochemistry, Southern Illinois University at Carbondale, Carbondale, Illinois

G. V. Smith
Molecular Science Program and Department of Chemistry and Biochemistry, Southern Illinois University at Carbondale, Carbondale, Illinois

I. INTRODUCTION

The decarbonylation of aldehydes catalyzed by metallic palladium (supported on charcoal or barium sulfate) has been studied extensively from the viewpoint of organic synthesis and has been reviewed by Tsuji and Ohno [1], but the reaction mechanism is still not fully understood. One mechanism, proposed by the same authors [2], involves acyl palladium intermediates that result from the oxidative addition of an aldehyde to a palladium atom. This mechanism has been suggested for decarbonylation (and in the opposite direction for hydrocarbonylation) of aldehydes catalyzed by both palladium complexes in solution and palladium metal crystallites. On the other hand, Smolik and Kraus [3] proposed a three-step mechanism for the decarbonylation of benzaldehyde and its derivatives. They pointed out that cleavage of the formyl C—H bond is rate determining for Pd catalysts, since substituent effects are very small and an isotope effect is significant. However, the effect of alkyl groups in the α-position of cinnamaldehydes on the rate of decarbonylation over a palladium catalyst is observed, cleavage of the C—C bond between the C=C and C=O bonds was proposed as the rate-determining step [3,4]. In a recent paper, Davis and Barteau [5] observed a large effect of alkyl substituents on the decarbonylation of aliphatic aldehydes on a Pd(111) surface. So they propose that decarbonylation proceeds through acyl and ketene intermediates, and that C—H bond cleavage of the acyl intermediates is rate determining.

Prompted by these different mechanistic viewpoints and our interest in decarbonylation as a preparative tool, we decided to reinvestigate the mechanism. A series of silica-supported palladium catalysts with a wide range of dispersions had been newly prepared and characterized in our laboratory. Also, a series of silica-supported platinum catalysts was available. Together these catalysts offered the additional ability to examine the structure sensitivity (or lack of it) of the reaction. Therefore, we selected for study several representative catalysts over a range of dispersions, to look for correlations between the active site(s) for decarbonylation and some feature of surface structure (corners, edges, planes).

Isobutyraldehyde was chosen as the substrate because it forms two products, propene and propane. These two products could arise from decarbonylation through either consecutive or parallel reactions, but the literature is not clear about which reaction path occurs. So isobutyraldehyde seemed to be a likely candidate to distinguish between these two mechanistic possibilities.

II. EXPERIMENTAL

A. Reactor

The reactor was a 0.4-mm i.d. glass tube filled with 0.025 g (in most cases) of catalyst, held in place by two small plugs of glass wool, and heated in a carefully controlled furnace (Lindberg Heviduty). The reactor effluent was sampled through a 1 cm^3 sample loop into a gas chromatograph. The reactant gas stream was formulated by passing the purified helium through a saturator filled with isobutyraldehyde. The saturator was immersed in ice water, which assured the partial pressure of isobutyraldehyde in helium to be 0.05 atm.

B. Reagents

The gases, hydrogen and helium, were high purity and were further purified by passing first through deoxygenating columns (BTS) and then through molecular sieve 13X and 5A columns before entering the reaction system. A final trap containing Mn/SiO_2 was used to indicate the possible presence of oxygen. These columns and the trap were regenerated periodically. Isobutyraldehyde was purchased from Matheson Coleman & Bell and was further purified by simple distillation under argon.

Table 28.1   Preparation and Characterization Data for Catalysts

Catalyst	Dispersion (%)	Method of preparation
Pt/SiO$_2$		
1.48%	21.5	ion exchange
1.48%	27.0	ion exchange
1.17%	40.7	impregnation
0.48%	63.0	ion exchange
0.83%	81.0	ion exchange
Pd/SiO$_2$		
1%	4.4	ion exchange
1%	36.0	ion exchange
1%	49.7	ion exchange
1%	63.1	ion exchange
1%	75.4	ion exchange
1%	84.0	ion exchange
1%	93.6	ion exchange
Pd black[a]	0.51	

[a]Donated by Engelhard Industries.

C. Catalysts

We used two types of catalyst, Pd/SiO$_2$ and Pt/SiO$_2$. The Pt/SiO$_2$ originated from the laboratories of Burwell and Butt. The methods of preparation and the characterization data are shown in Table 28.1; other details may be found in their article [6]. The Pd/SiO$_2$ catalysts were prepared by the ion-exchange method, and their dispersions were determined by static hydrogen chemisorption.

D. Analysis

Gas chromatographic (GC) analysis was carried out for all reactions using a 1/8 in × 3 m column packed with Porapak QS. The helium flow rate was 40 mL/min; the column temperature, which was programmed, was kept at 60°C for 6 minutes, then increased to 200°C at a rate of 10°C/min. The thermal conductivity detector was at 150°C and the filament current at 130 mA.

E. Catalyst Pretreatment

Both series of catalysts were pretreated by heating to 250°C in He, then flowing hydrogen overnight (10 h) at this temperature, and finally cooling in He to the reaction temperature.

III. RESULTS AND DISCUSSION

Preliminary experiments demonstrated that decarbonylation of iso-butyraldehyde does not occur below 400°C in the absence of the catalysts and that dehydrogenation of propane does not take place below 400°C in the presence of the catalysts. These experiments also showed that the catalysts lose their activity rapidly for the first few hours, and the rate of deactivation depends on the reaction temperatures (Table 28.2) and also, on the dispersions of the catalysts, a result that is discussed later. Because the rate of deactivation increases as reaction temperature increases, the following experiments were carried out at lower temperature (200°C). The initial activity of the deactivated catalyst can be recovered by either flowing hydrogen at 250°C or exposing to air at room temperature.

The activities of the two series of catalysts for decarbonylation of isobutyraldehyde in terms of turnover frequencies are shown in Table 28.3. The reactions were carried out at 200°C with the space time (W/F) of 7.8 g of catalyst per mole per hour. Because the activity of the catalyst declined in approximately linear fashion, initial activity levels were determined in terms of initial turnover frequencies by extrapolation to zero time. This was done by a least-squares fit computer program as shown in the results plotted in Figure 28.1. The reaction times range from 0.25 to 6 hours. The 4.4% dispersed catalyst cannot be treated in the same way because it lost its activity completely within 12 minutes, the turnover frequency (0.07) listed in Table 28.3 was obtained at a 6 minute reaction time.

The slopes of the lines in Figure 28.1 reflect the rates of deactivation of the catalysts. When the slope for each catalyst is plotted against the percentage dispersion in Figure 28.2, the rates of deactivation decrease as the dispersions increase for dispersions below 50%; for highly dispersed catalysts, however, the rates of deactivation are the same. This observation seems to be inconsistent with

Table 28.2 Activity Lost for 20 h Reaction at Different Temperatures[a]

Catalyst	Dispersion (%)	Temperature (°C)	Time	Activity lost (%)
1% Pd/SiO$_2$	84.0	200	20	12.3
1% Pd/SiO$_2$	84.0	280	20	46.1

[a]Reaction was carried out with 0.025 g of fresh catalyst in each case.

Table 28.3 Decarbonylation of Isobutyraldehyde on Pd/SiO$_2$ and Pt/SiO$_2$

Catalyst	Dispersion (%)	Turnover frequencies (s^{-1})			
		TOF$_t$[a]	TOF$_e$[b]	TOF$_a$[c]	ANE/ENE[d]
Pd black	0.51	0.100	0.090	0.010	0.11
1%Pd/SiO$_2$	4.4	0.070	0.070	0.0	0.0
1%Pd/SiO$_2$	36.0	0.168	0.168	0.0	0.0
1%Pd/SiO$_2$	49.7	0.197	0.179	0.019	0.10
1%Pd/SiO$_2$	63.1	0.284	0.207	0.077	0.37
1%Pd/SiO$_2$	75.4	0.268	0.168	0.100	0.60
1%Pd/SiO$_2$	84.0	0.267	0.169	0.107	0.63
1%Pd/SiO$_2$	93.6	0.260	0.140	0.120	0.86
1.48%Pt/SiO$_2$	21.5	0.094	0.068	0.026	0.38
1.48%Pt/SiO$_2$	27.0	0.084	0.061	0.023	0.38
1.17%Pt/SiO$_2$	40.0	0.077	0.062	0.017	0.28
0.48%Pt/SiO$_2$	63.0	0.152	0.113	0.039	0.35
0.83%Pt/SiO$_2$	81.0	0.063	0.046	0.017	0.37

[a]Number of isobutyraldehydes converted per surface site per second.
[b]Number of propenes formed per surface site per second.
[c]Number of propanes formed per surface site per second.
[d]Molar ratio of propane to propene formed.

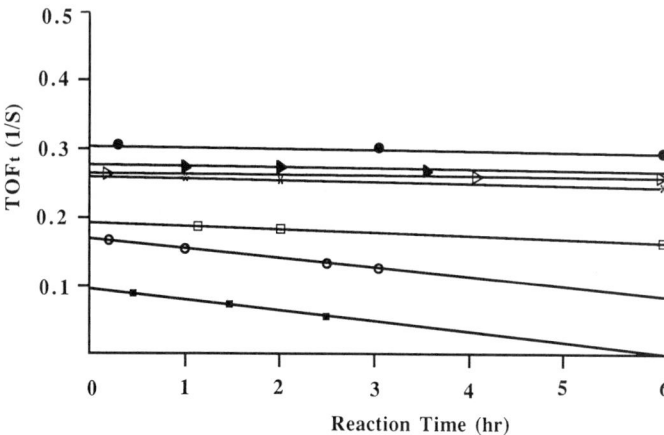

Figure 28.1 Plot of TOF$_t$ versus reaction time.

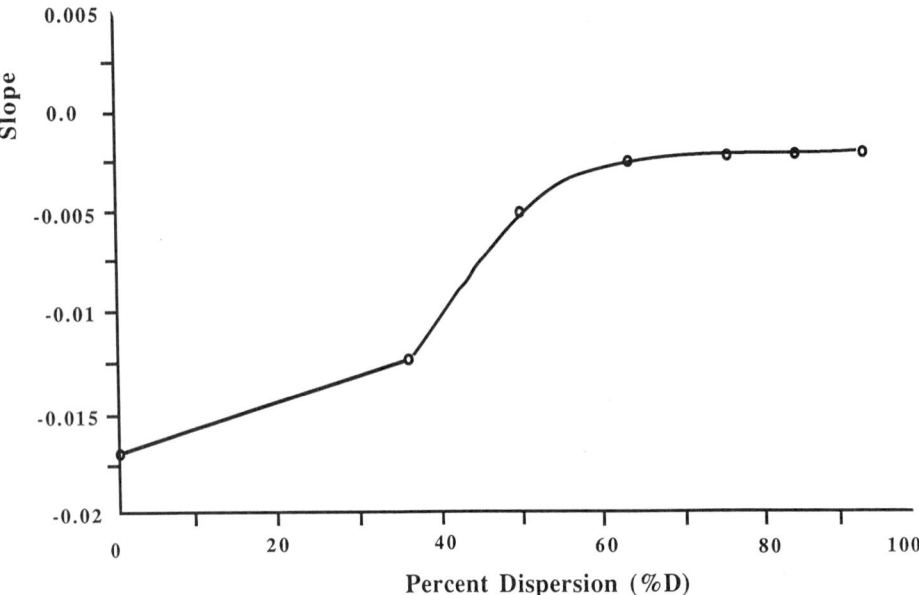

Figure 28.2   Slope versus percent dispersion.

literature data based on many studies devoted to the deactivation
of Pd/SiO_2 catalysts.

For example, it has been reported that the deactivation of carbon
monoxide adsorption sites could be caused either by decomposition
of hydrocarbons [7] or by disproportionation of carbon monoxide
to yield surface carbon [8]. And the active site density for both
reactions is particle size dependent, being greater on smaller particles. However, it seems that the disproportionation of CO is not
responsible for deactivation of the catalysts in the decarbonylation
of isobutyraldehyde under the conditions of this investigation because
the temperature is too low for the reaction to take place. So the
main reason for deactivation may be from carbon layer buildup resulting from decomposition of carbon-containing surface species, and
the rate of formation of this surface carbon may be controlled by
the concentration of surface hydrogen.

Support for this possibility is found in the observation that
since the lower dispersed catalysts have a low concentration of
surface hydrogen (they have lower activities than the higher dispersed
catalysts), the carbon-containing surface species have more opportunities to decompose to surface carbon. In fact, a comparison of
the turnover frequencies of the different dispersed catalysts with

their rates of deactivation reveals that they are parallel; the lower the turnover frequency, the greater the rate of deactivation.

The relation between the total turnover frequencies (TOF_t) and the dispersions is shown in Figure 28.3 for Pd/SiO_2 and Pt/SiO_2 catalysts. The initial turnover frequencies in terms of total conversion of isobutyraldehyde exhibits mild structure sensitivity on both series of catalysts. For example, on Pd/SiO_2 there is an initial increase to $D = 63.1\%$, then a slow decrease in the region of $D = 63.1-93.6\%$. For Pt/SiO_2 there is an apparent maximum in TOF_t in the vicinity of $D = 60\%$ such that the TOF_t values decrease with D beyond this point. These maxima for the two series of catalysts correspond approximately to the fraction of edge sites on regular octahedra and truncated octahedra crystallites of the metals [9].

Energies of activation (E_a) were evaluated for three catalysts in each series (Pd/SiO_2: 49.7%D, 63.1%D, 93.6%D; Pt/SiO_2: 40%D, 63%D, 81%D) in the temperature range 140-200°C for Pd/SiO_2 and 160-200°C for Pt/SiO_2. For both series, E_a is independent of dispersion within experimental error: 57.5 + 0.8 kJ/mol (Pd/SiO_2) and 57.3 + 1.4 kJ/mol (Pt/SiO_2). These numbers are close to that of furfural in the absence of hydrogen [10,11].

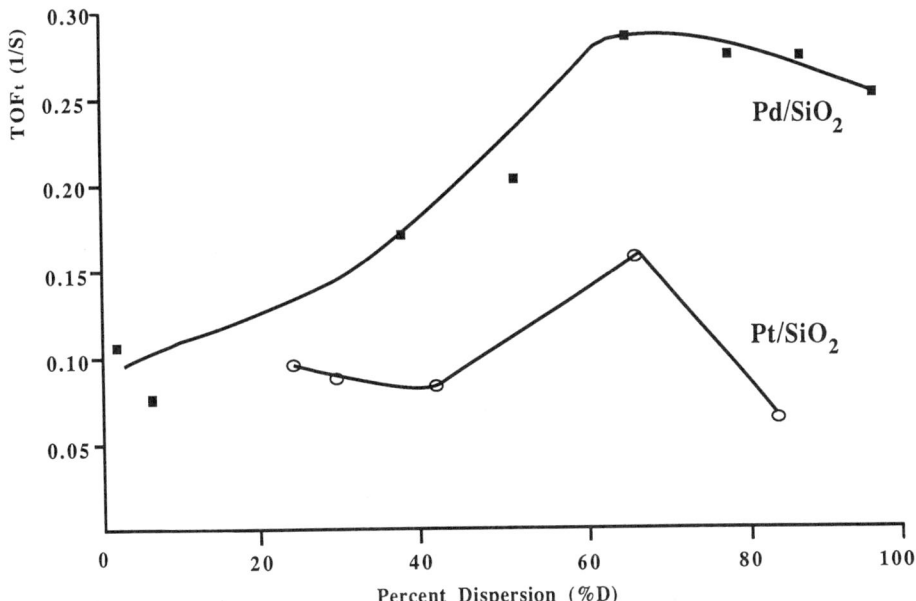

Figure 28.3 TOF_t versus percent dispersion.

The distribution of products with respect to space-time was
accomplished by changing the amount of catalyst and the flow rate
of the carrier gas. The results are shown in Figures 28.4-28.7.
Figures 28.4 and 28.5 show the reaction profiles for 0.025 g of Pd/SiO_2
(93.6%D) and Pt/SiO_2 (81%D), respectively. For Pt/SiO_2, the percent
yields of propene and propane increase linearly as the space-time
increases and the ratio of propane to propene does not vary as
shown in Figure 28.6. When a similar experiment was carried out
on another Pt catalyst (40%D), the same phenomenon was observed,
which indicates that both these compounds are primary products
on Pt. However, for Pd/SiO_2 the percent yields of propene and
propane change in such a way that the ratio of propane to propene
increases linearly as the space-time increases. The yield of propane
increases from 17.2 to 63.6% and the yield of propene decreases
from 42.6 to 32.1% (Fig. 28.4). When 0.005 g of 1% Pd/SiO_2 (84.0%D)
was used to carry out a similar study, only propene was formed
(Fig. 28.7). These results indicate that propane is formed from
the hydrogenation of propene on the Pd/SiO_2 catalysts.

Based on our results and others in the literature, we suggest
the reaction mechanisms for decarbonylation of aldehydes shown
in Figure 28.8 and explained as follows.

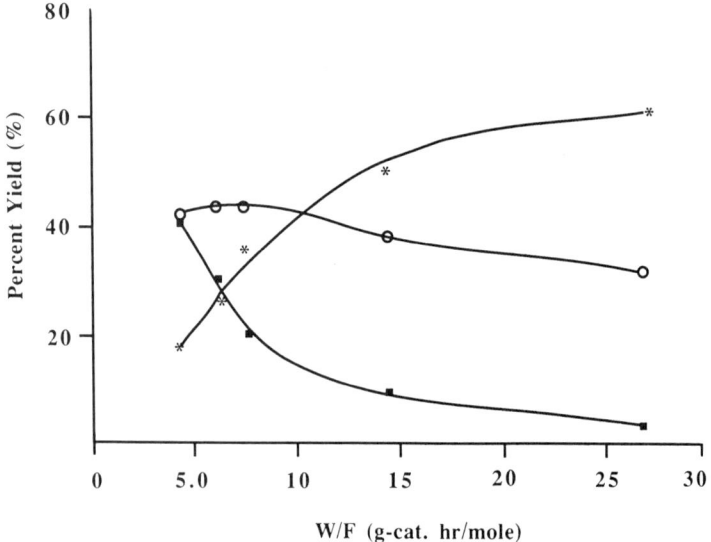

Figure 28.4 Plot of percent yield versus space time for the reaction
with 25 mg of Pd/SiO_2 (93.6%D) at 200°C.

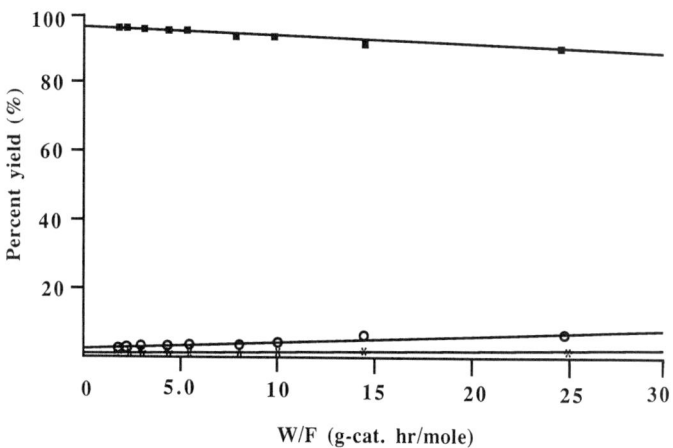

Figure 28.5 Plot of percent yield versus space time for the reaction
with 25 mg of Pt/SiO$_2$ (81%D) at 200°C.

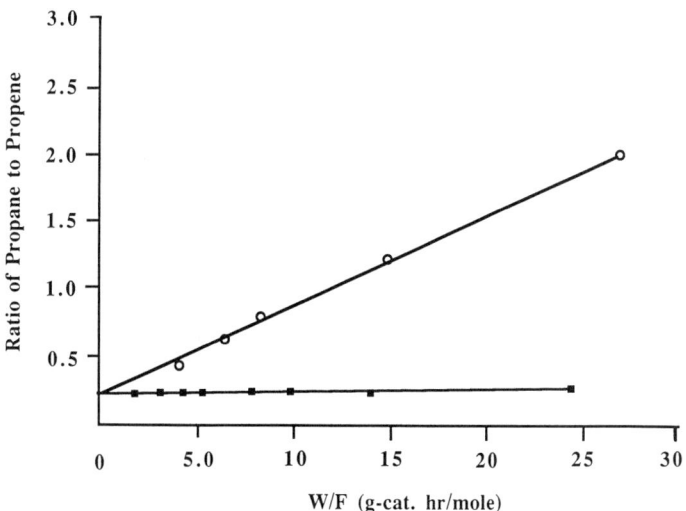

Figure 28.6 Plot of ratio of propane to propane versus space time
for Pd/SiO$_2$ (93.6%D) and Pt/SiO$_2$ (81%D).

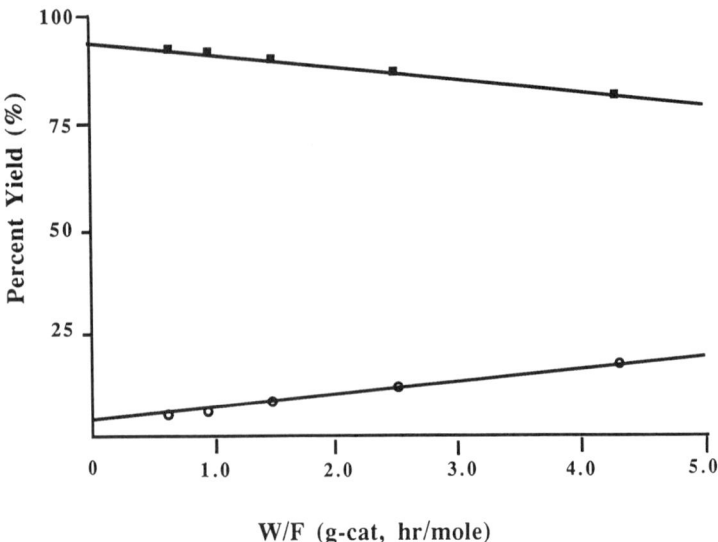

Figure 28.7   Percent yield versus space time for the reaction with
5 mg of Pd/SiO$_2$ (84.0%D) at 200°C.

Cleavage resulting from two types of adsorption of aldehydes
on palladium and platinum catalysts surface have been reported
[5,12] as shown in the first step of the mechanism in Figure 28.8.
Either cleavage of the C—H bond of the formyl group or cleavage
of the C—C bond between the R group and the C=O bond of the
adsorbed aldehyde must occur for the reaction to take place. Ease
of cleavage of the C—H and C—C bonds depends on the R group
of the aldehyde. When R is an aryl group, cleavage of the C—C
bond is easier than that of the C—H bond due to the strong interaction of the aryl group with the catalyst surface. This interaction
can reduce resonance effects between the aryl and carbonyl groups.
Moreover, the C—C bond between them may be weakened as a result
of coordination of these two groups to the same palladium atom.
Thereby, cleavage of the C—C bond is facilitated and precedes
cleavage of the C—H bond of the formyl group. The latter has
been proposed as the rate-determining step for decarbonylation
of benzaldehyde and its derivatives by Smolik and Kraus [3] based
on their experimental observations that K_H/K_D is 2.14 for the reaction of benzaldehyde and that substituents on the benzene ring
have no influence on the rate of reaction.
        Cleavage of the C—H bond of the formyl group in intermediate
I can proceed on the same site at which the C—C bond cleavage

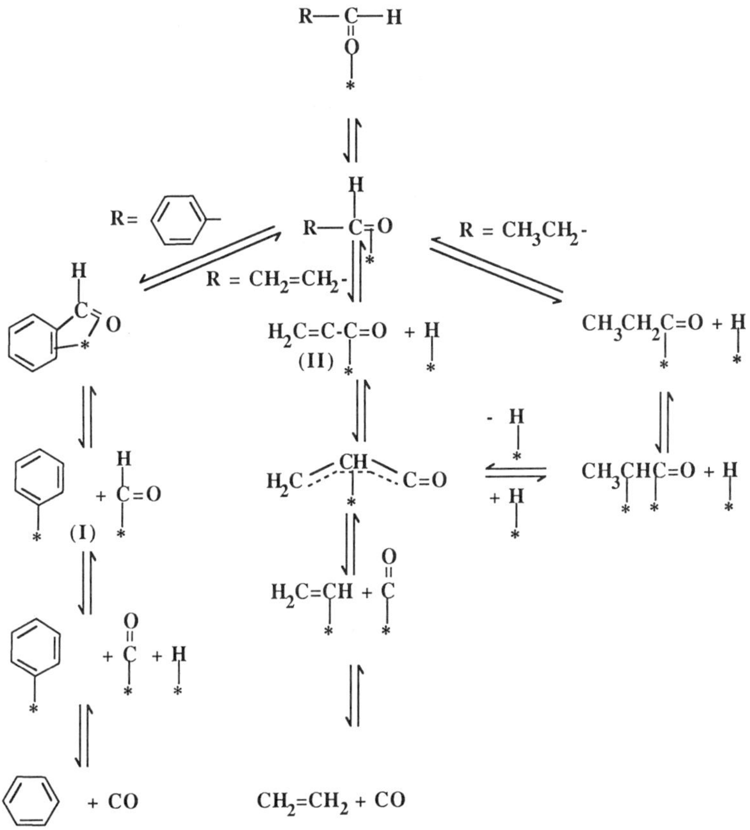

Figure 28.8   Reaction mechanisms for decarbonylation of aldehydes.

occurred if this site is a Siegel type [3]M (the active site is assumed to be [2]M in Fig. 28.8) or if there is assistance from other nearby active sites. The last step, then, is the combination of the newly formed aryl group with surface hydrogen, followed by desorption of CO from the surface to give the final products.

When the aldehyde to be decarbonylated is aliphatic or α,β-unsaturated, cleavage of the C—H bond of the formyl group precedes that of the C—C bond because of the lack of any strong interaction of the R group with the catalyst surface, and an acyl palladium intermediate (II) is formed. The C=O stretch of the intermediate (II) has been seen in high resolution electron energy loss (HREEL) spectra from decarbonylation of acetaldehyde, propanal, and acrolein [5], and the effect of an alkyl group in the α-position of cinnamaldehyde

on the rate of reaction was observed by Hoffman and Puthenpurackal
[4]. Cleavage of the C—C bond of the acyl palladium intermediate
(II) proceeds via an π-allylic intermediate if the C=C bond or the
β-hydrogen is available, as shown in Figure 28.8. This pathway
is able to explain why the reaction of acrolein is easier than that
of propanal and acetaldehyde and why no propane is formed initially
during the reaction of isobutyraldehyde on Pd catalysts. It leads
to the conclusion that the carbon–carbon double bond must be formed
before cleavage of the carbon–carbon single bond.

For aldehydes without a β-hydrogen, such as acetaldehyde,
cleavage of the C—C bond proceeds via a ketene intermediate as
suggested by Davis and Barteau [5]. For decarbonylation of isobutyraldehyde on platinum catalysts, the reaction may follow the
same pathway as that of acetaldehyde on Pd catalysts, which could
explain the formation of propane as a primary product on Pt.

IV. CONCLUSIONS

The decarbonylation of isobutyraldehyde on silica-supported palladium
and platinum catalysts is structure sensitive under the conditions
of investigation. The reaction occurs on edge sites via an π-allyl
intermediate.

REFERENCES

1. Tsuji, J., and Ohno, K., *Synthesis*, 157 (1969).
2. Tsuji, J., and Ohno, K., *J. Am. Chem. Soc.*, *90*, 94 (1968).
3. Smolik, J., and Kraus, M., *Collect. Czech. Chem. Commun.*,
   37, 3042 (1972).
4. Hoffman, N. E., and Puthenpurackal, T., *J. Org. Chem.*,
   30, 420 (1965).
5. Davis, J. L., and Barteau, M. A., *J. Am. Chem. Soc.*, *111*,
   1782 (1989).
6. Uchijima, T., Hermann, J. M., Inoue, Y., Burwell, R. L.,
   Jr., Butt, J. B., and Cohen, J. B., *J. Catal.*, *50*, 467 (1977).
7. Durrer, W. G., Poppa, H., and Dickinson, J. T., *J. Catal.*,
   115, 310 (1989).
8. Ichikawa, S., Poppa, H., and Boudart, M., *J. Catal.*, *91*,
   1 (1985).
9. Notheisz, F., Bartok, M., Ostgard, D., and Smith, G. V.,
   J. Catal., *101*, 212 (1986).
10. Srivastava, R. D., and Guha, A. K., *J. Catal.*, *91*, 254 (1985).

11. Singh, H., Prasad, M., and Srivastava, R. D., *J. Chem. Technol. Biotechnol.*, *30*, 293 (1980).
12. McCabe, R. W., DiMaggio, C. L., and Madix, R. J., *J. Phys. Chem.*, *89*, 854 (1985).

29

The Use of Dispersed Metal Catalysts for Organometallic Reactions

Shaun T. O'Leary and Robert L. Augustine

Department of Chemistry, Seton Hall University, South Orange, New Jersey

I. INTRODUCTION

The use of soluble transition metal complexes to catalyze synthetically useful organic reactions is well established [1]. While these reactions are generally rather selective because of the presence of a single catalytically active species, there are some problems associated with the use of these soluble monometallic species. Foremost among them is the difficulty that may be encountered in separating the metallic species from the desired organic material. In small-scale reactions chromatography can be used effectively for this separation, but in large-scale reactions this can be a significant problem. Furthermore, the recovery of the metallic species and the regeneration of the catalytically active species for further reactions is frequently noneconomical. One apparent way of solving these problems is to use an insoluble heterogeneous species to catalyze the reaction. The so-called heterogenized homogeneous catalysts, prepared by complexing the appropriate metallic species with an insoluble ligand, have been successfully used to catalyze a number of reactions and are readily removed from the reaction mixture. While such materials can be quite effective in synthetic reactions run in a batch reactor, their use in flow reactors leaves much to be desired because in these flow systems even weakly dissociated ligands can be lost from the catalyst and become mixed with the reaction products. This can result not only in the deactivation of the catalyst but also in the need to purify the product of some undesirable materials present in rather small amounts.

One obvious way of overcoming these difficulties would be to use a supported metal catalyst for these reactions, but this possibility does not seem to have been investigated to any extent. In a few of the reports concerning the homogeneously catalyzed reactions, there is some mention made of the use of supported metal catalysts. These are generally afterthoughts, however, and most of the information concerned with the use of such species is confined to tables describing the effect of reaction variables on product yield or selectivity. In most of those instances in which a supported metal was observed to promote the reaction under study, it was assumed that this reactivity was due to the dissolution of the metal to give a reactive soluble species.

As discussed previously [2], the surfaces of these dispersed metal catalysts are made up of a large number of surface atoms of different types, each with varying adsorption and reaction characteristics. It appeared, then, that with the complex nature of the metal surface, there should be present at least one type of site on which these synthetically useful C—C bond formation reactions should be promoted.

II. EXPERIMENTAL

The butyl vinyl ether was purchased from Aldrich and distilled using a short path apparatus immediately before use. The 4-nitrobenzoyl chloride and 4-nitroiodobenzene were obtained from Aldrich. The benzoyl chloride was purified using Kugelrohr distillation, while the iodobenzene was used as received. The N-ethylmorpholine, also from Aldrich, was distilled and stored over 3A molecular sieves. The solvents were Aldrich ACS grade and were distilled from sodium and benzophenone prior to use. The palladium-on-alumina and palladium-on-silica catalysts were prepared from palladium chloride by impregnation [3] and ion exchange [4], respectively.

The reactions were run in a three-necked 50 mL reaction flask equipped with a stirring motor, condenser, thermometer, and septum. The catalyst, internal standard, and 15 mL of the solvent were added to the reaction flask and it was purged twice with nitrogen. A mixture of the aryl halide or aroyl halide, vinyl ether, and amine in 10 mL of the solvent was added to the flask under nitrogen and the solution was brought to 100°C with vigorous stirring under a nitrogen atmosphere. The reaction was monitored by gas chromatography, and the reaction stirring was stopped for about 10 seconds to allow the catalyst to settle before sampling.

III. RESULTS AND DISCUSSION

Our initial work in this area involved a study of the Heck arylation [5] (Eq. 29.1).

(29.1)

This reaction can be used to prepare aryl enol ethers which, on hydrolysis, give the aldehyde or ketone, materials potentially useful themselves or as intermediates for further reactions. The reaction is commonly run using $Pd(OAc)_2$ as a soluble catalyst to promote the coupling of aryl halides with enol ethers [6,7]. A tertiary amine is used to neutralize the acid formed in the reaction. Aryl acid chlorides are also used, but the decarbonylation of the complexed intermediate requires the use of somewhat higher temperatures [8]. Reaction of p-nitrobenzoyl chloride with butyl vinyl ether over $Pd(OAc)_2$ in toluene in the presence of N-ethylmorpholine gives as the predominant products the *E* (1) and *Z* (2) isomers of the β-enol ether along with some of the α-isomer, 3.

It has been reported that Pd/C was also effective in promoting this reaction, but no further data concerning the use of this catalyst were reported, other than a few entries in a table [7,8]. When we repeated this reaction using toluene as the solvent and Pd/Al_2O_3 as the catalyst, extensive catalyst deactivation was observed because of the precipitation of the amine hydrochloride onto the catalyst. With dioxane as the solvent, though, the hydrochloride salt remained in solution and the reaction proceeded smoothly.

Table 29.1 lists the products obtained from this reaction run in dioxane over both soluble and supported palladium metal catalysts. With $Pd(OAc)_2$ essentially the same product composition is obtained in either toluene or dioxane. This soluble catalyst, however, was somewhat less reactive than the supported metals so, on a molar basis, four times more of the soluble Pd was needed to obtain yields similar to those observed with the supported Pd catalysts. With all catalysts, the *E/Z* β-isomer product ratio (1/2) was greater than 1, with the supported catalysts producing the largest ratios.

The Pd/Al_2O_3 catalyst was more reactive than Pd/SiO_2. A small
amount of the α-isomer, 3, was observed in each reaction along
with some of the butyl ester, 4.

To test whether the reaction was being promoted by some Pd
species solubilized into the reaction mixture, a reaction was run
in an inert atmosphere in a Schlenk tube over Pd/Al_2O_3 to about
10-20% completion. The reaction liquid was separated from the catalyst
by filtration and its composition determined by gas chromatography.
The liquid was then heated for an additional 2 hours, sufficient
time for further reaction to occur if a soluble catalyst were present,
and the composition again determined. There was no detectable
difference in the liquid composition between these two measurements,
showing that there was no soluble catalyst present in the reaction
system and that the process was, indeed, being promoted by the
metal surface of the dispersed catalysts.

To avoid ester formation, which we felt might have an influence
on the product selectivity in these reactions, the reaction of p-
nitroiodobenzene with butyl vinyl ether (eq. 29.2) was also run.

$$(29.2)$$

As the data in Table 29.2 show, this reaction proceeded at a slower
rate than that of the benzoyl chloride. The product composition
was also significantly different between the two reactions. The
iodobenzene reaction gave very little if any of the α-isomer, 3,
and β E/Z (1/2) ratios less than 1. The heterogeneous catalysts
produced ratios closer to 1 than did the soluble species. As in
the benzoyl chloride reaction, the Pd/SiO_2 was less active than
the Pd/Al_2O_3, but both were more active than the $Pd(OAc)_2$, which
required four times as much catalyst to give comparable results.
These results show that the supported Pd catalysts can be used
for organometallic reactions of this type.

While soluble catalysts have only one catalytically active species
present, there are surface atoms of a number of different types

Table 29.1 *p*-Nitrobenzoyl Chloride Reaction Run over Various Palladium Catalysts[a]

Catalyst	Yield (%)			Yield (%)	
	β-*E* (1)	β-*Z* (2)	*E/Z* ratio	α (3)	Ester (4)
Pd(OAc)$_2$ [b]	50.4	27.1	1.86	6.3	3.2
Pd/γ-Al$_2$O$_3$	42.9	15.6	2.75	5.8	4.4
Pd/SiO$_2$	30.7	15.4	1.99	4.5	7.8

[a]The reactions were run with 2.5 mmol of *p*-nitrobenzoyl chloride, 5.0 mmol of butyl vinyl ether, 3.75 mmol of *n*-ethylmorpholine, and catalyst in 0.25 mol% (based on *p*-NBC) in 25 mL of dioxane. Dodecane was used as an internal standard. The experiment was performed under a blanket of N$_2$ at the reflux temperature of the solvent.
[b]1 mol % (based on *p*-NBC); xylene was the solvent.

Table 29.2 *p*-Nitroiodobenzene Reaction Run over Various Palladium Catalysts[a]

Catalyst	Yield (%)			Yield of
	β-*E* (1)	β-*Z* (2)	*E/Z* ratio	α (3) (%)
Pd(OAc)$_2$ [b]	10.2	15.1	0.68	0.1
Pd/γ-Al$_2$O$_3$	14.9	15.2	0.98	0.7
Pd/SiO$_2$	5.0	6.8	0.74	0.0

[a]The reactions were run with 2.5 mmol of *p*-nitroiodobenzene, 5.0 mmol of butyl vinyl ether, 3.75 mmol of *n*-ethylmorpholine, and catalyst in 0.25 mol % (based on *p*-NIB) in 25 mL of dioxane. Dodecane was used as an internal standard. The experiment was performed under a blanket of N$_2$ at the reflux temperature of the solvent for 4 hours.
[b]1 mol % based on *p*-NIB.

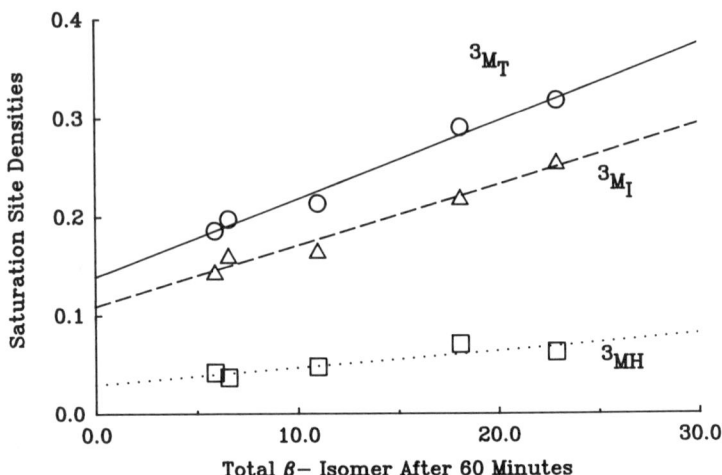

Figure 29.1   Saturation site densities versus total β-isomer after 60 minutes: (○) total saturation sites, (△) direct saturation sites, and (□) two-step saturation sites.

present on the metal crystallites of the supported catalyst [2]. It is important, then, to determine what type of surface site is responsible for promoting these reactions. To accomplish this, a series of single turnover (STO) characterized [3,4] Pd/Al_2O_3 catalysts, having differing STO reactive site densities, was used to promote the reaction shown in equation (29.1) and the product composition determined after a 60 minute reaction time. The results shown in Figure 29.1 establish that a direct relationship exists between the extent of β-isomer formation and the STO saturation site densities on these catalysts. This is true for both the direct (^3M_I) and two-step (^3MH) saturation sites, which have been shown to be corner or kink atoms on the surface [8]. Given the mechanism for the Heck arylation shown in equation (29.3) [5], it is not surprising that this reaction should take place on the more coordinately unsaturated atoms on the metal surface, since only these species have the ability to adsorb both reagents on the same atom in order to have them interact to give the product.

$$(2$$

IV. CONCLUSION

It has been shown that, at least for the Heck arylation reactic
supported metal catalysts can be used as effectively if not mo
so than the commonly used soluble transition metal complexes.
finding can be of importance to the specialty chemical and fin
chemical industries, since it may provide them with suitable a
tives to the use of the generally costly soluble transition met
catalysts to promote desired C—C bond forming reactions.

ACKNOWLEDGMENT

This research was supported by grant DE-FG02-84ER45120 f
the U.S. Department of Energy, Office of Basic Energy Sci
The metal salts were obtained through the Johnson-Matthey
Metal Loan Program.

REFERENCES

1. Collmann, J. P., Hegedus, L. S., Norton, J. R., and Finke, R. G., *Principles and Applications of Organotransition Metal Chemistry*, University Science Books, Mill Valley, CA, 1987.
2. Augustine, R. L., and O'Hagan, P. J., in *Catalysis in Organic Reactions* (D. W. Blackburn, ed.), Dekker, New York, 1990, p. 111.
3. Augustine, R. L., Baum, D. R., High, H. G., Szivos, L. S., and O'Leary, S. J., *J. Catal.*, *127*, 675 (1991).
4. Augustine, R. L., and Warner, R. W., *J. Catal.*, *80*, 358 (1983).
5. Heck, R. F., *Acc. Chem. Res.*, *12*, 146 (1979).
6. Hallberg, A., Westfelt, L., and Holm, B., *J. Org. Chem.*, *46*, 5414 (1981).
7. Andersen, C. M., Hallberg, A., and Daves, G. D., Jr., *J. Org. Chem.*, *52*, 3529 (1987).
8. Andersen, C. M., and Hallberg, A., Jr., *J. Org. Chem.*, *53*, 235 (1988).
9. Augustine, R. L., and Thompson, M. M., *J. Org. Chem.*, *52*, 1911 (1987).

30

Phase-Transfer-Catalyzed Aromatic Nitro Displacements

Mark D. Conner and Michael E. Ford

Air Products and Chemicals Inc., Allentown, Pennsylvania

I. INTRODUCTION

Nucleophilic aromatic displacements are well known and widely practiced reactions [1]. Such displacements provide facile routes to
ethers and thioethers—for example, from suitably activated aromatic
substrates. Among the more widely used substrates for such reactions
are o/p-fluoronitrobenzene, o/p-chloronitrobenzene, and o/p-fluoro(or
chloro)benzoates.

Some of the most industrially useful displacements are those
effected on o/p-dinitrobenzenes [2,3]. These displacements have
the advantages of facile displacement of either nitro group to provide
a single product, and lower cost of dinitrobenzene substrates relative
to the corresponding halonitrobenzene isomers. Additionally, products
of these displacements provide ready access to amines by simple
reduction. For example, diamines formed from reaction of dinitrobenzene with diols followed by reduction find uses in polyurethanes
and polyimides [4]. The positional isomerism of the nitro group
can also moderate the reactivity of the resultant amines for different
applications in the polymer systems.

Nucleophilic displacements from meta-substituted aromatic substrates proceed more slowly than those on ortho/para-substituted
aromatics and can be complicated by competitive substitution at
other positions.* Recently, however, phenoxides have been shown to
be effective nucleophiles for nitro displacements on m-dinitrobenzenes

---

*In our laboratory, for example, displacements run on 3-chloronitro-
benzene provided approximately equimolar amounts of products from
nitro and chloro displacement.

in the presence of added phase transfer catalysts [5,6]. Limited
literature reports on analogous reactions using aliphatic alkoxides
have appeared [7,8] (e.g., eq. 30.1).

$$(30.1)$$

Our primary objective was to extend current technology to
a wider range of alkoxides and phase transfer catalysts as a route
to soft-segment diamine monomers. A few such monomers are accessible via substitution of α,ω-dihalides with *m*-nitrophenol [9]. However,
this approach is limited by the relative unavailability of the required
dihalides and expense of the *m*-nitrophenol. Determination of the
scope of alkoxides suitable for nitro displacement from *m*-dinitro-
benzene was expected to lead to a family of new α,ω-di(*m*-aminophenyl)
ethers.

II. RESULTS AND DISCUSSION

A. Monohydric Alcohols

Our initial approach was to examine a range of aliphatic alkoxides
using representative monohydric alcohols as models. Dihydric analogues of promising candidates could then be evaluated as precursors
to the α,ω-di(*m*-nitrophenyl)ethers.

Accordingly benzyl, phenethyl, cyclohexyl, and butoxyethyl
alcohols were examined and found to react with *m*-dinitrobenzene
under basic, phase-transfer-catalyzed conditions with reasonable
facility. Cyclohexyl alcohol produced a significant amount of high
molecular weight by-products, which likely contributed to the lower
yield of product obtained in this case. Unoptimized product yields
for the above-mentioned alcohols are shown in Table 30.1.

B. Phase Transfer Catalysts

Four phase transfer catalysts (PTCs) were evaluated: tris-(3,6-
dioxaheptyl)amine (TDA-1), polyvinylpyrrolidone (average MW 4000),
polyethylene glycol monomethyl ether (average MW 350), and 18-
crown-6 (Table 30.2). Significantly, TDA-1 was the most effective
PTC, and 18-crown-6 exhibited similar activity.

Table 30.1   Alcohol Screening Results[a]

Alcohol (ROH)	Product	Yield (%)[b]
Benzyl		39
Phenethyl		42
Cyclohexyl		20
Butoxyethyl		47

[a]Conditions: *m*-dinitrobenzene/alcohol, 1.25:1; 140°C; 17 hours.; *N*,*N*-dimethylformamide solvent.
[b]Based on gas-liquid chromatographic area percent analysis, non-optimized.

Table 30.2   Relative Effectiveness of Phase Transfer Catalysts[a]

Phase transfer catalyst	Relative yield
tris-(1,3-Dioxaheptyl)amine	2.7
18-Crown-6	2.5
Polyvinylpyrrolidone[b]	1.6
Polyethylene glycol[c]	1.0

[a]Conditions: *m*-dinitrobenzene/2-butoxyethanol, 1.25:1; 140°C; 17 hours; *N*,*N*-dimethylformamide solvent. PTC loading 2 mole % (except PVP, which was screened at 0.2 mol % because of its high molecular weight).
[b]Average molecular weight 4000.
[c]Average molecular weight 350.

C.   Diols

Results obtained with diols were consistent with those observed in the model reactions. Unfortunately, very high loadings of PTC were required to obtain good selectivity to difunctionalized (vs. monofunctionalized) products. Thus, production of a 4:1 ratio of difunctional to monofunctional diethylene glycol to dinitrobenzene adducts required a loading of approximately 25 wt % polyvinyl-pyrrolidone (Figure 30.1).

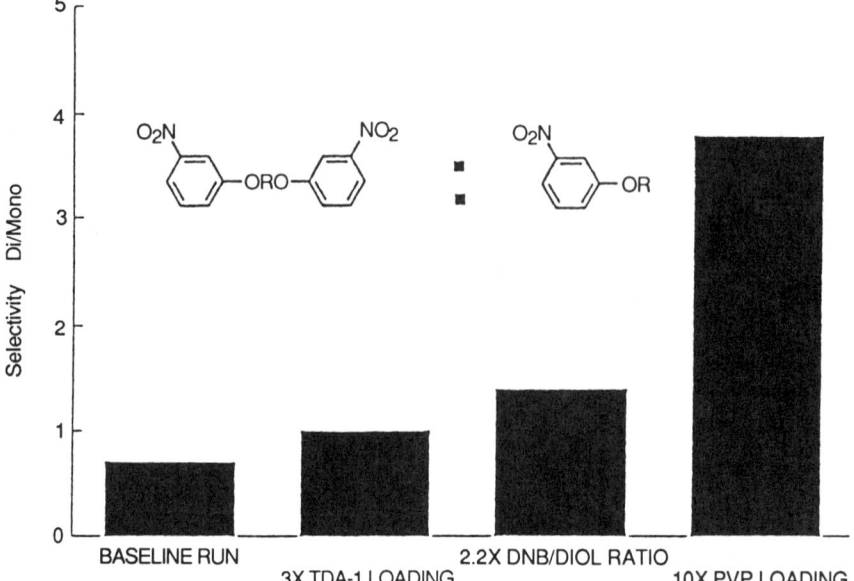

Figure 30.1 Selectivity to difunctional product. Conditions of base-
line run: *m*-dinitrobenzene/diethylene glycol, 2.5:1; 2 mol % TDA-1;
phase transfer catalyst, solvent, DMF; 17 hours; concentration
of DNB, 1.6 M.

III. CONCLUSION

Displacement of a nitro group from *m*-dinitrobenzene is shown to
provide a feasible route to aryl ethers using simple inexpensive
starting materials. Incomplete conversions, especially with diols,
may lead to limitations in processing, particularly involving starting
material separation/recycle.

REFERENCES

1. For a review on the subject see Zoltewicz, J. A., *Top. Curr.
   Chem.*, *59*, 33 (1975).
2. Shaffer, T. D., *J. Polym. Sci.*, *C, Polym. Lett.*, 27:11, 457
   (1989).
3. Takekoshi, T., Wirth, J. G., Heath, D. R., Kochanowski,
   J. E., Manello, J. S., and Webber, M. J., *J. Polym. Sci.*,
   Polym. Chem. Ed., *18*, 3069 (1980).

4. European Patent Application 268,849, to Dow Chemical Company (1987).
5. Sommes, P. G., Thetford, D., and Voyle, M., *J. Chem. Soc., Chem. Commun.*, 1373 (1987).
6. European Patent Application 192,480, to Mitsui Toatsu Chemicals, Inc. (1986).
7. Kornblum, N., Cheng, L., Kerber, R. C., Kester, M. M., Newton, B. N., Pinnick, H. W., Smith, R. G., and Wade, P. A., *J. Org. Chem.*, *41*, 1560 (1976).
8. Montanari, F., Pelosi, M., and Folla, F., *Chem. Ind.*, 412 (1982).
9. Harris, F. W., and Sridhar, K., *Polym. Prepr.*, *29*:2, 304 (1988).

31

Potential of ZSM-5 for Supplementing Clays in Pharmaceutical Applications

William E. Garwood and Pochen Chu

Mobil Research and Development Corporation, Paulsboro, New Jersey

I. INTRODUCTION

Certain clays have been used since ancient times [1] and marketed for years over the counter in pharmacies for gastrointestinal problems. Also a number of nonfibrous natural zeolites, principally clinoptilolite and phillipsite, have been researched as animal feed supplements, and the suggestion that they be tested in humans has been made [2]. Both the clays and natural zeolites reported in the literature have been mixtures varying in composition, structure, stability, and sorptive properties [2]. Synthetic ZSM-5, which is also non-fibrous, can be synthesized at desired silica/alumina ratios and particle size; it has good acid stability, assuring minimum change in structure during passage through the stomach with its low pH, and reproducible sorptive properties with a high selectivity for both amines and ammonia, with the latter believed to be a critical factor accounting for the beneficial effect of zeolites in animal feed. Of special interest is the potential of ZSM-5 due to its pore structure for adsorbing nitrosamines, in particular N-nitrosodimethylamine, a known carcinogen [3].

II. DATA AND DISCUSSION

A typical pharmaceutical clay contains about 67 wt % SiO_2, 20 wt % Al_2O_3, 1-2 wt % each Mg, Fe, and Ca, and 0.2-0.3 wt % each Na, K, and Ti, based on ash. HZSM-5 can be synthesized with Al_2O_3 contents varying from 50 ppm to about 8 wt %, with the remainder being SiO_2 and less than 0.1 wt % each of the remaining six elements

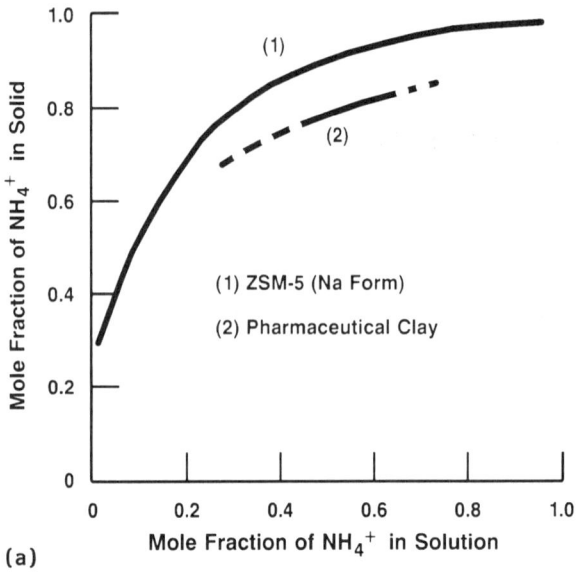

(a)

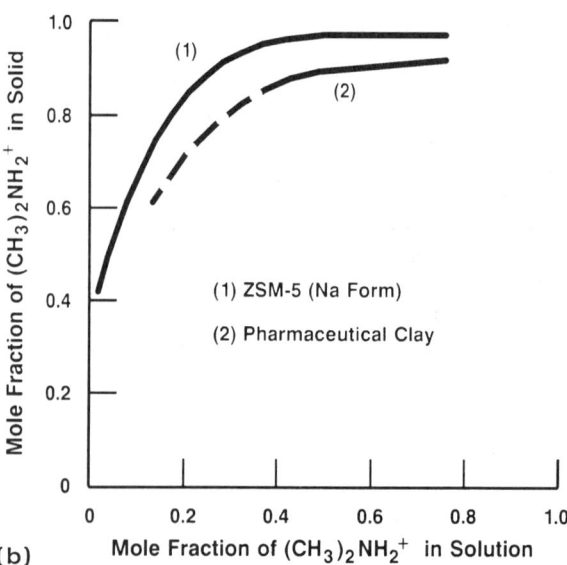

(b)

Figure 31.1 (a) $NH_4^+-Na^+$ and (b) $(CH_3)_2$ NH_2^+-Na exchange isotherms at 25°C.

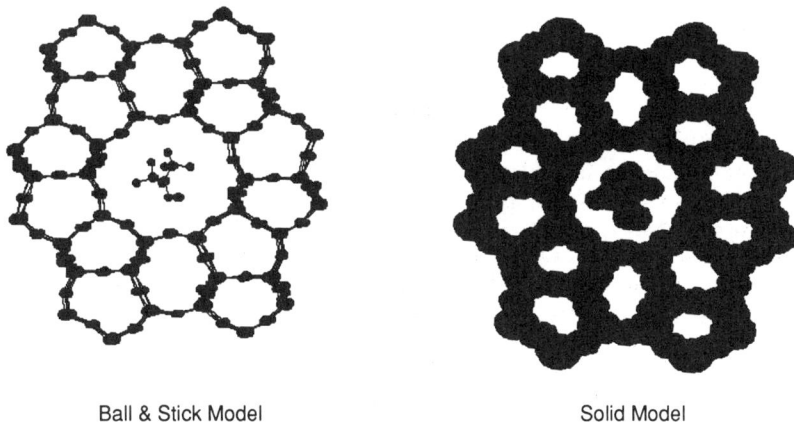

Ball & Stick Model   Solid Model

Figure 31.2   Computer-generated models of N-nitroso-dimethylamine in pores of ZSM-5.

found in the clay. The HZSM-5 used in this study contained 2.4 wt % Al_2O_3, corresponding to a SiO_2/Al_2O_3 ratio of 70:1.

The acid stability of ZSM-5 to dealumination is very high, even at approximately 100°C in solutions with pH as low as 1.4 [4]. Thus it should pass unchanged through the stomach, where the pH can be 2 or even lower [2]. Scanning electron micrographs of 2 μm ZSM-5 show it to be nonfibrous [5]. The particle size of ZSM-5 can be varied, from 0.02 to greater than 5 μm.

NH_4^+-Na^+ and $(CH_3)_2NH_2^+$-Na^+ exchange isotherms at 25°C were performed on both 70:1 SiO_2/Al_2O_3 NaZSM-5 and the typical pharmaceutical clay. ZSM-5 was more selective for the nitrogen compounds in both cases (Fig. 31.1).

The structure of ZSM-5 [6] is such that not only dimethylamine, but also its reaction product with a nitrosating agent such as nitrous acid, the carcinogen N-nitrosodimethylamine, fits easily in the pores (Fig. 31.2).

ACKNOWLEDGMENTS

The computer-generated models by N. Bhore are greatly appreciated, as are helpful discussions with J. B. Higgins, I. D. Johnson, T. P. McMahon, D. H. Olson, and J. L. Schlenker.

REFERENCES

1. *De re metallica*, translated from the first Latin edition of 1556, by Herbert C. and Lou H. Hoover, Dover Publications, New York, 1950. The translation was initially published in *Mining Magazine*, London, 1912.
2. Pond, W. G., and Mumpton, F. A., eds., *Zeo-Agriculture*, Westover Press, Boulder, CO, 1984.
3. Hotchkiss, J. H., 197th National Meeting of the American Chemical Society, April 9-14, 1989, Dallas, Division of Agricultural and Food Chemistry, Abstract 60; also Tannenbaum, S. R., Wishnok, J., and Leaf, C., Abstract 61.
4. Garwood, W. E., Chu, P., Chen, N. Y., and Bailar, J. C., Jr., *Inorg. Chem.*, *27*, 4331 (1988).
5. Stucky, G. D., and Dwyer, F. G., eds., *Intrazeolite Chemistry*, ACS Symposium Series, 218, American Chemical Society, Washington, DC, 1983.
6. Kokotailo, G. T., Lawton, S. L., Olson, D. H., and Meier, W. M., *Nature* (London), *272*, 437 (1978).

32

Process for the Production of 2,2,6,6-Tetraalky-4-piperidylamines

R. E. Malz, Jr., and C.-Y. Lin
Uniroyal Chemical Company, Inc., Naugatuck, Connecticut

Harold Greenfield
First Chemical Corporation, Pascagoula, Mississippi

I. INTRODUCTION

The reductive alkylation of 2,2,6,6-tetramethyl-4-piperidone (1) with hexamethylenediamine (2) for the production of N,N'-bis(2,2,6,6-tetramethyl-4-piperidyl)hexamethylenediamine (3) has been studied. In contrast to the conventional platinum catalyst used for such a reaction, better product yield has been achieved by the use of a palladium catalyst in a low pressure process. Also, a relatively substantial amount of unrecyclable by-products such as 2,2,6,6-tetramethyl-4-piperidinol (4: see Fig. 32.1), of the platinum catalyst under a higher pressure, can be eliminated by the use of the same palladium catalyst.

2,2,6,6-Tetraalkyl-4-piperidylamines, such as N,N'-bis-(2,2,6,6-tetramethyl-4-piperyl)hexamethylenediamines (3) are well-known ultraviolet stabilizers for polymeric materials. Conventional procedure for the production of these compounds has generally involved the use of a platinum catalyst in an aliphatic alcohol solvent [1].

The yield of the product, as shown in Table 32.1, is relatively low when a platinum catalyst is employed in a low pressure process (100-200 psig). While the product yield increases under the use of a higher pressure (600-800 psig) as indicated in Table 32.2, such high pressure platinum-catalyzed processes will also result in the production of a relatively substantial amount of unrecyclable by-products (such as 2,2,6,6-tetramethyl-4-piperidinol derivative, (4).

Since the mechanism of the reaction (Fig. 32.2) involves an equilibrium reaction between an alkanolamine (A) on the one hand and ketimine (B) and water on the other, it is worth noting that the use of a reaction medium comprising at least about 10% water

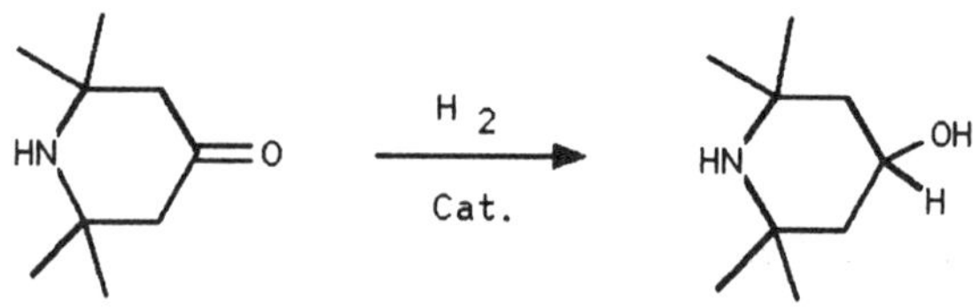

(1)   (4)

Figure 32.1   Piperidinol by-product formation.

Table 32.1   Yield of Product with Low Pressure (100-200 psig) Process

Example or comparative experiment	Reaction catalyst	Medium	Time at 80°C (h), Total	Reaction[a]	Bis product (mol %)[b]
1	Palladium	2-Propanol	5.0	3.5	97
2	Palladium	Water	6.2	5.2	96
3	Palladium	91% 2-Propanol,[c] 9% water	4.6	3.5	97
A	Platinum	Water	5.3	4.3	90

[a]Time at 80°C during apparent hydrogen absorption.
[b]Bis product = N,N'-bis(2,2,6,6-tetramethyl-4-piperidyl)hexamethylene-diamine.
[c]Percent by volume.

Table 32.2   Yield of Product with High Pressure Palladium Catalysis

Example number	Reaction catalyst	Solvent	Time (h)	Yield (mole %) Bis[a]	Mono[b]	TAA[c]	Alcohol[d]	HMDA[e]
4	Palladium	Water	6.7	68	10	0.16	0.07	N.D.
5	Palladium	Methanol	5.0	88	0.12	3.5	N.D.	N.D.
B	Platinum	Water	2.4	89	3.1	0.43	3.2	N.D.
C	Platinum	Methanol	4.8	93	0.47	2.9	0.57	N.D.

[a]N,N'-Bis(2,2,6,6-tetramethyl-4-piperidyl)hexamethylenediamine.
[b]N-(2,2,6,6-Tetramethyl-4-piperidyl)hexamethylenediamine.
[c]2,2,6,6-Tetramethyl-4-piperidone.
[d]2,2,6,6-Tetramethyl-4-piperidinol.
[e]Hexamethylenediamine; N.D. = none detected.

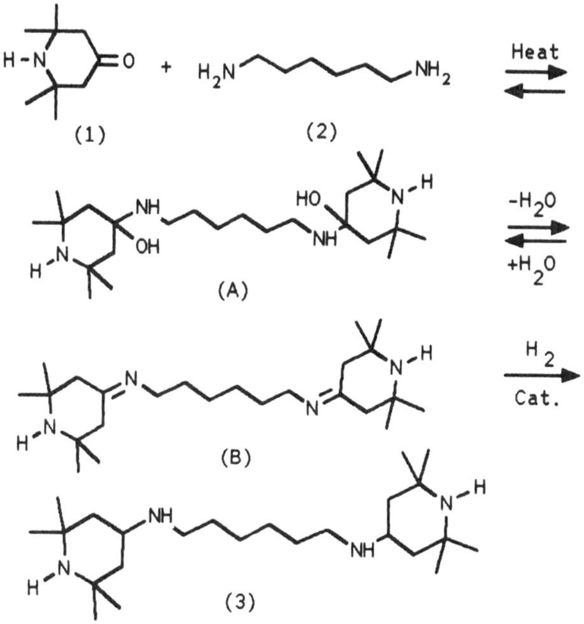

Figure 32.2 Synthesis of N,N'-bis(2,2,6,6-tetramethyl-4-piperidyl)hexamethylenediamine (3).

in the palladium-catalyzed process (cf. Table 32.1) would not materially affect the reaction rate and the yield of the product (4). The use of the water-methanol mixture would minimize the risk of methanol being ignited in the presence of the catalysts [2].

REFERENCES

1. (a) Cantatore, G., U.S. Patent 4,104,248 (1978). (b) Son, P. N., U.S. Patent 4,326,063 (1982). (c) Minagawa, M., et al., U.S. Patent 4,415,688 (1983). (d) DiBattista, P., et al., U.S. Patent 4,293,466 (1981).
2. Rylander, P. N., *Catalytic Hydrogenation over Platinum Metals*, Academic Press, New York, 1967, p. 12.

33

Raney Copper Catalyzed Hydration of a Hindered Nitrile

Mike G. Scaros, John P. Westrich,* Owen J. Goodmonson,†
and Michael L. Prunier

*Department of Chemical Development, Searle,
Skokie, Illinois*

I. INTRODUCTION

The use of Raney copper to effect the hydration of a nitrile to
an amide as in the case of commercially prepared acrylamide is
relatively well known. In our attempt to apply this chemistry to

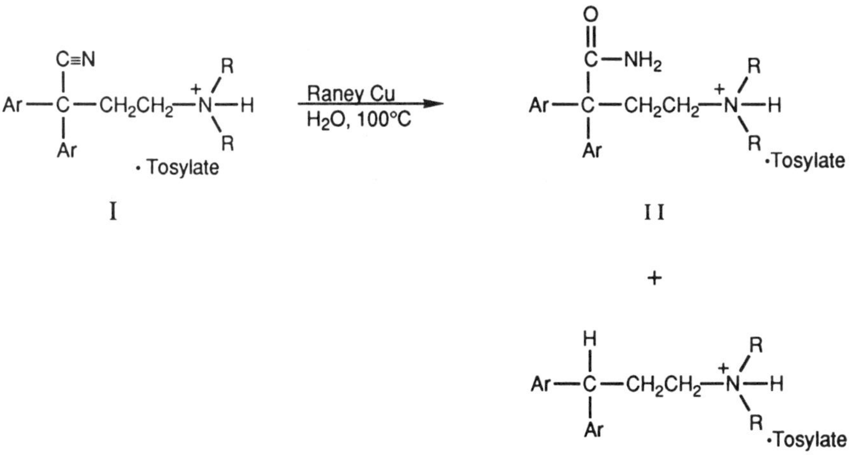

I

II

+

III

Decyanation Product

Current affiliations:
*S. C. Johnson and Son, Inc., Skokie, Illinois
†C.A.P.D., Abbott Laboratories, North Chicago, Illinois

Table 33.1   Alloy Digestion Temperature Results for Four Sources

Catalyst	Source	Digestion temperature (°C)	Conversion (%)	Comments
Archival	Kawaken	Unknown	>90	4-6 recycles
Commercial	W. R. Grace	100	50-60	No recycles
Laboratory preparation	W. R. Grace alloy	10-30	50-60	No recycles/ decyanation
Laboratory preparation	Kawaken Fine Chemical alloy	10-30	50-60	No recycles/ decyanation

a trisubstituted acetonitrile system (1), we encountered several unexpected problems. Once the sources of these problems had been discovered, a successful process was developed. The critical parameters that were identified included the preparation of the Raney copper from a copper aluminum alloy, the effective oxidation state of the catalyst, and the cause of catalyst deactivation. The elucidation of and solutions to these problems are described.

Table 33.1 contains the results of a study of both alloy digestion temperature and alloy/catalyst source. Based on these results, we assumed that alloy digestion temperature and alloy/catalyst source were not important and turned our attention to catalyst age.

II.   CATALYST OXIDATION STATE

The archival sample of Raney copper was charcoal gray, which is consistent with Cu^{2+} (cupric oxide), while the freshly prepared or purchased Raney copper was reddish, which is consistent with $Cu(0)$ and Cu^+ (copper metal and cuprous oxide). Consequently, atomic absorption data were obtained and the freshly prepared catalyst had a cupric-to-cuprous ratio of 1:9. Freshly prepared catalyst was then oxidized by bubbling air through an aqueous catalyst slurry, and the color of the slurry changed from reddish to gray. The cupric-to-cuprous ratio of this catalyst was determined to be 10:1. The catalytic activity of the oxidized catalyst (Fig. 33.1) was nearly identical to that obtained with the archival catalyst, including the catalyst's ability to be recycled. Thus we have successfully developed a catalyst preparation/activation sequence that gave a reliable Raney copper catalyst for this hydration. Controlled

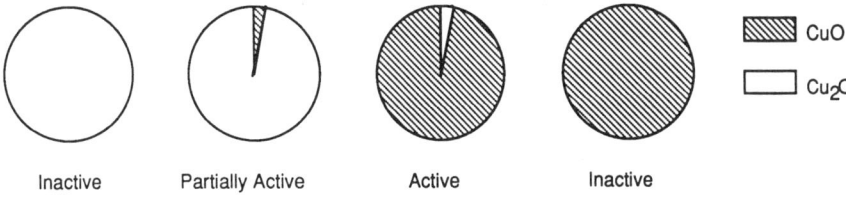

Figure 33.1   Catalytic activity with CuO and Cu_2O.

oxidation appeared to be important, since it was determined that exposing the catalyst sludge to air resulted in a strong exotherm (90°C) and an inactive catalyst, indicating that partial oxidation is required.

III. OXIDIZED CATALYST LIFE AND REACTIVATION

The oxidized catalyst was shown to function well for three or four hydrations without a noticeable decline in activity; however, after five or six hydrations, the reaction times became unacceptably long, with declining conversion rates. Various attempts to increase the number of recycles are listed in Table 33.2. Only the dilute NaOH washing of the Raney copper between hydrations provided enhanced longevity. As illustrated in Figure 33.2, the number of cycles could be almost doubled before significant increases in reaction times were observed.

Auger electron spectroscopic analysis of the Raney copper revealed that the ratio of aluminum to copper on the surface of the catalyst increased from 2:1 (fresh) to 6:1 (spent) after hydration.

Table 33.2   Methods for Increasing Number of Recycles

Oxidative approaches	Washes	
Oxidation between runs	Hot water	Prepare catalyst from $CuSO_4$/Al
Oxidation at higher pH	Dilute acid	Prepare catalyst from $CuSO_4$-Zn dust
Treatment with EDTA and oxidation	Dilute base	Addition of $NaNO_2$ promoter
Hydration in an atmosphere of air		

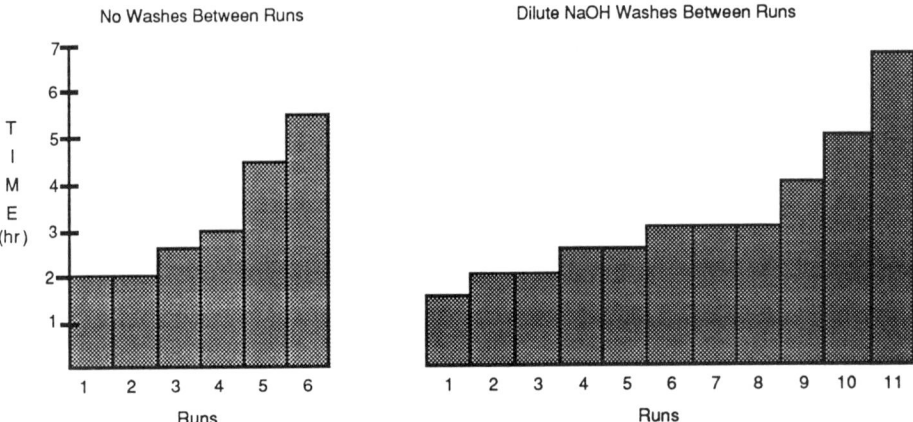

Figure 33.2   Number of cycles versus reaction time.

The dilute base wash of the catalyst between cycles in combination
with a dilute base storage of the oxidized catalyst between reactions
removed the aluminum that had migrated to the surface of the catalyst, and left the surface with a higher ratio of active copper.

IV.   MISCELLANEOUS OBSERVATIONS

The hydration must be performed under an inert atmosphere or
     unidentified green solids will form during the reaction.
Isolation of the product using this system was encumbered by the
     formation of an emulsion. Treatment of the filtered reaction
     mixture with EDTA solved this problem.
The Raney copper should be stored at a pH of 8-10 to maintain
     the catalyst maximum activity.
The catalyst must remain wet. Dried catalyst is inactive.
Uncontrolled exposure of the catalyst to air during recycle will
     decrease the catalyst's activity.

V.   CATALYST PREPARATION

Initial hydrations were run using an archival sample of Raney
copper* catalyst. The reactions resulted in greater than 90% con-

---

*Other than Kawaken Fine Chemical being the supplier, no informa-
tion on this catalyst was available; that is, it is not known whether
the alloy was purchased and the Raney copper prepared or the
Raney copper was purchased.

version to II and several catalyst recycles were possible. When
the archival sample was depleted, additional Raney copper was
purchased from W. R. Grace. Much to our surprise, the commercial
Raney copper resulted in only 50-60% conversion, with no recycle
possible. Also, when Raney copper prepared from a 1:1 copper
aluminum alloy (by digestion in water for 5 h at 10-30°C) was used
for this hydration, the reaction was incomplete and a significant
amount of the decyanation product III was observed.

Three major differences between the archival and commercial
Raney copper were identified.

1. *Alloy digestion temperature*. The archival temperature was unknown, versus 100°C for commercially prepared Raney copper.
2. *Alloy/catalyst source*. Archival Raney copper was purchased
   from Kawaken Fine Chemical, while commercial Raney copper
   and alloy was purchased from W. R. Grace.
3. *Catalyst age*. Archival Raney copper had been stored for several
   years, whereas the commercial catalyst/alloy had been recently
   purchased or prepared.

VI. CONCLUSION

Although Raney copper hydrations are well documented in the literature, we have determined that the success of the technique is
dependent on the preparation of the catalyst. We were successful
in identifying the requirements for obtaining a functional catalyst
for our particular substrate. The process affords product that
is comparable in yield and purity so that obtained from sulfuric
acid hydrations, and it eliminates the use of hot sulfuric acid and
provides a process that is not encumbered with the safety and
environmental issues associated with sulfuric acid.

ACKNOWLEDGMENTS

The authors acknowledges Chung Kim and Jerry Svarz for supplying
the various substrates. We also extend our appreciation to H. Dryden
for his suggestions and thoughts and to John Witt for his encourage-
ments and support. The authors are also indebted to Ann Worlatschek
for her diligent and conscientious help in the preparation of this
manuscript and to Searle for permission to publish this work.

REFERENCE

1. Svarz, J. J., Goretta, L. A., and Seale, V. L., U.S. Patent
   3,920,740 (Nov. 18, 1975).

Index